IMMUNOLOGY AND IMMUNE SYSTEM DISORDERS

IMMUNOGLOBULINS, MAGIC BULLETS AND THERAPEUTIC ANTIBODIES

IMMUNOLOGY AND IMMUNE SYSTEM DISORDERS

Additional books and e-books in this series can be found on Nova's website under the Series tab.

IMMUNOLOGY AND IMMUNE SYSTEM DISORDERS

IMMUNOGLOBULINS, MAGIC BULLETS AND THERAPEUTIC ANTIBODIES

MANZOOR AHMAD MIR
EDITOR

nova
Medicine & Health
New York

Library of Congress Cataloging-in-Publication Data

Names: Mir, Manzoor Ahmad, editor.
Title: Immunoglobulins, magic bullets and therapeutic antibodies / [edited by Manzoor Ahmad Mir].
Description: New York : Nova Science Publishers, Inc., [2019] | Series: Immunology and immune system disorders | Includes bibliographical references and index.
Identifiers: LCCN 2019056852 (print) | LCCN 2019056853 (ebook) | ISBN 9781536169034 (hardcover) | ISBN 9781536169041 (adobe pdf)
Subjects: LCSH: Immunoglobulins.
Classification: LCC QR186.7 .I474 2019 (print) | LCC QR186.7 (ebook) | DDC 615.3/7--dc23
LC record available at https://lccn.loc.gov/2019056852
LC ebook record available at https://lccn.loc.gov/2019056853

Published by Nova Science Publishers, Inc. † New York

CONTENTS

PREFACE

Immunology is a distinctive subject that rose in the mid- 20th century. The subject developed as scientists started to unravel the mysteries about the defence system against pathogens. Researchers started to understand the mechanisms employed by the innate and the adaptive immune system in defence against pathogens. During the last decade, the subject of immunology has been in sharp focus as the immunotherapies against diseases like cancer and AIDS seems last hope. Employing the body's own defence system against diseases like cancer and AIDS by activating specific cells of the immune system looks promising and therapies like CAR-T cell therapy have been approved. In the first edition of the book "Immunoglobulins, Magic Bullets and Therapeutic Antibodies" we have explained the Immunoglobulins, their genetic, production of Monoclonal antibodies, use of them in diagnostics and therapeutics and the role played by immunoglobulins in overall protection of our body.

The book is organised into four volumes. The first volume comprises of ten chapters and it describe the rise, history and scope of immunology and the building blocks of the immune system *viz.,* cells, molecules and organs of the immune system. The chapter second describes the cells of the innate and the adaptive immune system and how the granulocytes and macrophages employ defence mechanisms to protect the body against pathogenic invasions. In the chapter third of this book, we have described the organs if the immune systems and how different organs are involved in the differentiation and maturation of immune cells. The chapter has also focussed on the structure of lymph nodes and their function in concentrating the antigens. In the chapter four of this book we have described the terms like antigens, immunogens, antigenicity, immunogenicity and how immunogenicity of an antigen is affected and how antigenicity of an immunogens is related to the immune response. The innate and adaptive immune systems and the different types of cells and molecules employed by the two branches of immunity have been described in a separate chapter. The structure and biology of immunoglobulins, their types and function in antigen binding and antibody dependent cellular cytotoxicity (ADCC) have been described well in chapter six. Focus has been laid on the distinction between and antibody and an immunoglobulin. The structure and function and major histocompatibility complex (MHC) has been described well in the book. The education of cells about self and non-self during their maturation and the processing and presentation of antigens by MHC bearing cells and how MHC coordinates both humoral and cell-mediated immune responses has been explained well throughout the book. The book has explained the complement system and its components, mechanisms and functions in a separate chapter. At the end of the book, we have given an insight about the vaccines, their history, development and how they are useful and helpful in the defence against diseases. The book also discusses the immune disfunction and diseases associated with the dysregulation of immune responses.

The present book volume -2 entitles "Immunoglobulins Magic Bullets and Therapeutic Antibodies" comprises of ten chapters and it describe the Origin, production and scope of immunoglobulins which are the central molecules of immune system around which the field of immunology revolves. This book describes how the immune system, responds to injuries and insults by foreign antigens (bacteria, viruses etc) and produces immunoglobulins, which then through various immune response mechanisms protect the body against pathogenic invasions, how these antigens are involved in the differentiation and maturation of immune cells, how lymph nodes are involved in concentrating the antigens, how immunogenicity of an antigen is affected and how antigenicity of an immunogens is related to the immune response. Various types of immunoglobulins and the organization and expression of immunoglobulins genes are described separately. The antigen antibody interactions, antibody dependent cell mediated cytotoxicity are also discussed in separate chapters. The Magic bullets or monoclonal antibodies which has revolutionised the field of immunology, their production by hybridoma technology, phage display technology and the important role played by the monoclonal antibodies in research, diagnostics, in the treatment of cancer and autoimmune disorders is described separately. Lastly the Immunodiagnostic techniques like radioimmunoassay (RIA), Enzyme Linked Immunosorbent Assay (ELISA) and flow cytometry and their importance in the detection of various kinds of diseases like Cancer, HIV-AIDS, Tuberculosis, Malaria etc are discussed in detail separately.

Dr. Manzoor A. Mir

ACKNOWLEDGMENTS

First, I would like to thank Almighty for giving me strength, belief and good health. It is under his grace that we all live, learn and flourish.

Although it is difficult to thank everybody and interwove in words the genuine efforts made by the people directly or indirectly to make this book entitled "The Fundamentals of Immunology". But I would like to take this opportunity to pen down my appreciation for a number of people whose contribution in numerous ways immensely helped me to move towards my destination.

It gives me immense pleasure to express my sense of gratitude and respect to my mentors Prof Talat Ahmad, Dr. Javed N Agrewala, Prof Mohammad Afzal Zargar, Prof Raid Saleem Albaradie, and Dr Abid Hamid Dar whose constant support, utmost patience, invaluable advice and guidance rekindled my interest for science from time to time. I will benefit for a long time to come from their sincerity, originality and truthfulness which has nourished my intellectual maturity. I admire and respect them all for their sincerity, dedication, devotion, amazing memory, thorough knowledge and constructive criticism.

I owe deepest gratitude to my parents. A seed of loyalty, hard work, dedication and sincerity sown in me by my parents played a crucial role in completion of this work. They have been a source of great strength throughout my life. I would like to thank other family members for continuous and unconditional support of all my undertakings, scholastic and otherwise. Without their affection, support, best wishes, love and care this work would not have been completed

I would like to cordially thank Miss Najmu Nisa Research Scholar BGSBU for her support and co-operation. Mr. Aariz Manzoor Mir and Mr. Umar Mehraj deserves a special mention. I would like to thank everybody who was important to the successful realization of this book, as well as expressing my apology that I could not mention personally one by one.

Finally, I would like to thank again the Prime Mover Almighty in whom I have great faith

Dr. Manzoor A. Mir

In: Immunoglobulins …
Editor: Manzoor Ahmad Mir
ISBN: 978-1-53616-903-4

Chapter 1

INTRODUCTION TO IMMUNOGLOBULINS

Manzoor A. Mir**, *Umar Mehraj, Safura Nisar and Hina Qayoom
Department of Bioresources, School of Biological Sciences,
University of Kashmir, Srinagar J&K, India

ABSTRACT

Immunoglobulin's or antibodies are glycosylated, special types of proteins, produced in response to antigens. Immunoglobulin's specifically interact with antigens which induced their production. They are exclusively synthesized by B-cells (Plasma Cells) when a immunogen/antigen interact specifically with a B-cells receptors (BCR) present on the surface of B lymphocyte. This interaction induces a signal that directs the activation of transcription factors and the activation of concerned genes and finally the production of immunoglobulin's, specific for each immunogen (antigen) which stimulated the B-cell. A part from this immune system develops memory B-cells to remember the antigens that caused previous reaction. These memory cells produce the secondary immune response which is always quick and robust. There are five major types of immunoglobulins in humans: IgM, IgG, IgA, IgE, and IgD. Immunoglobulin's are composed of two heavy (H) and two light (L) chains. These are functionally divided into a variable (V) region that binds specifically to the antigen and constant (C) that has an effector function in activation of complement. The antibody has specific regions for interaction with the antigen known as paratopes which interact with the epitopes of the antigen.

Keywords: antibody, heavy chain, light chain, variable region, constant region.

* Corresponding Author: Dr Manzoor Ahmad Mir (Ph. D), Department of Bioresources, School of Biological Sciences, University of Kashmir, Srinagar J&K, India-190006, Contact:-+919622901319. Email: drmanzoor@kashmiruniversity.ac.in.

Objectives

- Introduction to immunoglobulins.
- Describe the structure of immunoglobulins.
- Describe the general characteristics of different immunoglobulins
- Difference between immunoglobulin's and immunogens.

Introduction

Immunoglobulin's also called as antibodies are glycoprotein's having specific amino acid sequences by which they can interact with specific antigens. Immunoglobulin's and antibodies are both interchangeable, although all antibodies are immunoglobulins, but all immunoglobulins are not antibodies. The main difference between immunoglobulin and antibody is that immunoglobulin has a transmembrane domain in order to be attached to the plasma membrane whereas antibody does not have a transmembrane domain. Immunoglobulin refers to any class of structurally related proteins in the serum and the cells of the immune system which function as antibodies. It is produced in response to an antigen. The name immunoglobulin is derived from the fact that they migrate with globular proteins when antibody-containing serum is placed in an electrical field. Antibodies are immunoglobulin's (Igs) which are produced in the body in response to the antigen or foreign bodies. Antibodies form 20% of the plasma proteins. These antibodies, the antigen-binding glycoproteins are synthesized exclusively by B-cells and in billions of forms, each with a different amino acid sequence and a different antigen binding site. Immunoglobulin's are the most abundant protein components in the blood, consisting about 20% of the total protein components in the blood plasma. These antibodies are present on B-cell membrane and also secreted by the plasma cells. Antibodies belong to a class of proteins called globulins and are multifunctional glycoprotein's produced in direct response against antigens by vertebrates and essential for the deterrence and resolution of infection by various pathogens. They initiate their biological effect by binding non-covalently to antigen without modifying the covalent structure of the antigen. Antibody molecules are referred to as multifunctional molecules as they perform more than one function. One part of the immunoglobulin is involved in the binding of the antigen while the other part of the molecule may be involved in binding to receptors on phagocytes, the activation of complement pathways and the activation of the NK cells of the immune system. The antibodies produced in response to a particular antigen are heterogeneous. Most antigens are complex and contain many different antigenic determinants, and the immune system usually responds by producing antibodies to several epitopes on the antigen. This response requires the recruitment of several clones of B cells. Their outputs are monoclonal antibodies, each of which specifically binds a single antigenic determinant. Together, these monoclonal antibodies make up the polyclonal and heterogeneous serum antibody response to an immunizing antigen.

CLASSES OF IMMUNOGLOBULINS

Immunoglobulin's (antibodies) are assembled into five major classes; IgM, IgG, IgA, IgD and IgE with unique antigenic properties present on their heavy chains (constant region) each designated with its own class of heavy chain-μ, γ, α, δ and ε respectively (Figure 1.1).

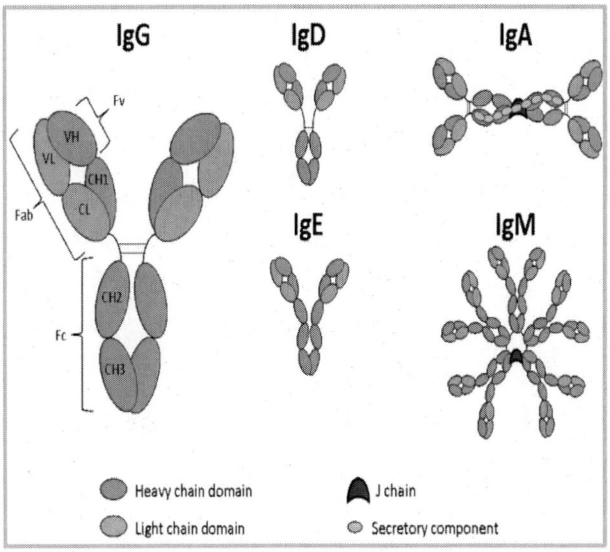

Figure 1.1. Schematic representation of the five immunoglobulin classes or isotopes in mammals.

In addition, immunoglobulin IgG and IgA are further divided into subclasses based on their antigenic differences within each of the five classes. In humans, IgG has four subclasses; IgG1, IgG2, IgG3, IgG4 having $^{\gamma 1, \gamma 2, \gamma 3}$ and $^{\gamma 4}$ heavy chains, respectively while as IgA has two subclasses; IgA1, IgA2. IgM, IgD and IgE have no subclasses. Each class and subclass has its own characteristic properties. Antibodies are antigen binding proteins that are present on the B-cell membrane and secreted by plasma cells. Membrane-bound antibody confers antigenic specificity to the B cells; antigen-specific proliferation of B-cell clones is elicited by the interaction of membrane antibody with antigen. Changes in the isotype expressed, which produce changes in the biological effector functions of Immune system is the ability to respond to a limitless array of foreign antigens. As immunoglobulin (Ig) sequence data accumulated is virtual, every antibody molecule studied is found to contain a unique amino acid sequence in its variable region but only one of a limited number of invariant sequences in its constant region. The genetic basis for this combination of constancy and tremendous variation in a single protein molecule lies in the organization of the immunoglobulin genes. In germ-line DNA, multiple gene segments encode portions of a single immunoglobulin heavy or light chain. These gene segments are carried in the germ cells but cannot be transcribed and translated into complete chains until they are rearranged into functional genes. During B-cell maturation in the bone marrow, certain gene segments are randomly shuffled by a dynamic genetic system capable of generating more than 106 combinations. The processes of B cell development are carefully regulated, the maturation of a progenitor B cell progresses through an ordered sequence of Ig-gene rearrangements coupled with modifications to the gene that contribute to the diversity of the final product. The individual B cell is thus antigenically committed to a specific epitope. After antigenic stimulation of a mature B cell in peripheral lymphoid organs, further rearrangement of constant-region gene segments can generate the immunoglobulin molecule without changing its specificity. By the end of this process, a mature, immune competent B cell will contain coding sequences for one functional heavy chain variable-region and one light-chain variable-region. Thus, mature B cells contain chromosomal DNA that is no longer identical to germ-line DNA. Ag-Ab interaction is highly specific and occurs in a similar way as a bimolecular association of an enzyme-substrate.

Characteristics of Various Types of Immunoglobulins

The characteristic features of various classes of antibodies are listed below:

- **Immunoglobulin Type G (IgG)**: Its Molecular weight- 180 kDa. It can cross placenta. It is a monomer form of immunoglobulin. Total percentage of this Ab in serum is 80%. Made up of heavy chain of ϒ and can fixes complement. Function is present during secondary immune response. Number of binding sites on this antibody are 2.
- **Immunoglobulin Type M (IgM):** Molecular weight- 900 kDa. It can't cross placenta. It is pentamer form of immunoglobulin with a total percentage of this Ab in serum is 6%. It is made up of heavy chain of μ and can fixes complement. Function is, present during primary immune response and number of binding sites on this antibody are 10.
- **Immunoglobulin Type A (IgA):** Molecular weight- 385 kDa. It can't cross placenta. It is a dimer form of immunoglobulin with a total percentage of this Ab in serum is 13% (monomer). It is made up of heavy chain of α. It can't fix complement. The function of this IgA Ab is to provide protection on Respiratory and Urogenital surface. The number of binding sites on this antibody are 4.

Table 1.1. Showing types of immunoglobulins and their characteristics

Types of Immunoglobulins

Properties	IgG	IgM	Secretory IgA	IgD	IgE
Molecular weight	180 kDa	900 kDa	385 kDa	180 kDa	200 kDa
Crosses placenta	Yes	No	No	No	No
Form of immunoglobulin	Monomer	Pentamer	Dimer	Monomer	Monomer
Total percentage of Ab in serum	80%	6%	13% (monomer)	< 1%	< 1%
Heavy chain	ϒ	μ	α	δ	ε
Fixes complement	Yes	Yes	No	No	No
Function	Present during secondary immune response	Present during primary immune response	Provide protection on Respiratory and Urogenital surface	B- cell receptor binding	Provide immunity against parasite and responsible for allergic condition
Number of binding site	2	10	4	2	2

- **Immunoglobulin Type D (IgD):** Molecular weight- 180 kDa. It can't cross placenta. Monomer form of immunoglobulin with a total percentage of this Ab in serum is <1%. It is made up of heavy chain of δ. It can't fixes complement and its function is to act as B- cell receptor (BCR). The number of binding sites on this antibody are 2.
- **Immunoglobulin Type E (IgE):** Molecular weight- 200 kDa. It can't cross placenta. It is monomer form of immunoglobulin with a total percentage of this Ab in serum is <1%. It is made up of heavy chain of ε. It can't fix complement. The functions of this antibody are to provide immunity against parasite and responsible for allergic condition. The number of binding sites on this antibody are 2.

Affinity and Avidity of Immunoglobulins

The binding between antigens and immune components involves weak non-covalent interactions. The binding forces are relatively weak and *reversible* and consist mainly of van der Waals forces, electrostatic forces, and hydrophobic forces. The combined strength of all non-covalent interactions between a single antigen-binding site of an antibody and a single epitope of the antigen is called *affinity* of that antibody for the concerned *epitope*. A low affinity antibody binds the concerned antigen weakly, and it tends to dissociate readily from the Ag-Ab complex. Many antigens are complex antigens in that they have multiple, repeating epitopes. Such antigens usually bind an antibody molecule having multiple binding sites at more than one binding site. Further, interaction between such an antibody molecule and a complex antigen molecule at one binding site will increase the probability of reaction between them at a second binding site. The cumulative strength of the multiple interactions between a multivalent antibody and antigen is called *avidity* (Figure 1.2).

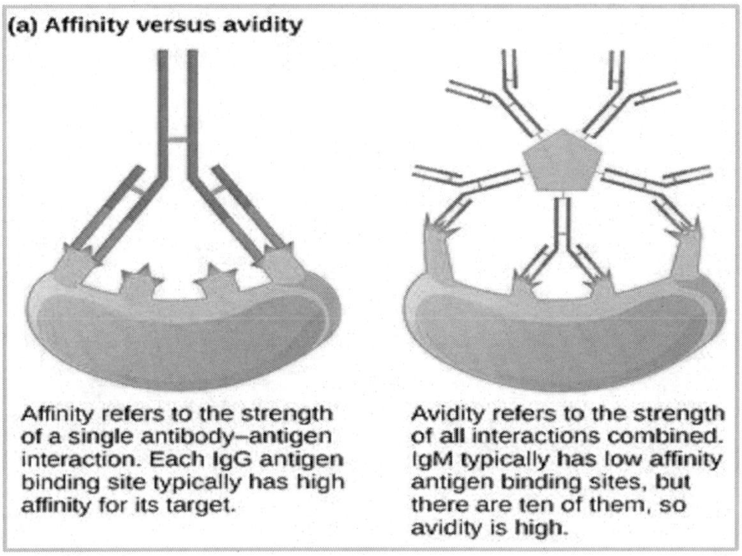

Figure 1.2. Shows affinity vs avidity of antibody molecule.

When an antibody elicited by one antigen (more precisely, an immunogen) reacts with an unrelated antigen, it is called *cross-reaction*. Cross reaction occurs when two unrelated antigens have an identical or a very similar epitope.

Antibody-Dependent Cell-Mediated Cytotoxicity (ADCC)

The antibody-dependent cell-mediated cytotoxicity also known as antibody-dependent cellular cytotoxicity (ADCC), is a mechanism of cell-mediated immune defense whereby an effector cell of the immune system actively lyses a target cell, whose membrane-surface antigens have been bound by specific antibodies. It can also be determined a lytic process in which antibody coated target cells are killed by the cells of the immune system such as NK cells and macrophages. The significance of this ADCC is that virus-infected or other diseased cells are targeted for destruction by components of the cell-mediated immune system, such as natural killer cells. ADCC is independent of the immune complement system that also lyses targets but does not require any other cell. ADCC requires an effector cell which classically is known to be natural killer (NK) cells that typically interact with IgG antibodies. However, macrophages, neutrophils and eosinophils can also mediate ADCC. ADCC is part of the adaptive immune response due to its dependence on a prior antibody response.

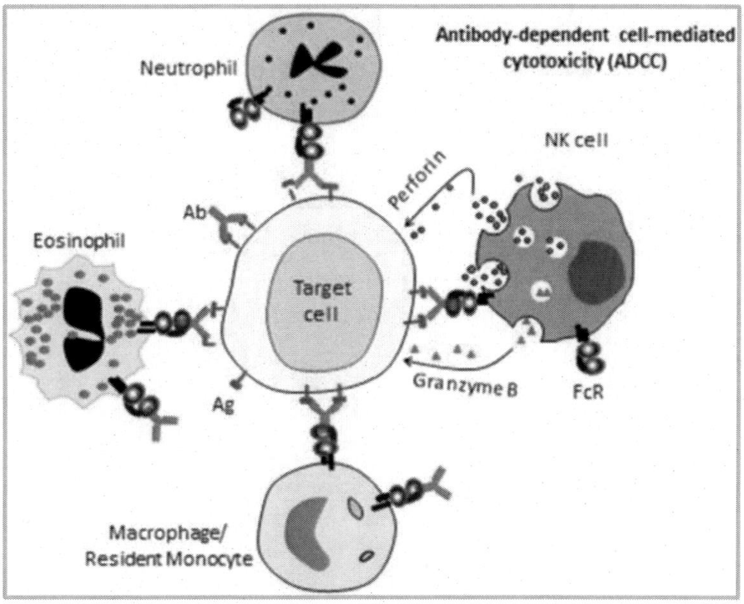

Figure 1.3. ADCC by four major immune effector cells: macrophages/resident monocytes, NK cells, neutrophils, and eosinophils. (Image source Jorge Raúl Geffner, in Encyclopedia of Immunology (Second Edition), 1998).

ADCC was initially described by Moller in 1967 as the ability of natural killer (NK) cells – at the time identified only as lymphocytes – to destroy antibody-coated target cells. A non-phagocytic mechanism, whereby most leucocytes can kill target cells in the absence of complement and without major histo-compatibility complex (MHC) restriction thus became apparent. The specificity of ADCC is conferred by target-bound antibody and can be induced by antibody concentrations well below those needed to activate complement-mediated lysis. Fc receptors (FcR) on the plasma membrane of the effector cell are necessary but not sufficient for ADCC activity. Once antibody has bound to the target cell, its interaction with either activating or inhibitory FcR on the cytotoxic effector cell appears to dictate the precise signalling pathways that are stimulated. When target-bound antibodies cross-link activating FcR, a cytotoxic pathway is initiated that results in the release of mediators such as perforin, tumor necrosis factor (TNFa), and reactive oxygen intermediates (ROI) – which induce target cell death. Conversely, ligation of inhibitory FcR prevents effector cell activation, resulting in survival of the target cell. Mechanism of ADCC depicted by four major immune effector cells: macrophages/resident monocytes,

NK cells, neutrophils, and eosinophils in the Figure 1.3. The target cell may either be microbe infected or a tumor cell that expresses the antigen recognized and bound by the antibody. Target cell killing is achieved by several cytolytic processes that also cause limitation of microbial or tumor growth.

Immunoglobulin Labelling and Flowcytometry

Flow cytometry is a popular cell biology technique that utilizes laser-based technology to count, sort, and profile cells in a heterogeneous fluid mixture. Using a flow cytometry machine, cells or other particles suspended in a liquid stream are passed through a laser light beam in single file fashion, and interaction with the light is measured by an electronic detection apparatus as light scatter and fluorescence intensity. If a fluorescent label, or fluorochrome, is specifically and stoichiometrically bound to a cellular component, the fluorescence intensity will ideally represent the amount of that particular cell component. The Alexa Fluor family of fluorescent dyes is a series of dyes invented by Molecular Probes, now a part of Thermo Fisher Scientific, and sold under the Invitrogen brand name. Alexa Fluor dyes are frequently used as cell and tissue labels in fluorescence microscopy and cell biology. Alexa Fluor dyes can be conjugated directly to primary antibodies or to secondary antibodies to amplify signal and sensitivity or other biomolecules. Flow cytometry is a powerful tool because it allows simultaneous multiparametric analysis of the physical and chemical characteristics of up to thousands of particles per second. This makes it a rapid and quantitative method for analysis and purification of cells in suspension. Using flow cytometry, we can determine the phenotype and function of the cells and even sort live cells. The use of flow cytometry in the clinical laboratory has grown substantially in the past decade. This is an attribute to the development of smaller, user-friendly, less-expensive instruments and a continuous increase in a number of clinical applications.

Production of Immunoglobulins by Hybridoma Technology

Hybridomas are cells that have been engineered to produce a desired antibody in large amounts. To produce monoclonal antibodies, B-cells are removed from the spleen of an animal that has been challenged with the relevant antigen. These B-cells are then fused with myeloma tumor cells that can grow indefinitely in culture (myeloma is a B-cell cancer). This fusion is performed by making the cell membranes more permeable. The fused hybrid cells (called Hybridomas), being cancer cells, will multiply rapidly and indefinitely and will produce large amounts of the desired antibodies. They have to be selected and subsequently cloned by limiting dilution. Supplemental media containing Interleukin-6 (such as briclone) are essential for this step. Monoclonal antibodies (MAbs) are an old immunological tool with applications in the fields of immunology, biotechnology, biochemistry, and applied biology.

Production of monoclonal antibodies using hybridoma technology was discovered in 1975 by Georges Kohler of West Germany and Cesar Milstein of Argentina. Modern-day research on MAbs from laboratories worldwide is revealing additional applications in diverse branches of sciences.

Monoclonal antibodies (mAb or moAb) are antibodies that are made by identical immune cells that are all clones of a unique parent cell. Monoclonal antibodies can have monovalent affinity, in that they bind to the same epitope. In contrast, polyclonal antibodies bind to multiple epitopes and are usually made by several different plasma cell lineages (Figure 1.4). Bispecific monoclonal antibodies can also be engineered, by increasing the therapeutic targets of one single monoclonal antibody to two epitopes. Monoclonal antibodies are also known as magic bullets and are produced by hybridoma technology (Figure 1.5) which involves identifying antigen-specific plasma/plasmablast cells that produce antibodies specific to an antigen of interest and fusing these cells with myeloma cells using polyethylene glycol as fusinogen and selecting the hybrid cells in HAT media followed by selection, purification, screening and identification.

These monoclonal antibodies have diagnostic, analytical and therapeutic uses. Recently, MAbs have been widely applied in the field of clinical medicine. Currently, MAbs account for one-third of all the new therapeutic treatments for breast cancer, leukemia, arthritis, transplant rejection, asthma, and psoriasis, with many more late-stage clinical trials being conducted.

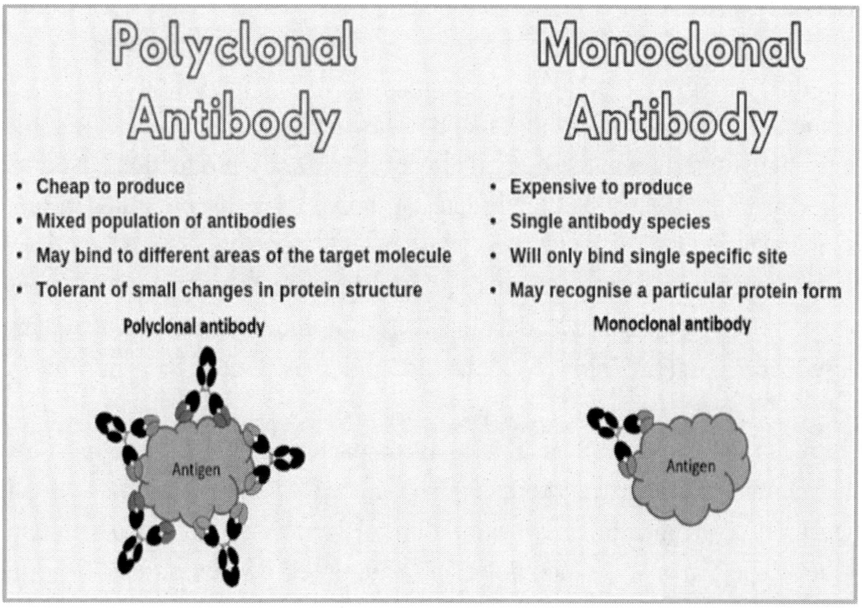

Figure 1.4. Schematic representation of polyclonal and monoclonal antibodies.

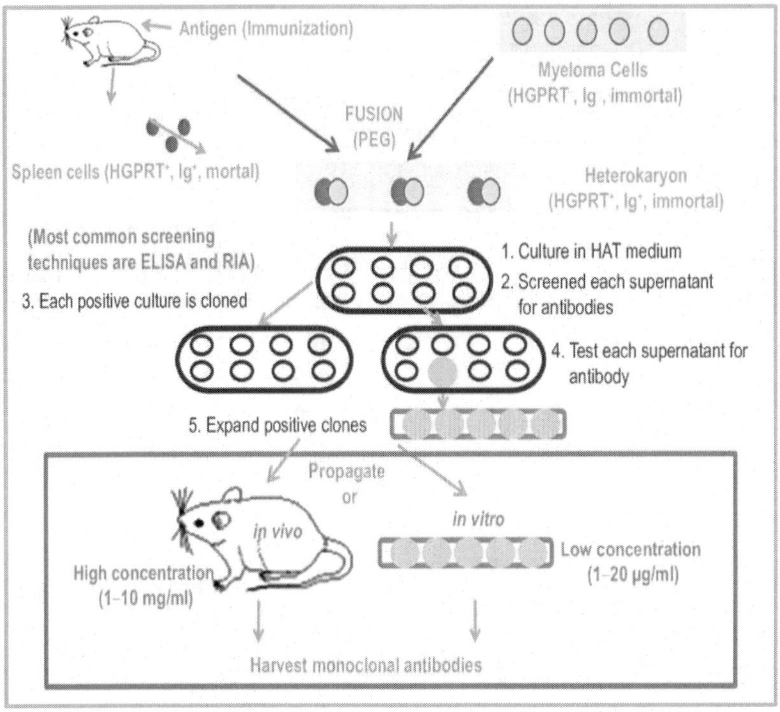

Image source: Journal, Indian Academy of Clinical Medicine 2013(4):9-21.

Figure 1.5: Showing the generation of monoclonal antibodies using hybridoma technology.

RIA and ELISA as Diagnostic Tests for Qualitative and Quantitative Antibody Detection

RIA, highly sensitive laboratory technique used to measure minute amounts of substances including antigens, hormones, and drugs present in the body. The substance or antigen (a foreign substance in the body that causes antibody production) to be measured is injected into an animal, causing it to produce antibodies. Serum containing the antibodies is withdrawn and treated with a radioactive antigen and later with a nonradioactive antigen. Measurements of the amount of radioactivity are then used to determine the amount of antigen present. The technique was developed by Solomon Berson and Rosalyn Yalow. For this discovery Rosalyn Yalow was awarded the 1977 Nobel Prize in Physiology or Medicine.

RIA is an in vitro assay. It calculates the presence of an antigen with very high sensitivity. Mainly, any biological product or substance for which a specific antibody exists can be measured, even in minute concentrations. RIA has been the first immune-assay procedure developed to analyse picomolar and nanomolar concentrations of hormones in biological fluids. The radioimmunoassay is perhaps *the oldest types of immunoassays.* Here, a radioisotope is attached to an antigen of interest and bound with its complementary antibody. Then a sample with the antigen to be measured is added. It competes with the radioactive antigen, kicks it out of the binding spot and replaces it. After washing away unbound antigens the radioactivity of the sample is measured. The amount of radioactive signal is inversely related to the amount of target antigen. ELISA (enzyme-linked immunosorbent assay) is a plate-based assay technique designed for detecting and quantifying peptides, proteins, antibodies and hormones. In an ELISA, an antigen must be immobilized to a solid surface and then complexed with an antibody that is linked to an enzyme. Detection is accomplished by assessing the conjugated enzyme activity via incubation with a substrate to produce a measurable product. The most crucial element of the detection strategy is a highly specific antibody-antigen interaction. ELISAs are typically performed in 96-well (or 384-well) polystyrene plates, which will passively bind antibodies and proteins. It is this binding and immobilization of reagents that makes ELISAs so easy to design and perform. Having the reactants of the ELISA immobilized to the microplate surface makes it easy to separate bound from non-bound material during the assay. This ability to wash away nonspecifically bound materials makes the ELISA a powerful tool for measuring specific analytes within a crude preparation.

Structure of Immunoglobulin

An immunoglobulin (Ig) is a Y shaped molecule consisting of 4-polypeptide chains with two identical *light (L)* chains, consisting of about 220 amino acid residues (25000 MW) and two identical *heavy (H) chains,* consisting of about 440 amino acid residues (50,000MW) held together by disulphide linkages and non-covalent interactions. Thus, an immunoglobulin is a dimer of H-L chain.

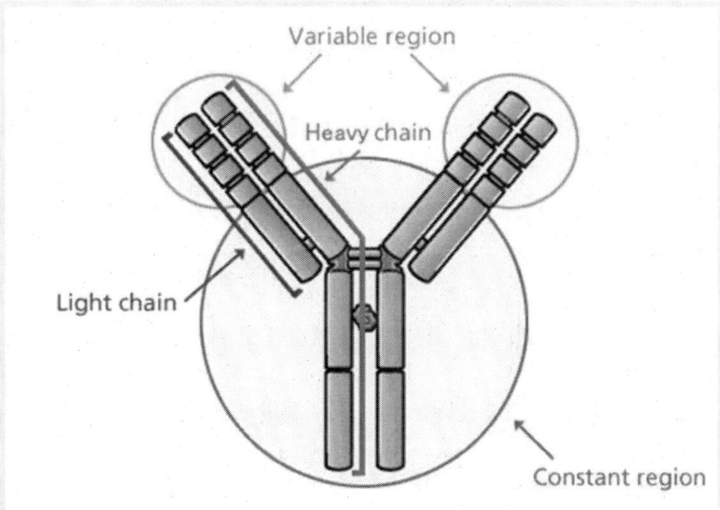

Image source: https://microbiologyinfo.com/antibody-structure-classes-and-functions/.

Figure 1.6. Immunoglobulin structure.

Constant and Variable Region

Each light and heavy chain (L or H) of immunoglobulin has two regions or domains: the constant and the variable. The amino terminal half of the light chain is the variable region (V_L) and the car boxy terminal half is the constant region (C_L) almost 110 amino acid long. In contrast, the heavy chain has one-quarter of the amino terminal region variable (V_H) almost 110 amino acid long and the remaining three-quarters are constant (C_{H1}, C_{H2}, C_{H3}) almost 330 or 440 amino acid long. The specific binding of an immunoglobulin (antibody) with an antigen is due to the amino acid sequence of variable region of both light and heavy chain.

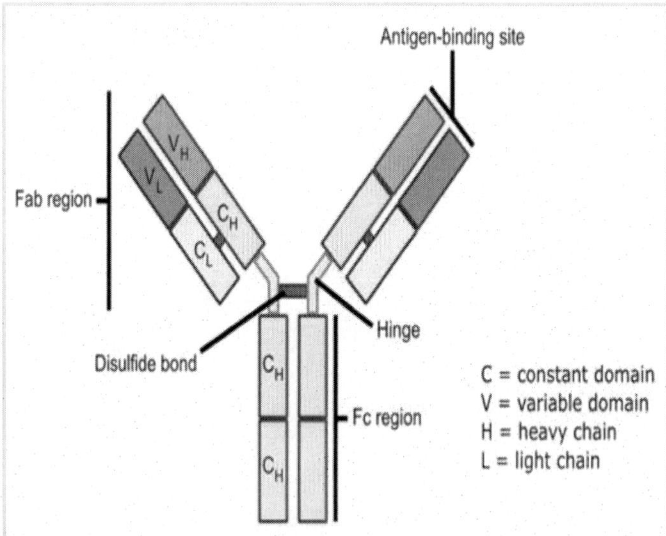

Image source; https://step1.medbullets.com/immunology/105060/antibodies.

Figure 1.7. Basic structure of immunoglobulin (antibody).

The basic structure of immunoglobulin molecule illustrated by Porter and Edelman with a different approach found that treatment with the enzyme *papain* split the immunoglobulin molecule into two fragments: one that binds with the antigen designated as *Fab; fragment antigen binding* and the other *Fc; fragment crystallisable* which does not bind to antigen. Each immunoglobulin molecule has two antigen-binding sites or domains, each being constituted of the variable region of one light chain and one heavy chain. The constant regions of two heavy chains of an immunoglobulin (antibody) molecule form its effector functional domain that determines its interaction with the other components of the immune system. As studied, there are two major classes of light chains: *kappa* (κ) and *lambda* (λ). A species can produce both types of light chains, however the light chains are either both *kappa* (κ) or both *lambda* (λ) and never one of each. The occurrence of *kappa* (κ) chain is more common in human immunoglobulins than *lambda* (λ) chain. Since there are two types of light chains, it has been shown that heavy chains consist of five different types: α, γ, δ, ε and μ. These five types of heavy chains are called isotypes. The type of heavy chain ultimately determines the class and the function of the given immunoglobulin: IgM (μ), IgG (γ), IgA (α), IgD (δ), and IgE (ε). Each class can have either *kappa* (κ) or *lambda* (λ) light chains and an individual of a species can make all heavy chains but a single immunoglobulin molecule has both identical heavy chains. Thus, an immunoglobulin molecule can have the structure κ2γ2 with two identical light chains and two identical heavy chains. Similarly, it could have the structure of λ2γ2. The variable regions of both light and heavy chains are diversified for most part to the three small hyper-variable regions (each almost 10 amino acid long) called *complementary determining regions (CDR)* in each chain; the remaining parts of the variable region, known as *framework regions,* are relatively constant. Proceeding from either V_L or V_H amino-terminal, these *complementary determining regions* are denoted as: CDR1, CDR2 and CDR3, CDR3 being most variable. The *framework regions* are designated as: FR1, FR2, FR3 and FR4.

The structure of the immunoglobulin molecule is determined by the primary, secondary, tertiary, and quaternary organization of the protein. The primary structure, the amino acid sequence, accounts for the variable and constant regions of the heavy and light chains. The secondary structure is formed by folding of the extended polypeptide chain back and forth upon itself into an antiparallel β-pleated sheet. The chains are then folded into a tertiary structure of compact globular domains, which are connected to neighbouring domains by continuations of the polypeptide chain that lie outside the β-pleated sheets. Finally, the globular domains of adjacent heavy and light polypeptide chains interact in the quaternary structure forming functional domains that enable the molecule to specifically bind antigen and, at the same time, perform a number of biological effector functions.

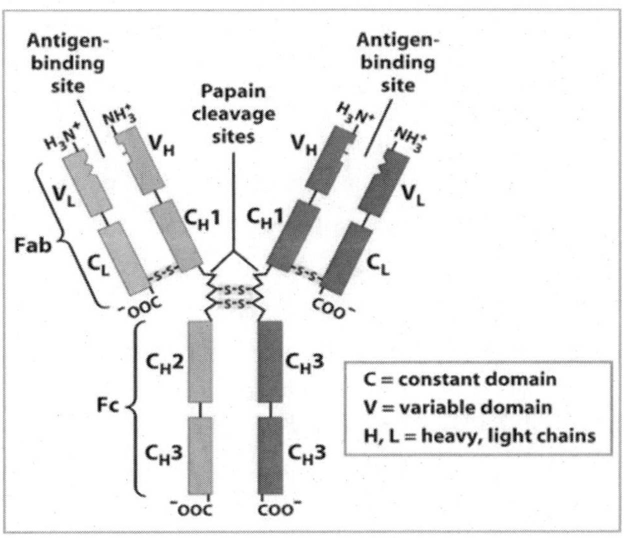

Image source; Dr. Anaya Mandal, MD.

Figure 1.8. Showing the variable and constant region of an antibody and the antigen binding sites.

Table 1.2. Difference between immunoglobulin's and immunogens

	Immunoglobulin's (Antibodies)	Immunogens (Antigens)
1.	Antibody is a protein molecule.	Antigen is a protein or polysaccharide molecule.
2.	It is synthesized to combat foreign material.	It is usually a foreign material that stimulates antibody formation.
3.	Antibody occurs on the surface of a plasma cell and also in body fluids.	Antigen may occur on the surface of a microbe or as a free molecule.
4.	Antibody directly joins an antigen to destroy the latter.	Antigen binds to a macrophage to reach a helper T-cell to initiate immune response.

FREQUENTLY ASKED QUESTIONS

Q1. What are antibodies?

Answer: Antibody, also called immunoglobulin, a protective protein produced by the immune system in response to the presence of a foreign substance, called an antigen. Antibodies recognize and latch onto antigens in order to remove them from the body. A wide range of substances are regarded by the body as antigens, including disease-causing organisms and toxic materials such as insect venom. Antibodies have a Y shaped structure made up of two heavy and two light chains. Antibodies have set of complimentary determining regions (CDRs) in the variable region which form paratope and binds specifically with the epitope of the antigen during antibody antigen interaction. Antibody can be polyclonal or monoclonal. There are five classes of antibodies namely IgA, IgD, IgE, IgG and IgM depending upon the presence of heavy chain in the antibody molecule.

Q2. Are antibodies and immunoglobulins the same?

Answer: Immunoglobulin and antibody are disease-fighting proteins developed by most vertebrates in response to a particular antigen. Both immunoglobulin and antibody are glycoproteins. Both of them contain similar regions in their molecules. The main difference between immunoglobulin and antibody is that immunoglobulin has a transmembrane domain in order to be attached to the plasma membrane whereas antibody does not have a transmembrane domain.

Q3. What is an immunoglobulin superfamily?

Answer: A superfamily is a group of proteins that share similar amino acid sequences where sequences are less than 50% identical. The immunoglobulin superfamily is a large and diverse group of proteins comprising more than 100 different polypeptides that contain segments of conserved sequence of 100 amino acids, an Ig homology unit which have significant similarities to those of immunoglobulin domains.

Q4. What do you understand by antibody dependent cell mediated cytotoxicity?

Answer: The antibody-dependent cell-mediated cytotoxicity also known as antibody-dependent cellular cytotoxicity (ADCC), is a mechanism of cell-mediated immune defense whereby an effector cell of the immune system actively lyses a target cell, whose membrane-surface antigens have been bound by specific antibodies. It can also be determined a lytic process in which antibody coated target cells are killed by the cells of the immune system such as NK cells and macrophages. The significance of this ADCC is that virus-infected or other diseased cells are targeted for destruction by components of the cell-mediated immune system, such as natural killer cells. ADCC is independent of the immune complement system that also lyses targets but does not require any other cell. ADCC requires an effector cell which classically is known to be natural killer (NK) cells that typically interact with IgG antibodies. However, macrophages, neutrophils and eosinophils can also mediate ADCC. ADCC is part of the adaptive immune response due to its dependence on a prior antibody response.

Q5. What does the term opsonisation mean?

Answer: Opsonisation is a process whereby a particle often a bacteria is rendered more attractive for ingestion by phagocytes by coating it with a specific antibody or by complement components. Opsonization is the molecular mechanism whereby molecules, microbes, or apoptotic cells are chemically modified to have a stronger attraction to the cell surface receptors on phagocytes and NK cells. An opsonin (from the Greek opsōneîn, to prepare for eating) is any molecule that enhances phagocytosis by marking an antigen for an immune response or marking dead cells for recycling. With the antigen coated in opsonins, binding to immune cells is greatly enhanced. Opsonization also mediates phagocytosis via signal cascades from cell surface receptors. Significance of the opsonisation is that in a healthy individual, they mark dead and dying self-cells for clearance by macrophages and neutrophils, activate complement proteins, and target cells for destruction through the action of natural killer (NK) cells.

Q6. What is a complement system?

Answer: The complement system is a part of the immune system that enhances the ability of antibodies and phagocytic cells to clear microbes and damaged cells from an organism, promote inflammation, and attack the pathogen's cell membrane. The activation of the complement system is one of the most important effector mechanisms of immunoglobulins. The complement system is a group of approximately 20 proteins/enzymes present in plasma. The complement system has four major function, including lysis of infectious organisms, activation of inflammation, opsonization and immune clearance. There are three different complement pathways, the classical complement pathway, the alternative complement pathway, and the mannose-binding lectin pathway. The complement system is also known as Cascade system because it is complementary to the antibody response of the adaptive immune system and after the first few complement proteins bind, a cascade of sequential binding events follows in which the pathogen rapidly becomes coated in complement proteins.

Q7. What are agglutinins and agglutinogens?

Answer: Antibodies that cause agglutination or aggregation of cells or bacteria are called as agglutinins. An agglutinin is a substance in the blood that causes particles to coagulate and aggregate; that is, to change from fluid-like state to a thickened-mass (solid) state. Agglutinins can be antibodies that cause antigens to aggregate by binding to the antigen-binding sites of antibodies. Agglutinogen on the other hand is any antigen, or foreign cell, toxin, bacteria, or anything else that gets the immune system reacting, that makes your body generate agglutinins. Agglutinins have multiple arms that can bind onto agglutinogens. This means one agglutinin can hold lots of invading pathogens together. Clumping the bad guys together helps soldier immune cells find them and get rid of them. This clumping process is called agglutination.

Q8. Define the framework region.

Answer: In molecular biology, a framework region is a subdivision of the variable region (Fab) of the antibody. The variable region is composed of seven amino acid regions, four of which are framework regions and three of which are hypervariable regions. The framework region makes up about 85% of the variable region. The invariant peptide sequence within the variable region that supports the CDRs is termed as the framework region. It acts as the scaffolding to which CDRs are fixed.

Q9. What are CDRs?

Answer: Complementarity-determining regions (CDRs) are part of the variable chains in immunoglobulins (antibodies) and T cell receptors, where these molecules bind to their specific antigen. A set of CDRs constitutes a paratope which binds with the epitope of the antigen during antibody-antigen interaction. The region within the variable region that generates the antigen binding site are called the complementarity determining regions or CDRs. Being the most variable parts of the immunoglobulin molecules, CDRs are crucial to the diversity of antigen specificities generated by lymphocytes. There are three CDRs (CDR1, CDR2 and CDR3), arranged non-consecutively, on the amino acid sequence of a variable domain of an antigen receptor. Since the antigen receptors are typically composed of two variable domains (on two different polypeptide chains, heavy and light chain), there are six CDRs for each antigen receptor that can collectively come into contact with the antigen. A single antibody molecule has two antigen receptors and therefore contains twelve CDRs total. There are three CDR loops per variable domain in antibodies. Sixty CDRs can be found on a pentameric IgM molecule.

Q10. What do you understand by an immunoglobulin fold?

Answer: Two β sheets linked by a disulphide bond and tightly packed against each other comprise the immunoglobulin fold. The immunoglobulin fold consists of a pair of β sheets, each built of antiparallel β strands, that surround a central hydrophobic core. This motif is found in the immunoglobulin domain. An immunoglobulin domain consists of a pair of β-sheets linked by a disulfide bond and hydrophobic interactions. Three hypervariable loops lie at one end of the structure. The immunoglobulin domain is a type of protein domain that consists of a 2-layer sandwich of 7-9 antiparallel β-strands arranged in two β-sheets with a Greek key topology, consisting of about 125 amino acids. The backbone switches repeatedly between the two β-sheets. Typically, the pattern is (N-terminal β-hairpin in sheet 1)-(β-hairpin in sheet 2)-(β-strand in sheet 1)-(C-terminal β-hairpin in sheet 2). The cross-overs between sheets form an "X", so that the N- and C-terminal hairpins are facing each other.

Q11. How many types of antibodies are there and how do they differ from one another?

Answer: There are five immunoglobulin classes (isotypes) of antibody molecules found in serum: IgG, IgM, IgA, IgE and IgD. They are distinguished by the type of heavy chain they contain. IgG molecules possess heavy chains known as γ-chains; IgMs have μ-chains; IgAs have α-chains; IgEs have ε-chains; and IgDs have δ-chains. The variation in heavy chain polypeptides allows each immunoglobulin class to function in a different type of immune response or during a different stage of the body's defence. The classes or isotypes of antibodies along with their characteristics are depicted in Table below:

Q12. What is meant by therapeutic antibodies?

Answer: Development of therapeutic antibodies boomed in the 1980's, and the first therapeutic antibody, a mouse antibody, was launched in 1986 as an immunosuppressive agent used during organ transplantation. Antibodies can swiftly provide therapeutics to target the disease-related molecules that have been discovered in genomic research because 1) the high level of specificity and affinity to the target molecule or antigen achieves a high level of efficacy and fewer adverse events, 2) their ability to target diverse molecules and the modes of action of the antibodies allow them to be applied to a wide range of therapeutic targets, and 3) modification and refinement by genetic engineering technology and the establishment of recombinant manufacturing technology has made industrial manufacturing possible.

Class	Sub-class	Description	Antibody complexes
IgA	2	Found in mucosal areas, such as the gut, respiratory tract and urogenital tract, and prevents colonization by pathogens. Also found in saliva, tears, and breast milk.	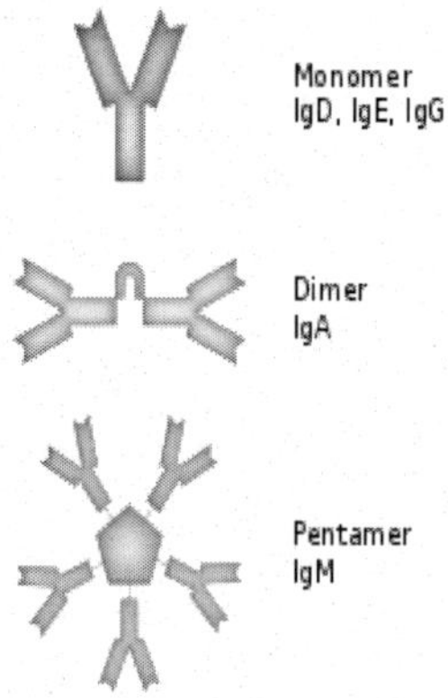
IgD	1	Functions mainly as an antigen receptor on B cells that have not been exposed to antigens. It has been shown to activate basophils and mast cells to produce antimicrobial factors.	
IgE	1	Binds to allergens and triggers histamine release from mast cells and basophils, and is involved in allergy. Also protects against parasitic worms.	
IgG	4	In its four forms, provides the majority of antibody-based immunity against invading pathogens. The only antibody capable of crossing the placenta to give passive immunity to the fetus.	
IgM	1	Expressed on the surface of B cells (monomer) and in a secreted form (pentamer) with very high avidity. Eliminates pathogens in the early stages of B cell-mediated (humoral) immunity before there is sufficient IgG.	

Q13. What are monoclonal antibodies?

Answer: Monoclonal antibodies (mAb or moAb) are antibodies that are made by identical immune cells that are all clones of a unique parent cell. Monoclonal antibodies can have monovalent affinity, in that they bind to the same epitope. In contrast, polyclonal antibodies bind to multiple epitopes and are usually made by several different plasma cell lineages. Bispecific monoclonal antibodies can also be engineered, by increasing the therapeutic targets of one single monoclonal antibody to two epitopes. Monoclonal antibodies are also known as magic bullets and are produced by hybridoma technology which involves identifying antigen-specific plasma/plasmablast cells that produce antibodies specific to an antigen of interest and fusing these cells with myeloma cells using polyethylene glycol as fusinogen and selecting the hybrid cells in HAT media followed by selection, purification, screening and identification. These monoclonal antibodies have diagnostic, analytical and therapeutic uses.

Q14. How are Monoclonal Antibodies (Mabs) produced?

Answer: Mabs are produced through hybridoma technology. The production of monoclonal antibodies involves several steps. In the first instance a laboratory animal is injected with a desired target to stimulate their immune system. Following this, antibody producing cells, B lymphocytes, are harvested from the animal's spleen and fused with an immortal myeloma cell line to create hybrid cells, or hybridomas. The hybrid cells are then screened to find those that secrete antibodies with the desired specificity for a particular target. Once identified the hybrid cell is cloned to establish a hybridoma colony. This colony is then maintained in a culture medium to provide a continual supply of monoclonal antibodies. Each of the antibodies secreted by the hybrid cell is identical (monoclonal) and has the capacity to bind to a specific receptor found on the surface of a cell.

Q15. How are Monoclonal Antibodies (Mabs) useful?

Answer: Monoclonal antibodies (mAb) are important reagents used in biomedical research, in diagnosis of diseases, and in treatment of such diseases as infections and cancer. The use of monoclonal antibodies (mAb) in biomedical research has been and will continue to be important for the identification of proteins, carbohydrates, and nucleic acids. Their use has led to the elucidation of many molecules that control cell replication and differentiation, advancing our knowledge of the relationship between molecular structure and function. These advances is basic biologic sciences have improved our understanding of the host response to infectious-disease agents and toxins produced by these agents, to transplanted organs and tissues, to spontaneously transformed cells (tumours), and to endogenous antigens (involved in autoimmunity). In addition, the exquisite specificity of mAb allows them to be used in humans and animals for disease diagnosis and treatment. Under the appropriate conditions, mAb-producing hybridomas survive indefinitely, so continued production of mAb is associated with the use of fewer animals, especially when production involves the use of in vitro methods.

MULTIPLE CHOICE QUESTIONS

Q1. Antibodies are produced by which cells in the human body?

A. Lymphocytes
B. Phagocytes
C. Monocytes
D. Thrombocytes

Q2. Antigen binding sites of an immunoglobulin are located in

A. Light chain alone
B. Heavy chain alone
C. Fc region of the antibody
D. Fab regions of the antibody

Q3. The class of immunoglobulins which can cross the placenta is

A. Immunoglobulin IgM
B. Immunoglobulin IgG
C. Immunoglobulin IgA
D. Immunoglobulin IgD

Q4. Self-reactive antibodies are not found in significant concentration in normal serum because of:

A. It is impossible to generate a self-reactive antibody
B. Self-reactive b cells are killed by cd8 t cells
C. B cells that are stimulated via their surface bound antibody in the absence of t cell help to commit suicide
D. Self-reactive b cells switch to iga which is all secreted and not present in serum

Q5. Simulation of which of the following will lead to appearance of first antibody to appear following stimulation by an antigen?

A. Immunoglobulin IgM
B. Immunoglobulin IgA
C. Immunoglobulin IgE
D. Immunoglobulin IgG

Q6. Which is the most efficient complement fixing class of antibody?

A. Immunoglobulin IgE
B. Immunoglobulin IgA
C. Immunoglobulin IgM
D. Immunoglobulin IgG

Q7. The class of immunoglobulin, most abundant in the human body is:

A. Immunoglobulin IgD
B. Immunoglobulin IgE
C. Immunoglobulin IgG
D. Immunoglobulin IgM

Q8. J chain is present in which of the following Immunoglobulins?

A. IgA Immunoglobulin
B. IgM Immunoglobulin
C. Both (a) and (b) Immunoglobulin
D. IgE Immunoglobulin

Q9. Which is not a function of IgG Immunoglobulin?

A. Major antibody in serum
B. First antibody type produced against an antigen during the primary antibody response
C. Activates or fixes complement
D. Involved in opsonisation

Q10. Which of the following (s) is/are function (s) of Fc fragment of IgG?

A. Immunoglobulin?
B. It determines catabolic rate
C. It binds complement
D. It is related to passage of IgG across the placental barrier
E. All of these

Q11. Which is not a function of IgA Immunoglobulin?

A. Protect mucosal surfaces
B. Fix complement
C. Protect eyes
D. Agglutinate antigen

Q12. Which of the following is not true about antibody structure?

A. Antibodies have multiple identical antigen binding sites.
B. Antibodies are built from equal numbers of large (heavy) & small (light) peptide chains.
C. Antibodies are secreted and function away from the cell. They are not attached to the cell membrane.
D. The class of the antibody molecule is determined solely by its heavy chain.

Q13. Where do precursor T-lymphocytes develop into fully competent but not yet activated T-cells?

A. The thymus gland
B. The bone marrow
C. The lymph nodes
D. The spleen

Q14. Which of the following is not a feature of a secondary immune response to an antigen, when compared to the first response to the same antigen?

A. The antibody is generated faster.
B. More antibody is produced.
C. The antibody produced has greater affinity for the antigen.
D. Antibody is generated without T-cell help.

Q15. Cancer cells often have reduced amounts of cell surface proteins, including class I MHC antigens. Which of the following cells of the immune system can exploit this property to kill cancer cells?

A. Cytotoxic T-cells
B. Natural killer cells
C. Helper T-cells
D. Macrophages

Answer Key

1(A), 2(D), 3(B), 4(C), 5(A), 6(C), 7(C), 8(C), 9(B), 10(D), 11(B), 12(C), 13(A), 14(D), 15(B)

Assignments

Q1. If the variable region of the antibody is localised towards the c terminal region, do you think it is going to affect antibody function in any way?
Q2. Do you think having two dissimilar antigen binding sites on antibody molecules will be advantageous to human beings or not?
Q3. Why do you think the body needs five different classes of antibodies?
Q4. How do you believe that monoclonal antibodies will change the map of treatment and diagnostics in future?
Q5. What are few of the areas of current research in therapeutic antibodies?

References

Aisenberg, AC. and Bloch, KJ., 1972. Immunoglobulins on the surface of neoplastic lymphocytes. *New England Journal of Medicine*, *287*(6), pp. 272-276.

Akita EM, Nakai S. Immunoglobulins from egg yolk: isolation and purification. *Journal of Food Science*. 1992 May; 57(3):629-34.

Amzel LM, Poljak RJ. Three-dimensional structure of immunoglobulins. *Annual Review of Biochemistry*. 1979 Jul; 48(1):961-97.

Arnold JN, Wormald MR, Sim RB, Rudd PM, Dwek RA. The impact of glycosylation on the biological function and structure of human immunoglobulins. *Annu. Rev. Immunol.*. 2007 Apr 23; 25:21-50.

Beale D, Feinstein A. Structure and function of the constant regions of immunoglobulins. *Quarterly Reviews of Biophysics*. 1976 May; 9(2):135-80.

Capon DJ, Lasky LA, inventors; Genentech Inc, assignee. *Hybrid immunoglobulins*. United States patent US 5,428,130. 1995 Jun 27.

Dorrington KJ, Tanford C. Molecular size and conformation of immunoglobulins. In *Advances in Immunology* 1970 Jan 1 (Vol. 12, pp. 333-381). Academic Press.

Eikelenboom P, Stam FC. Immunoglobulins and complement factors in senile plaques. *Acta neuropathologica*. 1982 Jun 1; 57(2-3):239-42.

Ey PL, Prowse SJ, Jenkin CR. Isolation of pure IgG1, IgG2a and IgG2b immunoglobulins from mouse serum using protein *A-sepharose. Immunochemistry.* 1978 Jul 1; 15(7):429-36.

Frangione B. *Immunoglobulins and Immunogenetics and Immunodeficiency*. 1975:1.

Goldstein IM, Roos D, Kaplan HB, Weissmann G. Complement and immunoglobulins stimulate superoxide production by human leukocytes independently of phagocytosis. *The Journal of Clinical Investigation.* 1975 Nov 1; 56(5):1155-63.

Grabar P. Autoantibodies and the physiological role of immunoglobulins. *Immunology Today*. 1983 Dec 1; 4(12):337-40.

Green NM. Electron Mircroscopy of the Immunoglobulins. In *Advances in Immunology* 1969 Jan 1 (Vol. 11, pp. 1-30). Academic Press.

Harboe NM, Ingild A. Immunization, isolation of immunoglobulins and antibody titre determination. *Scandinavian Journal of Immunology.* 1983 Dec; 17:345-51.

Hijmans, W, Schuit, HRE. and Klein, F., 1969. An immunofluorescence procedure for the detection of intracellular immunoglobulins. *Clinical and Experimental Immunology*, *4*(4), p. 457.

Hill, RL, Delaney, R, Fellows Jr, RE. and Lebovitz, HE, 1966. The evolutionary origins of the immunoglobulins. *Proceedings of the National Academy of Sciences of the United States of America*, *56*(6), p. 1762.

Hurley WL. Immunoglobulins in mammary secretions. In *Advanced Dairy Chemistry—1 Proteins* 2003 (pp. 421-447). Springer, Boston, MA.

Hurley, WL. and Theil, PK, 2011. Perspectives on immunoglobulins in colostrum and milk. *Nutrients*, *3*(4), pp. 442-474.

Johansson SG, Bennich H, Wide L. A new class of immunoglobulins in human serum. *Immunology*. 1968 Feb; 14(2):265.

Korhonen H, Marnila P, Gill HS. Milk immunoglobulins and complement factors. *British Journal of Nutrition*. 2000 Nov; 84(S1):75-80.

Lemieux R, Bazin R, Néron S. Therapeutic intravenous immunoglobulins. *Molecular Immunology*. 2005 May 1; 42(7):839-48.

Lennox ES, Cohn M. Immunoglobulins. *Annual Review of Biochemistry*. 1967 Jul; 36(1):365-406.

Leslie, GA and Clem, LW. 1969. Phylogeny of immunoglobulin structure and function: III. Immunoglobulins of the chicken. *Journal of Experimental Medicine*, *130*(6), pp. 1337-1352.

Malek A, Sager R, Kuhn P, Nicolaides KH, Schneider H. Evolution of maternofetal transport of immunoglobulins during human pregnancy. *American Journal of Reproductive Immunology*. 1996 Nov; 36(5):248-55.

McClellan BH, Whitney CR, Newman LP, Allansmith M. R. Immunoglobulins in tears. *American Journal of Ophthalmology*. 1973 Jul 1; 76(1):89-101.

Mellors RC. Korngold L. The cellular origin of human immunoglobulins (γ2, γ1M, γ1A). *Journal of Experimental Medicine*. 1963 Sep 1; 118(3):387-96.

Mestecky J, Moro I, Kerr MA, Woof JM. Mucosal immunoglobulins. In *Mucosal Immunology* 2005 (pp. 153-181). Academic Press.

Milstein C, Pink JR. Structure and evolution of immunoglobulins. *Progress in Biophysics and Molecular Biology*. 1970 Jan 1; 21:209-63.

Muyldermans S, Baral TN, Retamozzo VC, De Baetselier P, De Genst E, Kinne J, Leonhardt H, Magez S, Nguyen VK, Revets H, Rothbauer U. Camelid immunoglobulins and nanobody technology. *Veterinary Immunology and Immunopathology*. 2009 Mar 15; 128(1-3):178-83.

Nezlin R. *The Immunoglobulins: Structure and Function*. Academic Press; 1998 May 6.

Papamichail M, Brown JC, Holborow EJ. Immunoglobulins on the surface of human lymphocytes. *The Lancet*. 1971 Oct 16; 298(7729):850-2.

Pernis B, Forni L, Amante L. Immunoglobulins as cell receptors. *Annals of the New York Academy of Sciences*. 1971 Dec; 190(1):420-31.

Porter RR. Structural studies of immunoglobulins. *Science*. 1973 May 18; 180(4087):713-6.

Preud'Homme JL, Seligmann M. Surface bound immunoglobulins as a cell marker in human lymphoproliferative diseases. *Blood*. 1972 Dec 1; 40(6):777-94.

Rabellino E, Colon S, Grey HM, Unanue ER. Immunoglobulins on the surface of lymphocytes: I. Distribution and quantitation. *Journal of Experimental Medicine*. 1971 Jan 1; 133(1):156-67.

Radaev S, Sun P. Recognition of immunoglobulins by Fcγ receptors. *Molecular Immunology*. 2002 May 1; 38(14):1073-83.

Raghavan M, Bjorkman PJ. Fc receptors and their interactions with immunoglobulins. *Annual Review of Cell and Developmental Biology*. 1996 Nov; 12(1):181-220.

Rowe DS, Fahey JL. A new class of human immunoglobulins: I. A unique myeloma protein. *Journal of Experimental Medicine*. 1965 Jan 1; 121(1):171-84.

Rümke P. The origin of immunoglobulins in semen. *Clinical and Experimental Immunology*. 1974 Jun; 17(2):287.

Saji F, Samejima Y, Kamiura S, Koyama M. Dynamics of immunoglobulins at the feto-maternal interface. *Reviews of Reproduction*. 1999 May 1; 4(2):81-9.

Schroeder Jr HW, Cavacini L. Structure and function of immunoglobulins. *Journal of Allergy and Clinical Immunology*. 2010 Feb 1; 125(2):S41-52.

Soothill JF, Hayes K, Dudgeon JA. The immunoglobulins in congenital rubella. *Lancet*. 1966:1385-8.

Spiegelberg HL. Biological activities of immunoglobulins of different classes and subclasses. In *Advances in Immunology* 1974 Jan 1 (Vol. 19, pp. 259-294). Academic Press.

Tomasi Jr TB. Secretory immunoglobulins. *New England Journal of Medicine*. 1972 Sep 7; 287(10):500-6.

Warner NL. Membrane immunoglobulins and antigen receptors on B and T lymphocytes. In *Advances in Immunology* 1974 Jan 1 (Vol. 19, pp. 67-216). Academic Press.

Weaver DM, Tyler JW, Van Metre DC, Hostetler DE, Barrington GM. Passive transfer of colostral immunoglobulins in calves. *Journal of Veterinary Internal Medicine*. 2000 Nov 1; 14(6):569-77.

In: Immunoglobulins …
Editor: Manzoor Ahmad Mir

ISBN: 978-1-53616-903-4

Chapter 2

TYPES OF IMMUNOGLOBULINS

Manzoor A Mir*[*], Umar Mehraj, Safura Nisar and Hina Qayoom
Department of Bioresources, School of Biological Sciences,
University of Kashmir, India

ABSTRACT

The immune system of our body relies on two arms, the cell mediated immunity and the humoral immune response. The cell mediated immunity comprises the cells of immune system particularly T cells, while the humoral arm comprises the antibody secreting B cells. B cells express and secrete antibodies of different types and grouped into classes. Immunoglobulin's (antibodies) are grouped into five classes: IgM, IgG, IgA, IgE and IgD distinguished on the basis of heavy chain found in each molecule. Immunoglobulin IgG is separated into four subclasses: IgG1, IgG2, IgG3 and IgG4 while as IgA is separated into two classes: IgA1 and IgA2. IgG is the most abundant almost 80% present in blood circulation. IgM is the largest (Pentamer) and the first immunoglobulin produced. IgA is present in mucosal secretions such as: saliva, tears etc, and most abundantly present in colostrum of milk providing first immunity to the infants after birth. IgD is a trace antibody that is present on the surface of B-cells involved in stimulating and suppressing B-cell formation. IgE is found in very low concentration in human serum and it increases during allergic reactions.

Keywords: immunoglobulin, Pentamer, antibody, saliva, maternal antibody, colostrum

---

[*] Corresponding Author: Dr Manzoor Ahmad Mir (Ph. D), Department of Bioresources, School of Biological Sciences, University of Kashmir, Srinagar J&K, India-190006, Contact:-+919622901319. Email: drmanzoor@kashmiruniversity.ac.in.

OBJECTIVES

- Discuss different types of immunoglobulins.
- Explain the functions of different immunoglobulins.
- Describe the structure of immunoglobulins.
- Distinguish between different types of immunoglobulins.
- Give an outline of the antigenic determinants

INTRODUCTION

Immunoglobulin's (antibodies) are assembled into five major classes; IgM, IgG, IgA, IgD and IgE with unique antigenic properties present on their heavy chains (constant region) each designated with its own class of heavy chain-μ, γ, α, δ and ε respectively. In addition, immunoglobulin IgG and IgA are further divided into subclasses based on their antigenic differences within each of the five classes. In humans, IgG has four subclasses; IgG1, IgG2, IgG3, IgG4 having $^{\gamma1, \gamma2, \gamma3}$ and $^{\gamma4}$ heavy chains, respectively while as IgA has two subclasses; IgA1, IgA2. IgM, IgD and IgE have no subclasses. Each class and subclass has its own characteristic properties. Antibodies are antigen binding proteins that are present on the B-cell membrane and secreted by plasma cells. Membrane-bound antibody confers antigenic specificity to the B cells; antigen-specific proliferation of B-cell clones is elicted by the interaction of membrane antibody with antigen. Secreted antibodies circulate in the blood, where they serve as the effectors of humoral immunity by searching out and neutralizing antigens or marking them for elimination. All antibodies have same structural features that bind to antigen, and participate in a limited number of effector functions. The antibodies produced in response to a particular antigen are thus heterogeneous. Most antigens are complex and contain many different antigenic determinants. The immune system usually responds by producing antibodies to several epitopes on the antigen. This response requires the recruitment of several clones of B cells. Their outputs are monoclonal antibodies, each of which specifically binds a single antigenic determinant. Together, these monoclonal antibodies make up the polyclonal and heterogeneous serum antibody response to an immunizing antigen. In 1890, von Behring and Kitasato reported the existence of an agent in the blood that could neutralize diphtheria toxin. The following year, reference was made to antibodies, in studying the ability of an agent to discriminate between two immune substances. The substance which induces the production of an antibody is referred to as antigen. Thus, an antibody and its antigen represent a classic of reiteration.

In 1939, Tiselius and Kabat used electrophoresis to separate immunized serum into albumin, alpha-globulin, beta-globulin, and gamma-globulin fractions. Absorption of serum against the antigen depleted the gamma-globulin fraction, yielding gamma globulin, immunoglobulin (Ig) and columns for sizing were used to separate immunoglobulins into heavy (IgM) regular (IgA, IgE, IgD, IgG), and light (light chain dimmers).

Many years of investigations into the structure and function of immunoglobulin only served to emphasize the complex nature of a protein, were receptors bind to a limited and defined set of ligands. An individual immunoglobulin also binds to a limited and defined set of ligands, immunoglobulins as a population can bind to a virtually unlimited array of antigens sharing little or no similarity. This property of adjustable binding depends on a complex array of mechanisms that alter the DNA of individual B cells. Immunoglobulins also serve two purposes: that of cell-surface receptors for antigen which permit cell signaling and cell activation and that of soluble effector molecules which can individually bind and neutralize antigens at a distance.

Classes of Immunoglobulin's

IgM

IgM is the largest and first antibody produced by B-cells. It accounts for almost 5-10% of total antibodies in human serum with an average serum concentration of 1.5 mg/ml. IgM occurs as a pentamer, composed of 5 Y-shaped units linked by disulphide bonds. It appears to be the first type of immunoglobulin to arise during evolution, being most abundant in primitive vertebrates. IgM is the first antibody produced in response to an antigen in the body and is most effective against microorganisms which enter the blood and activate phagocytosis. It cannot traverse the blood vessels, so it's only restricted to the blood stream. It's most efficient in agglutination and complement fixation. It's also called as *"natural antibody"* as it increases the production of IgG. It does not cross placenta.

Monomeric IgM, with a molecular weight of 180,000, is expressed as membrane-bound antibody on B cells. IgM as secreted from plasma cells as pentamer which have five monomeric units are held together by disulphide bonds that link their carboxyl-terminal heavy chain domains (C_H4/C_H4) and their C_H3/C_H3 domains. The five monomer subunits are arranged with their Fc regions in the centre of the pentamer and the ten antigen-binding sites on the periphery of the molecule. Each pentamer contains an additional Fc-linked polypeptide called the *J (joining) chain,* which is a disulphide-bonded to the carboxyl-terminal cysteine residue of two of the ten chains. The J chain appears to be required for polymerization of the monomers to form pentameric IgM, and it is added just before the secretion of the pentamer. The presence of the J chain allows IgM to bind to receptors on secretory cells, which transport it across epithelial linings to enter the external secretions that bath mucosal surfaces. Although IgA is the major isotype found in these secretions, IgM plays an important role as a secretory immunoglobulin. It's the first immunoglobulin class produced in a primary response to an antigen, and it is also the first immunoglobulin to be synthesized by the neonate. Because of its pentameric structure with 10 antigen-binding sites, serum this immunoglobulin has a higher valency than the other isotypes. IgM molecule can bind 10 small hapten molecules; however, because of steric hindrance, only 5 or fewer molecules of larger antigens can be bound simultaneously. Because of its high valency, pentameric IgM is more efficient than other isotypes in binding antigens with many repeating epitopes such as viral particles and red blood cells (RBCs). For example, when RBCs are incubated with specific antibody, they clump together into large aggregates in a process called agglutination. It takes 100 to 1000 times more molecules of IgG than of IgM to achieve the same level of agglutination. A similar phenomenon occurs with viral particles, less IgM than IgG is required to neutralize viral infectivity. IgM is more effective than IgG at activating the complement. Complement activation requires 2 Fc regions in close proximity, and the pentameric structure of a single molecule of IgM fulfils this requirement. Because of its large size, IgM does not diffuse well and therefore is found in very low concentrations in the intercellular tissue fluids.

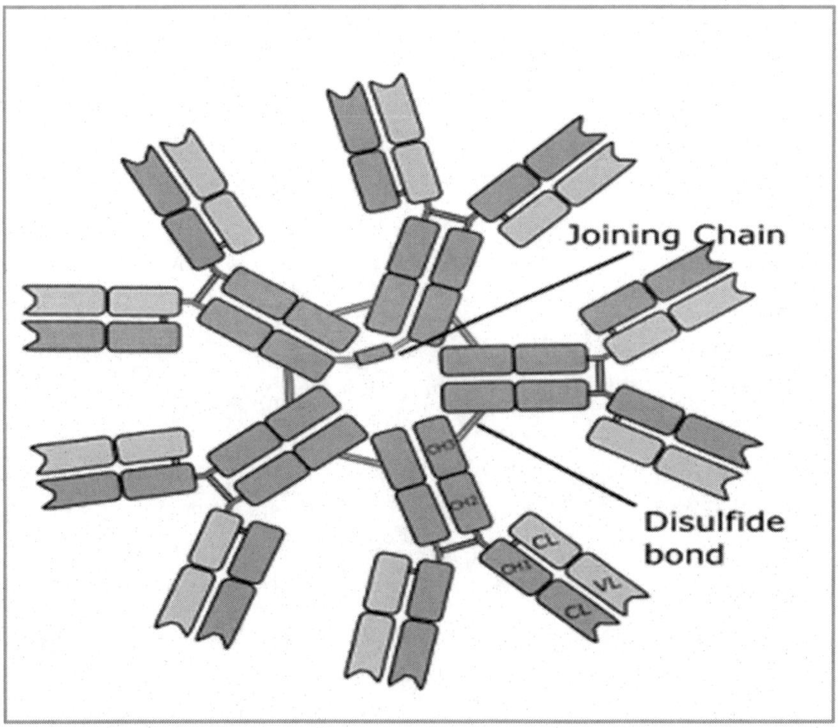

Image source: https://www.antibodies-online.com/.

Figure 2.1. Immunoglobulin IgM Pentamer structure.

IgM is the first immunoglobulin expressed during B cell development. Naïve B cells express monomeric IgM on their surface and associate with CD79a and CD79b, polypeptide chains that participate in IgM cell signaling. Upon maturation and antigenic stimulation, multimeric (usually pentameric, rarely hexameric) IgM, in which single IgM units link to each other by disulfide bonds in the CH4 region, is secreted. The pentamer also contains a polypeptide chain, the J-chain, which is bound to two of the monomers by means of a disulfide bond. The J-chain facilitates secretion at mucosal surfaces. Generally, while monomeric IgM molecules have low affinity due to their immaturity; high avidity can be attained by means of multimeric interactions between the pentameric secreted antibody and the antigen, especially if that antigen contains multiple repeating epitopes itself. IgM functions by opsonizing (coating) antigen for destruction and fixing complement. The pentameric nature of the antibody renders it very efficient in this process. IgM antibodies are associated with a primary immune response and are frequently used to diagnose acute exposure to an immunogen or pathogen. IgM is expressed early in B cell development and the μ heavy chain associates with VH and VL that have not undergone much somatic mutation in response to antigen. As a result, IgM antibodies tend to be more poly-reactive than other isotypes, which allows IgM-bearing B-cells to respond quickly to a variety of antigens. These relatively low-affinity IgM antibodies are also called natural antibodies. Some of these natural antibodies not only participate as a first line of defense, but also play a role in immunoregulation. Natural antibodies may react with auto antigens, but are rarely responsible for autoimmune disease or pathogenesis. Pathogenic auto antibodies tend to drawn away the somatically mutated, high affinity IgG population.

IgG

IgG is the most abundant antibody constituting almost 80% in human serum. A predominant antibody in the secondary response. It's also called as *"maternal antibody"* because it's the only antibody that crosses the placenta and confers immunity to the foetus for the first month after birth. It prevents infection spread in the body and opsonises i.e., enhances phagocytosis. It fixes complement and binds to soluble antigens, exotoxins and viruses most actively. It is composed of a single Y-shaped monomer divided into four subclasses: IgG1, IgG2, IgG3 and IgG4 distinguished by differences in γ-chain sequence. The amino acid sequences that distinguish the four IgG subclasses are encoded by different germ-line CH genes, whose DNA sequences are 90%–95% homologous. The structural characteristics that distinguish these subclasses from one another are the size of the hinge region and the number and position of the interchain disulphide bonds between the heavy chains. The subtle amino acid differences between subclasses of IgG affect the biological activity of the molecule: IgG1, IgG3, and IgG4 readily cross the placenta and play an important role in protecting the developing foetus. IgG3 is the most effective complement activator, followed by IgG1; IgG2 is less efficient, and IgG4 is not able to activate complement at all. IgG1 and IgG3 bind with high affinity to Fc receptors on phagocytic cells and thus mediate opsonization. IgG4 has an intermediate affinity for Fc receptors, and IgG2 has an extremely low affinity.

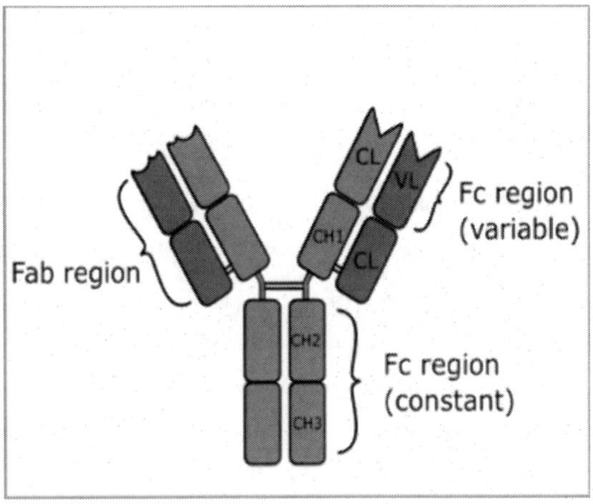

Image source: https://www.antibodies-online.com/.

Figure 2.2. Immunoglobulin IgG.

IgG is the predominant isotype found in the body. It has the longest serum half-life of all immunoglobulin isotypes. It is also the most extensively studied class of immunoglobulins. Based on structural, antigenic and functional differences in the constant region of the heavy chain, and CH1 and CH3 in particular, four IgG subclasses (IgG1, IgG2, IgG3 and IgG4) were identified. These IgG subclasses were numbered in reference to the rank order (IgG1>IgG2>IgG3>IgG4) of the serum levels of these antibodies in the blood of normal, healthy individuals living in an affluent western European environment. The differences in the CH domains affect antibody flexibility and functional affinity, some of which facilitate cooperative interactions with multivalent antigens. The mobility or flexibility of the F (ab) and the Fv portions of the antibody are primarily controlled by the CH1 domain and hinge region. The IgG subclasses exhibit different functional activities. Activation of the complement cascade is an important means of clearance of opsonized pathogens. While IgG4 is the only subclass that fails to fix complement; affinity for C1q, which is the first component of the complement pathway and which binds to the CH2 domain of IgG, differs between members of the other three IgG subclasses, with IgG3>IgG1>IgG2. There are also defined on the differences in the affinity to the three classes of FcγR (I, II and III). IgG1 and IgG3 bind to all three FcγR classes. IgG4 binds only FcγRII and III, albeit significantly weaker than the binding of IgG1. And IgG2 binds only to FcγRII.

There are also similarities within the subclasses such as trans placental transport and participation in the secondary immune response. Within the secondary antibody response, there is skewing in the predominant subclass that is induced. For example, IgG1 and IgG3 antibodies are generally induced in response to protein antigens whereas IgG2 and IgG4 are associated with polysaccharide antigens. The response to a given antigen can also result in a skewed IgG subclass response and this is frequently as a source of investigation as to correlate the protection or for the design of vaccines.

Specific subclasses may be associated with individual disease processes. For example, in pemphigus vulgaris, a mucocutaneous blistering disease, IgG4 antibodies to desmoglein 3 are pathogenic whereas first degree relatives with IgG1 auto antibodies to the same protein show no evidence of the disease.

IgG antibodies also contribute directly to an immune response including neutralization of toxins and viruses. Here again, IgG subclass impacts on the outcome of this interaction. In HIV, it has been shown that IgG3 antibodies may be more effective at neutralizing virus than IgG1 antibodies, presumably through an increase in antibody flexibility improving antibody access or inducing changes in the oligomer structure of the virus.

IgA

IgA is the heaviest and second most abundant antibody found in body secretions including breast milk, saliva, tears and mucus of the bronchial, genitourinary and digestive tracts. IgA occurs as a single monomer predominantly in serum or double unit (dimer) present in external secretions, called secretory IgA, held together by j-chain polypeptide. The IgA class of immunoglobulin contains two subclasses; IgA1 (93%) and IgA2 (7%). It constitutes almost 10-15% of total Ig in serum which is most abundant in new born. It does not fix complement and binds with the bacterial antigens present on their surface and remove them. IgA prevents attachment of microorganisms and fight against any foreign substances entering into the body cells. The production of secretory IgA is maximum among other immunoglobulin's (Igs). It's the major immunoglobulin found in the colostrum of milk in the nursing mothers and provides a major source of protection against pathogens to the new born after birth. Although IgA constitutes only 10%–15% of the total immunoglobulin in serum, it is the predominant immunoglobulin class in external secretions such as breast milk, saliva, tears, and mucus of the bronchial, genitourinary, and digestive tracts. In serum, IgA exists primarily as a monomer, but polymeric forms (dimers, trimers, and some tetramers) are sometimes seen, all containing a J-chain polypeptide. The IgA of external secretions, called secretory IgA, consists of a dimer or tetramer, a J-chain polypeptide, and a polypeptide chain called secretory component. This secretory component is derived from the receptor that is responsible for transporting polymeric IgA across cell membranes. The J-chain polypeptide in IgA is identical to that found in pentameric IgM and serves a similar function in facilitating the polymerization of both serum IgA and secretory IgA. The secretory component is a 70,000-MW polypeptide produced by epithelial cells of mucous membranes. It consists of five immunoglobulin-like domains that bind to the Fc region domains of the IgA dimer. This interaction is stabilized by a disulphide bond between the fifth domain of the secretory component and one of the chains of the dimeric IgA. The daily production of secretory IgA is greater than that of any other immunoglobulin class. IgA-secreting plasma cells are concentrated along mucous membrane surfaces. Along the jejunum of the small intestine, for example, there are more than 2.5 -10^{10} IgA-secreting plasma cells—a number that surpasses the total plasma cell population of the bone marrow, lymph, and spleen combined. Every day, a human secretes from 5 g to 15 g of secretory IgA into mucous secretions. The plasma cells that produce IgA preferentially move to sub epithelial tissue, where the secreted IgA binds tightly to the receptor for polymeric immunoglobulin molecules. This *poly-Ig receptor* is expressed on the basolateral surface of most mucosal epithelia like; the lining of the digestive, respiratory, and genital tracts and on glandular epithelia in the mammary, salivary, and lacrimal glands. After polymeric IgA binds to the poly-Ig receptor, the receptor-IgA complex is transported across the epithelial barrier to the lumen. Transport of the receptor-IgA complex involves receptor-mediated endocytosis into coated pits and directed transport of the vesicle across the epithelial cell to the luminal membrane, where the vesicle fuses with the plasma membrane. The poly-Ig receptor is then cleaved enzymatically from the membrane and becomes the secretory component, which is bound to and released together with polymeric IgA into the mucous secretions. The secretory component masks sites susceptible to protease cleavage in the hinge region of secretory IgA, allowing the polymeric molecule to exist longer in the protease-rich mucosal environment. Pentameric IgM is also transported into mucous secretions by this mechanism, although it accounts for a much lower percentage of antibody in the mucous secretions than does IgA. The poly-Ig receptor interacts with the J chain of both polymeric IgA and IgM antibodies.

Secretory IgA serves an important effector function at mucous membrane surfaces, which are the main entry sites for most pathogenic organisms. Because it is polymeric, secretory IgA can cross-link large antigens with multiple epitopes. Binding of secretory IgA to bacterial and viral surface antigens prevents attachment of the pathogens to the mucosal cells, thus inhibiting viral infection and bacterial colonization. Complexes of secretory IgA and antigen are easily entrapped in mucus and then eliminated by the ciliated epithelial cells of the respiratory tract or by peristalsis of the gut. Secretory IgA has been shown to provide important lines of defense against bacteria such as *Salmonella, Vibrio cholerae,* and *Neisseria gonorrhoeae* and viruses such as polio, influenza, and reovirus. Breast milk contains secretory IgA and many other molecules that help protect the new-born against infection during the first month of life. Because the immune system of infants is not fully functional, breast-feeding plays an important role in maintaining the health of new-born.

Generally, a monomer in the serum, IgA at the mucosa, termed as secretory IgA (IgA), is a dimer (sometimes trimer and tetramer) associated with a J-chain and another polypeptide chain, the secretory component. Similar to IgM, the CH3 domains of IgA have short tail pieces to which the J-chain binds via disulfide bonds whereas the secretory component is disulfide bonded to one of the CH2 domains of the dimer. There are two subclasses of IgA, IgA1 and IgA2, whose structures differ mainly in their hinge regions. IgA1 has a longer hinge region with a duplicated stretch of amino acids that is lacking in IgA2. This elongated hinge region increases the sensitivity of IgA1 to bacterial proteases in spite of partial protection by glycans. Such increased protection against protease digestion may explain why IgA2 predominates in the many mucosal secretions, such as the genital tract, whereas more than 90% of serum IgA is in the form of IgA1.

IgA is critical at protecting mucosal surfaces from toxins, virus and bacteria by direct neutralization or by prevention of binding to the mucosal surface. Intracellular IgA may also be important in preventing bacterial or viral infection or pathogenesis. The polymeric nature of secretory IgA may be particularly important. For example, polymeric IgA is more effective than monomeric IgA at preventing *Clostridium difficile* toxin A damage to epithelial cells. While complement fixation by IgA does not appear to be a major effector mechanism at the mucosal surface, the IgA receptor is expressed on neutrophils which may be activated to mediate ADCC locally. As described above, specific bacteria may be trapped by the glycans on IgA. Finally, it has been proposed that IgA may also act as a potentiator of the immune response in intestinal tissue by uptake of antigen to dendritic cells.

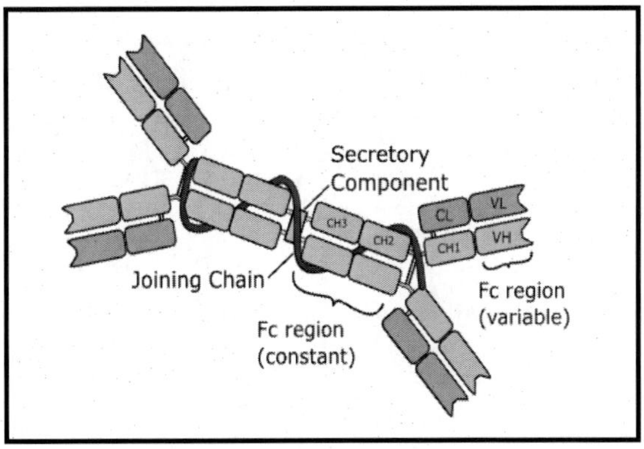

Image source: https://www.antibodies-online.com/.

Figure 2.3. Immunoglobulin IgA, Dimer-Secretory component.

IgE

IgE is composed of a single Y-shaped monomer present in minute concentration in blood. This type of immunoglobulin is concentrated in mucous membranes, skin and lungs. IgE is associated with the body's allergic responses and mediate immediate hypersensitivity. It binds tightly with tissue mast cells and blood basophils which release histamine and cause allergy. It does not fix complement. It's a main host defence against helminths infection. The potent biological activity of IgE allowed it to be identified in serum despite its extremely low average serum concentration (0.3-g/ml). IgE antibodies mediate the immediate hypersensitivity reactions that are responsible for the symptoms of hay fever, asthma, hives, and anaphylactic shock. The presence of a serum component responsible for allergic reactions was first demonstrated in 1921 by K. Prausnitz and H. Kustner, who injected serum from an allergic person intra-dermally into a non-allergic individual. When the appropriate antigen was later injected at the same site, a wheal and flare reaction (analogous to hives) developed. This reaction, called the *P-K reaction* (named for its originators, Prausnitz and Kustner), was the basis for the first biological assay for IgE activity. Actual identification of IgE was accomplished by K. And T. Ishizaka in 1966. They obtained serum from an allergic individual and immunized rabbits with it to prepare anti-isotype anti-serum. The rabbit anti-serum was then allowed to react with each class of human antibody known at that time (i.e., IgG, IgA, IgM, and IgD). In this way, each of the known anti-isotype antibodies was precipitated and removed from the rabbit anti-serum. What remained was an anti-isotype antibody specific for an unidentified class of antibody. This antibody turned out to completely block the P-K reaction. The new antibody was called IgE. IgE binds to Fc receptors on the membranes of blood basophils and tissue mast cells. Cross-linkage of receptor bound IgE molecules by antigen (allergen) induces basophils and mast cells to translocate their granules to the plasma membrane and release their contents to the extracellular environment, a process known as degranulation. As a result, varieties of pharmacologically active mediators are released and give rise to allergic manifestations. Localized mast-cell degranulation induced by IgE also may release mediators that facilitate a buildup of various cells necessary for antiparasitic defense.

Though it is present at the lowest serum concentration with the shortest half-life, IgE is a very potent immunoglobulin. It is associated with hypersensitivity and allergic reactions as well as the response to parasitic worm infections. IgE binds with extremely high affinity to the FcεRI which is expressed on mast cells, basophils, Langerhans cells and eosinophils. Circulating IgE up regulates FcεR expression on these cells. The combination of strong binding and up regulation of FcεR expression contributes to the remarkable potency of this immunoglobulin.

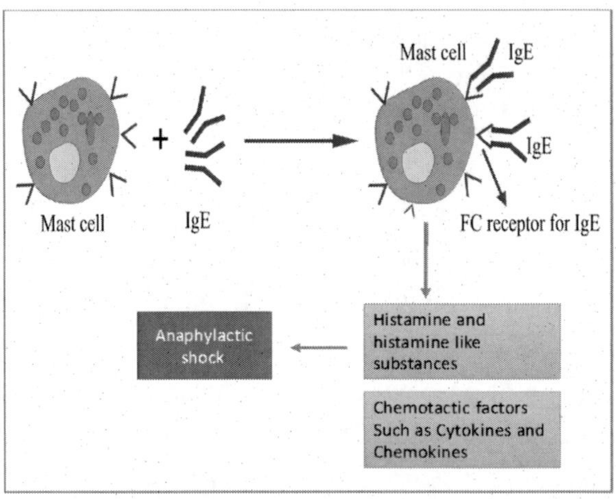

Image source, https://www.antibodies-online.com/.

Figure 2.4. Immunoglobulin IgE.

Recently, there has been the development of anti-IgE antibodies as therapy for allergy and asthma. Antibodies are designed to target free IgE as well as B cells with membrane bound IgE but not IgE bound to FcεR as the latter would stimulate degranulation and the release of inflammatory mediators. IgE has a much lower affinity for FcεRII or CD23, which is expressed both on the same cells as FcεRI as well as B cells, NK cells and platelets.

IgD

IgD is composed of a single Y-shaped monomer, present in lowest concentration in blood circulation. It constitutes almost 0.2% of the total Ig in serum. Together with the IgM, it's the major membrane bound immunoglobulin expressed by B-cells. This functions as antigen receptors present on the surface of B-cells and in serum. Help in activating B-cells. No other biological function has been identified for IgD. IgD was first discovered when a patient developed a multiple myeloma whose myeloma protein failed to react with anti-isotype antisera against the then-known isotypes: IgA, IgM, and IgG. When rabbits were immunized with this myeloma protein, the resulting antisera were used to identify the same class of antibody at low levels in normal human serum. The new class, called IgD, has a serum concentration of 30 g/ml and constitutes about 0.2% of the total immunoglobulin in serum. IgD, together with IgM, is the major membrane bound immunoglobulin expressed by mature B cells, and its role in the physiology of B cells is still under study. For IgD there is no functional biological effector.

Circulating IgD is found at very low levels in the serum with a short serum half-life which may be attributed to sensitivity of the molecule, with the hinge region in particular, to proteolysis. The function of circulating IgD is unclear, as it is not known to participate in the major antibody effector mechanisms. Circulating IgD can react with specific bacterial proteins, such as the IgD binding protein of *Moraxella catarrhalis*, independent of the variable regions of the antibody. Binding of these bacterial proteins to the constant region of IgD results in B cell stimulation and activation.

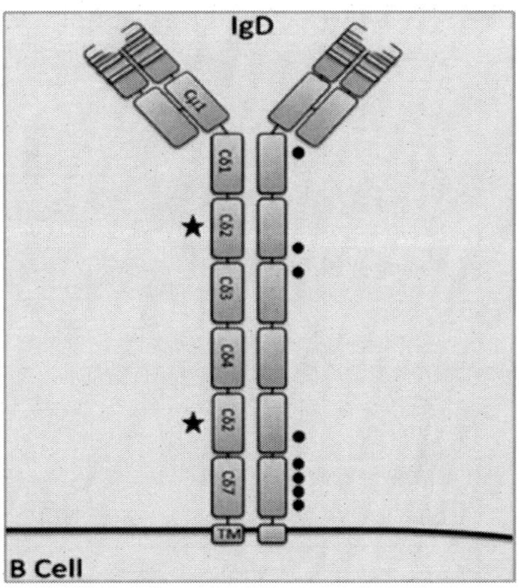

Image source: https://www.antibodies-online.com/.

Figure 2.5. Immunoglobulin IgD.

Table 2.1. Different classes of immunoglobulins

Characteristics	IgM	IgG	IgA	IgE	IgD
Structure	Pentamer	Monomer	Monomer or Dimer	Monomer	Monomer
Number of antigen binding sites	10	2	4	2	2
Percentage of total antibody in serum	5-10%	80%	10-15%	0.002%	0.2-1%
Crosses placenta	No	Yes	No	No	No
Fixes complement	Yes	Yes	No	No	No
Function	Main antibody of primary responses, serves as first line of defence.	Main blood antibody of secondary responses neutralizes toxins, opsonisation.	Secreted into mucus, tears, saliva, and colostrum.	Antibody of allergy and antiparasitic activity.	B-cell receptor.
Location	Blood, lymph, B-cell surface (as monomer).	Blood, lymph, intestine.	Secretions (saliva, tears, mucus, colostrum), blood, lymph.	Bound to mast cells and basophil cells, blood.	B-cell surface, blood, lymph.
Heavy chain	μ	Γ	α	ε	δ
Light chain	κ or λ	κ or λ	κ or λ	κ or λ	κ or λ

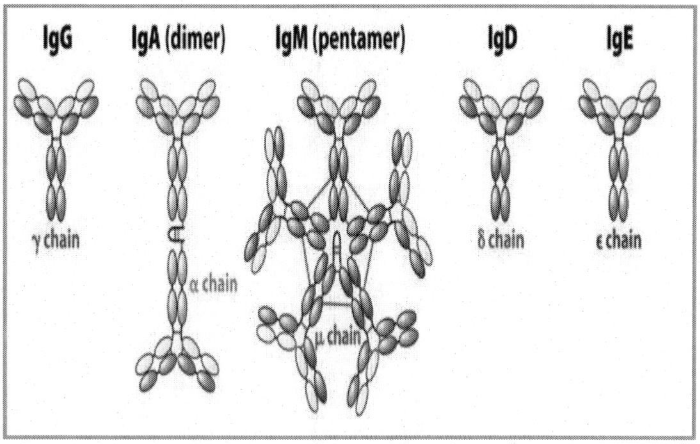

Image source: Immunoglobulin's-properties, structure and functions by Arthur westmann.

Figure 2.6. Molecular depiction of different types of antibodies.

While the membrane-bound form of IgD that has been more extensively studied remains poorly understood in function. Similar to IgM, membrane bound IgD is associated with CD79a and CD79b for signaling. IgD is expressed on the membranes of B cells when they leave the bone marrow and populate secondary lymphoid organs. Most IgD^+ B cells also co-express IgM and both participate in B cell receptor signaling through CD79a and CD79b. IgD can replace IgM and vice versa on IgD^+IgM^+ B cells. It has been proposed that membrane bound IgD regulates B cell fate at specific developmental stages through changes in activation status.

ANTIBODIES AS ANTIGENS

Immunoglobulins are protein molecules and therefore can be Immunogens too. Immunologists have exploited this property to generate anti-immunoglobulins and this anti-antibody is used to characterize epitopes present on antibody. Immunoglobulin-antigen interactions typically take place between the *paratope*, the site on the Ig at which the antigen binds, and the *epitope*, which is the site on the antigen that is bound. *In vivo*, immunoglobulins tend to be produced against intact antigens in soluble form, and thus preferentially identify surface epitopes that can represent conformational structures that are noncontiguous in the antigen's primary sequence. This ability to identify component parts of the antigen independently of the rest makes it possible for the B cell to discriminate between two closely related antigens, each of which can be viewed as a collection of epitopes. It also permits the same antibody to bind divergent antigens that share equivalent or similar epitopes, a phenomenon referred to as *cross-reactivity*.

Immunization of heterologous species with monoclonal antibodies (or a restricted set of immunoglobulins) allowed the identification of both common and individual immunoglobulin antigenic determinants. Individual determinant(s), termed *idiotype(s)*, are contained within V domains. Common determinants, termed *isotypes*, are specific for the constant portion of the antibody and allow grouping of immunoglobulins into recognized classes, with each class defining an individual type of C domain. Determinants common to subsets of individuals within a species, yet differing between other members of that species, are termed *allotypes* and define inherited polymorphisms that result from gene alleles

The antigenic determinants or epitopes on immunoglobulins have been classified into three major types:

- Isotypic

The prefix 'Iso' means same in all members of the same species. Isotypic determi-nant is present in the constant region of the heavy chain. Isotypic determinant is characteristic for a particular species and is present in all members of the same species.

Therefore, all individuals of the same species have the same isotype. But the isotypic determinants between different species are not the same. Consequently, when an antibody from one species is injected into another species, the injected antibodies are recognized as foreign, resulting in the induction of antibodies (anti-antibodies) against the isotypic determinants of the injected antibodies. The antiantibodies react with isotypic determinants in the constant region of the injected antibodies.

- Allotypic

The prefix 'Alio' means that different in individuals of the same species. Within a species multiple allele exist for certain isotypic genes; consequently, some members within the same species have different antigenic determinants called allotypic determinants.

Therefore, injection of antibodies from one animal to another animal of the same species induces antiantibodies (anti-allotypic antibodies) formation against the allotypic determinants of the injected antibodies, provided that the two animals differ in their allotypic determinants. Genes in the constant region of immunoglobulin code the allotypic determinants. In humans allotypic determinants are found in the four IgG subclasses, IgA2 subclass, and k light chain.

- Idiotypic

Antigen-binding site in antibody molecule is formed by the amino acids of the VH and VL chains. The amino acids of the antigen binding site also act as immunogen. The antigenic determinants of the VH and VL region are called idiotypic determinants or idiotopes. One antibody molecule has many idiotopes in the antigen-binding site. The sum of the individual idiotopes in an antibody molecule is called the idiotype of the antibody. The idiotopes are further designated alpha, beta, and gamma idiotopes.

FREQUENTLY ASKED QUESTIONS

Q1. What are antibodies?

Answer: Antibody, also called immunoglobulin, a protective protein produced by the immune system in response to the presence of a foreign substance, called an antigen. Antibodies recognize and latch onto antigens in order to remove them from the body. A wide range of substances are regarded by the body as antigens, including disease-causing organisms and toxic materials such as insect venom. Antibodies have a Y shaped structure made up of two heavy and two light chains. Antibodies have set of complimentary determining regions (CDRs)in the variable region which form paratope and binds specifically with the epitope of the antigen during antibody antigen interaction. Antibody can be polyclonal or monoclonal. There are five classes of antibodies namely IgA, IgD, IgE, IgG and IgM depending upon the presence of heavy chain in the antibody molecule.

Q2 Are antibodies and immunoglobulins the same?

Answer: Immunoglobulin and antibody are disease-fighting proteins developed by most vertebrates in response to a particular antigen. Both immunoglobulin and antibody are glycoproteins. Both of them contain similar regions in their molecules. The main difference between immunoglobulin and antibody is that immunoglobulin has a transmembrane domain in order to be attached to the plasma membrane whereas antibody does not have a transmembrane domain.

Q3. What do you understand by idiotypic determinants?

Answer: Antigen-binding site in antibody molecule is formed by the amino acids of the VH and VL chains. The amino acids of the antigen binding site also act as immunogen. The antigenic determinants of the VH and VL region are called idiotypic determinants or idiotopes. One antibody molecule has many idiotopes in the antigen-binding site. The sum of the individual idiotopes in an antibody molecule is called the idiotype of the antibody. The idiotopes are further designated alpha, beta, and gamma idiotopes.

Q4. What do you understand by allotypic determinants?

Answer: The prefix 'Alio' means that different in individuals of the same species. Within a species multiple allele exist for certain isotypic genes; consequently, some members within the same species have different antigenic determinants called allotypic determinants.

Therefore, injection of antibodies from one animal to another animal of the same species induces antiantibodies (anti-allotypic antibodies) formation against the allotypic determinants of the injected antibodies, provided that the two animals differ in their allotypic determinants. Genes in the constant region of immunoglobulin code the allotypic determinants. In humans allotypic determinants are found in the four IgG subclasses, IgA2 subclass, and k light chain.

Q5. What do you understand by isotypic determinants?

Answer: The prefix 'Iso' means same in all members of the same species. Isotypic determi-nant is present in the constant region of the heavy chain. Isotypic determinant is characteristic for a particular species and is present in all members of the same species.

Therefore, all individuals of the same species have the same isotype. But the isotypic determinants between different species are not the same. Consequently, when an antibody from one species is injected into another species, the injected antibodies are recognized as foreign, resulting in the induction of antibodies (anti-antibodies) against the isotypic determinants of the injected antibodies. The antiantibodies react with isotypic determinants in the constant region of the injected antibodies.

Q6. Give an account of IgE antibody?

Answer: IgE is composed of a single Y-shaped monomer present in minute concentration in blood. This type of immunoglobulin is concentrated in mucous membranes, skin and lungs. IgE is associated with the body's allergic responses and mediate immediate hypersensitivity. It binds tightly with tissue mast cells and blood basophils which release histamine and cause allergy. It does not fix complement. It's a main host defence against helminths infection. The potent biological activity of IgE allowed it to be identified in serum despite its extremely low average serum concentration (0.3-g/ml). IgE antibodies mediate the immediate hypersensitivity reactions that are responsible for the symptoms of hay fever, asthma, hives, and anaphylactic shock.

Q7. Give an account of IgD immunoglobulin?

Answer: IgD is composed of a single Y-shaped monomer, present in lowest concentration in blood circulation. It constitutes almost 0.2% of the total Ig in serum. Together with the IgM, it's the major membrane bound immunoglobulin expressed by B-cells. This functions as antigen receptors present on the surface of B-cells and in serum. Help in activating B-cells. No other biological function has been identified for IgD. IgD was first discovered when a patient developed a multiple myeloma whose myeloma protein failed to react with anti-isotype antisera against the then-known isotypes: IgA, IgM, and

IgG. When rabbits were immunized with this myeloma protein, the resulting antisera were used to identify the same class of antibody at low levels in normal human serum.

Q8. Give an account of IgG antibody?

Answer: IgG is the most abundant antibody constituting almost 80% in human serum. A predominant antibody in the secondary response. It's also called as "maternal antibody" because it's the only antibody that crosses the placenta and confers immunity to the foetus for the first month after birth. It prevents infection spread in the body and opsonises i.e., enhances phagocytosis. It fixes complement and binds to soluble antigens, exotoxins and viruses most actively. It is composed of a single Y-shaped monomer divided into four subclasses: IgG1, IgG2, IgG3 and IgG4 distinguished by differences in γ-chain sequence. The amino acid sequences that distinguish the four IgG subclasses are encoded by different germ-line CH genes, whose DNA sequences are 90%–95% homologous. The structural characteristics that distinguish these subclasses from one another are the size of the hinge region and the number and position of the interchain disulfide bonds between the heavy chains. The subtle amino acid differences between subclasses of IgG affect the biological activity of the molecule: IgG1, IgG3, and IgG4 readily cross the placenta and play an important role in protecting the developing foetus. IgG3 is the most effective complement activator, followed by IgG1; IgG2 is less efficient, and IgG4 is not able to activate complement at all. IgG1 and IgG3 bind with high affinity to Fc receptors on phagocytic cells and thus mediate opsonization. IgG4 has an intermediate affinity for Fc receptors, and IgG2 has an extremely low affinity.

Q9. Give an account of IgM?

Answer: IgM is the largest and first antibody produced by B-cells. It accounts for almost 5-10% of total antibodies in human serum with an average serum concentration of 1.5 mg/ml. IgM occurs as a pentamer, composed of 5 Y-shaped units linked by disulphide bonds. It appears to be the first type of immunoglobulin to arise during evolution, being most abundant in primitive vertebrates. IgM is the first antibody produced in response to an antigen in the body and is most effective against microorganisms which enter the blood and activate phagocytosis. It cannot traverse the blood vessels, so it's only restricted to the blood stream. It's most efficient in agglutination and complement fixation. It's also called as "natural antibody" as it increases the production of IgG. It does not cross placenta.

Q10. What do you understand by antibody as antigen concept?

Answer: Immunoglobulins are protein molecules and therefore can be Immunogens too. Immunologists have exploited this property to generate anti-immunoglobulins and this anti-antibody is used to characterize epitopes present on antibody. Immunoglobulin-antigen interactions typically take place between the *paratope*, the site on the Ig at which the antigen binds, and the *epitope*, which is the site on the antigen that is bound. *In vivo*, immunoglobulins tend to be produced against intact antigens in soluble form, and thus preferentially identify surface epitopes that can represent conformational structures that are noncontiguous in the antigen's primary sequence. This ability to identify component parts of the antigen independently of the rest makes it possible for the B cell to discriminate between two closely related antigens, each of which can be viewed as a collection of epitopes. It also permits the same antibody to bind divergent antigens that share equivalent or similar epitopes, a phenomenon referred to as *cross-reactivity*.

Q11. Describe the structure of IgA?

Answer: IgA is the heaviest and second most abundant antibody found in body secretions including breast milk, saliva, tears and mucus of the bronchial, genitourinary and digestive tracts. IgA occurs as a single monomer predominantly in serum or double unit (dimer) present in external secretions, called secretory IgA, held together by j-chain polypeptide. The IgA class of immunoglobulin contains two subclasses; IgA1 (93%) and IgA2 (7%). It constitutes almost 10-15% of total Ig in serum which is most abundant in new born. It does not fix complement and binds with the bacterial antigens present on their surface and remove them. IgA prevents attachment of microorganisms and fight against any foreign substances entering into the body cells. The production of secretory IgA is maximum among other immunoglobulin's (Igs). It's the major immunoglobulin found in the colostrum of milk in the nursing mothers and provides a major source of protection against pathogens to the new born after birth. Although IgA constitutes only 10%–15% of the total immunoglobulin in serum, it is the predominant immunoglobulin class in external secretions such as breast milk, saliva, tears, and mucus of the bronchial, genitourinary, and digestive tracts. In serum, IgA exists primarily as a monomer, but polymeric forms (dimers, trimers, and some tetramers) are sometimes seen, all containing a J-chain polypeptide.

Q12. What is meant by therapeutic antibodies?

Answer: Development of therapeutic antibodies boomed in the 1980's, and the first therapeutic antibody, a mouse antibody, was launched in 1986 as an immunosuppressive agent used during organ transplantation. Antibodies can swiftly provide therapeutics to target the disease-related molecules that have been discovered in genomic research because 1) the high level of specificity and affinity to the target molecule or antigen achieves a high level of efficacy and fewer adverse events, 2) their ability to target diverse molecules and the modes of action of the antibodies allow them to be applied to a wide range of therapeutic targets, and 3) modification and refinement by genetic engineering technology and the establishment of recombinant manufacturing technology has made industrial manufacturing possible.

Q13. What are monoclonal antibodies?

Answer: Monoclonal antibodies are laboratory produced antibodies designed to recognise and bind to specific receptors found on the surface of cells. They are derived from natural antibodies; complex proteins derived from a single B cell made by the body's immunological defence system to recognise and fight foreign invaders such as bacteria and viruses.

Q14. Discuss the function of IgA and its secretory component?

Answer: Secretory IgA serves an important effector function at mucous membrane surfaces, which are the main entry sites for most pathogenic organisms. Because it is polymeric, secretory IgA can cross-link large antigens with multiple epitopes. Binding of secretory IgA to bacterial and viral surface antigens prevents attachment of the pathogens to the mucosal cells, thus inhibiting viral infection and bacterial colonization. Complexes of secretory IgA and antigen are easily entrapped in mucus and then eliminated by the ciliated epithelial cells of the respiratory tract or by peristalsis of the gut. Secretory IgA has been shown to provide important lines of defense against bacteria such as *Salmonella, Vibrio cholerae,* and *Neisseria gonorrhoeae* and viruses such as polio, influenza, and reovirus. Breast milk contains secretory IgA and many other molecules that help protect the new-born against infection during the first month of life. Because the immune system of infants is not fully functional, breast-feeding plays an important role in maintaining the health of new-born.

Q15. How are mAbs useful?

Answer: Monoclonal antibodies (mAb) are important reagents used in biomedical research, in diagnosis of diseases, and in treatment of such diseases as infections and cancer. The use of monoclonal antibodies (mAb) in biomedical research has been and will continue to be important for the identification of proteins, carbohydrates, and nucleic acids. Their use has led to the elucidation of many molecules that control cell replication and differentiation, advancing our knowledge of the relationship between molecular structure and function. These advances is basic biologic sciences have improved our understanding of the host response to infectious-disease agents and toxins produced by these agents, to transplanted organs and tissues, to spontaneously transformed cells (tumours), and to endogenous antigens (involved in autoimmunity). In addition, the exquisite specificity of mAb allows them to be used in humans and animals for disease diagnosis and treatment. Under the appropriate conditions, mAb-producing hybridomas survive indefinitely, so continued production of mAb is associated with the use of fewer animals, especially when production involves the use of in vitro methods.

MULTIPLE CHOICE QUESTIONS

Q1. Antibodies are produced by which cells in the human body?

A. Lymphocytes
B. Phagocytes
C. Monocytes
D. Thrombocytes

Q2. Antigen binding sites of an immunoglobulin are located in

A. Light chain alone
B. Heavy chain alone
C. Fc region of the antibody
D. Fab regions of the antibody

Q3. The class of immunoglobulins which can cross the placenta is

A. Immunoglobulin IgM
B. Immunoglobulin IgG
C. Immunoglobulin IgA
D. Immunoglobulin IgD

Q4. Self-reactive antibodies are not found in significant concentration in normal serum because of:

A. It is impossible to generate a self-reactive antibody
B. Self-reactive b cells are killed by cd8 t cells
C. B cells that are stimulated via their surface bound antibody in the absence of t cell help to commit suicide
D. Self-reactive b cells switch to iga which is all secreted and not present in serum

Q5. Simulation of which of the following will lead to appearance of first antibody to appear following stimulation by an antigen?

A. Immunoglobulin IgM
B. Immunoglobulin IgA
C. Immunoglobulin IgE
D. Immunoglobulin IgG

Q6. Which is the most efficient complement fixing class of antibody?

A. Immunoglobulin IgE
B. Immunoglobulin IgA
C. Immunoglobulin IgM
D. Immunoglobulin IgG

Q7. The class of immunoglobulin, most abundant in the human body is:

A. Immunoglobulin IgD
B. Immunoglobulin IgE
C. Immunoglobulin IgG
D. Immunoglobulin IgM

Q8. J chain is present in which of the following Immunoglobulins?

A. IgA Immunoglobulin
B. IgM Immunoglobulin
C. Both (a) and (b) Immunoglobulin
D. IgE Immunoglobulin

Q9. Which is not a function of IgG Immunoglobulin?

A. Major antibody in serum
B. First antibody type produced against an antigen during the primary antibody response
C. Activates or fixes complement
D. Involved in opsonisation

Q10. Which of the following (s) is/are function (s) of Fc fragment of IgG Immunoglobulin?

A. It determines catabolic rate
B. It binds complement
C. It is related to passage of IgG across the placental barrier
D. All of these

Q11. Which is not a function of IgA Immunoglobulin?

A. Protect mucosal surfaces
B. Fix complement
C. Protect eyes
D. Agglutinate antigen

Q12. Which of the following is not true about antibody structure?

A. Antibodies have multiple identical antigen binding sites.
B. Antibodies are built from equal numbers of large (heavy) & small (light) peptide chains.
C. Antibodies are secreted and function away from the cell. They are not attached to the cell membrane.
D. The class of the antibody molecule is determined solely by its heavy chain.

Q13. Where do precursor T-lymphocytes develop into fully competent but not yet activated T-cells?

A. The thymus gland
B. The bone marrow
C. The lymph nodes
D. The spleen

Q14. Which of the following is not a feature of a secondary immune response to an antigen, when compared to the first response to the same antigen?

A. The antibody is generated faster.
B. More antibody is produced.
C. The antibody produced has greater affinity for the antigen.
D. Antibody is generated without T-cell help.

Q15. Cancer cells often have reduced amounts of cell surface proteins, including class I MHC antigens. Which of the following cells of the immune system can exploit this property to kill cancer cells?

A. Cytotoxic T-cells
B. Natural killer cells
C. Helper T-cells
D. Macrophages

Answer Key

1(A), 2(D), 3(B), 4(C), 5(A), 6(C), 7(C), 8(C), 9(B), 10(D), 11(B), 12(C), 13(A), 14(D), 15(B)

References

Biggar RJ, Christiansen M, Rostgaard K, Smedby KE, Adami HO, Glimelius B, Hjalgrim H, Melbye M. Immunoglobulin subclass levels in patients with non-Hodgkin lymphoma. *International Journal of Cancer*. 2009 Jun 1; 124(11):2616-20.

Boyle MD, Reis KJ. Antigenic relationships among bacterial immunoglobulin-binding protein. *Bacterial Immunoglobulin-Binding Proteins*. 1990 Jan 1; 1:175-86.

Cathou RE, Kulczycki A, Haber E. Structural features of γ-immunoglobulin, antibody, and their fragments. Circular dichroism studies. *Biochemistry*. 1968 Nov 1; 7(11):3958-64.

Cotton RG, Milstein C. Fusion of two immunoglobulin-producing myeloma cells. *Nature.* 1973 Jul; 244(5410):42.

Deans RJ, Denis KA, Taylor A, Wall R. Expression of an immunoglobulin heavy chain gene transfected into lymphocytes. *Proceedings of the National Academy of Sciences.* 1984 Mar 1; 81(5):1292-6.

Dingli D, Kyle RA, Rajkumar SV, Nowakowski GS, Larson DR, Bida JP, Gertz MA, Therneau TM, Melton LJ, Dispenzieri A, Katzmann JA. Immunoglobulin free light chains and solitary plasmacytoma of bone. *Blood.* 2006 Sep 15; 108(6):1979-83.

Elgersma AV, Zsom RL, Norde W, Lyklema J. The adsorption of different types of monoclonal immunoglobulin on positively and negatively charged polystyrene latices. *Colloids and Surfaces.* 1991 Jan 1; 54:89-101.

Fox SM, Stavely-Haiber LM. Immunoglobulin A (IgA) levels in blood products and plasma derivatives. *Immunohematology.* 1988; 4(1):5-9.

Fridman WH. Fc receptors and immunoglobulin binding factors. *The FASEB Journal.* 1991 Sep; 5(12):2684-90.

Fröland S, Natvig JB, Berdal P. Surface-bound immunoglobulin as a marker of B lymphocytes in man. *Nature New Biology.* 1971 Dec 22; 234(51):251.

Garweg JG, Garweg SD, Flueckiger F, Jacquier P, Boehnke M. Aqueous humor and serum immunoblotting for immunoglobulin types G, A, M, and E in cases of human ocular toxoplasmosis. *Journal of Clinical Microbiology.* 2004 Oct 1; 42(10):4593-8.

Gerrard JW, Rao DC, Morton NE. A genetic study of immunoglobulin E. American journal of human genetics. 1978 Jan;30(1):46. Korsrud FR, Brandtzaeg P. Quantitative immunohistochemistry of immunoglobulin-and J-chain-producing cells in human parotid and submandibular salivary glands. *Immunology.* 1980 Feb; 39(2):129.

Haas IG, Wabl M. Immunoglobulin heavy chain binding protein. *Nature.* 1983 Nov; 306(5941):387.

Hagiya M, Davis DD, Takahashi T, Okuda K, Raschke WC, Sakano H. Two types of immunoglobulin-negative Abelson murine leukemia virus-transformed cells: implications for B-lymphocyte differentiation. *Proceedings of the National Academy of Sciences.* 1986 Jan 1; 83(1):145-9.

Heremans JF. Immunoglobulin formation and function in different tissues. In *Current Topics in Microbiology and Immunology* 1968 (pp. 131-203). Springer, Berlin, Heidelberg.

Hood L, Davis M, Early P, Calame K, Kim S, Crews S, Huang H. Two types of DNA rearrangements in immunoglobulin genes. In *Cold Spring Harbor Symposia on Quantitative Biology* 1981 Jan 1 (Vol. 45, pp. 887-898). Cold Spring Harbor Laboratory Press.

Johansson SG. Raised levels of a new immunoglobulin class (IgND) in asthma. *The Lancet.* 1967 Nov 4; 290(7523):951-3.

Kamerzell TJ, Ramsey JD, Middaugh CR. Immunoglobulin dynamics, conformational fluctuations, and nonlinear elasticity and their effects on stability. *The Journal of Physical Chemistry B*. 2008 Mar 13; 112(10):3240-50.

Kumar S, Dispenzieri A, Katzmann JA, Larson DR, Colby CL, Lacy MQ, Hayman SR, Buadi FK, Leung N, Zeldenrust SR, Ramirez-Alvarado M. Serum immunoglobulin free light-chain measurement in primary amyloidosis: prognostic value and correlations with clinical features. *Blood*. 2010 Dec 9; 116(24):5126-9.

Kurosawa YO, Tonegawa SU. Organization, structure, and assembly of immunoglobulin heavy chain diversity DNA segments. *Journal of Experimental Medicine*. 1982 Jan 1; 155(1):201-18.

Lefranc MP, Lefranc G. *The Immunoglobulin Factsbook*. Academic press; 2001 May 23.

Lin J, Sampson HA. The role of immunoglobulin E-binding epitopes in the characterization of food allergy. *Current Opinion in Allergy and Clinical Immunology*. 2009 Aug 1; 9(4):357-63.

Lobb N, Curtain CC, Kidson C. Regulatory genes controlling the synthesis of heavy chains of human immunoglobulin-G. *Nature*. 1967 May; 214(5090):783.

Mai HL, Sheikh-Hamad D, Herrera GA, Gu X, Truong LD. Immunoglobulin heavy chain can be amyloidogenic: morphologic characterization including immunoelectron microscopy. *The American Journal of Surgical Pathology*. 2003 Apr 1; 27(4):541-5.

Marsh DG, Bias WB, Ishizaka K. Genetic control of basal serum immunoglobulin E level and its effect on specific reaginic sensitivity. *Proceedings of the National Academy of Sciences*. 1974 Sep 1; 71(9):3588-92.

Mepham BL, Frater W, Mitchell BS. The use of proteolytic enzymes to improve immunoglobulin staining by the PAP technique. *The Histochemical Journal*. 1979 May 1; 11(3):345-57.

Milstein C, Clegg JB, Jarvis JM. C-terminal half of immunoglobulin λ chains. *Nature*. 1967 Apr; 214(5085):270.

Monteiro RC, Hostoffer RW, Cooper MD, Bonner JR, Gartland GL, Kubagawa H. Definition of immunoglobulin A receptors on eosinophils and their enhanced expression in allergic individuals. *The Journal of Clinical Investigation*. 1993 Oct 1; 92(4):1681-5.

Moroz C, Amir J, De Vries A. A hereditary immunoglobulin A abnormality: absence of light-heavy—chain assembly: Study of immunoglobulin synthesis in tonsillar cells. *The Journal of Clinical Investigation*. 1971 Dec 1; 50(12):2726-33.

Myhre EB, Kronvall G. Specific binding of bovine, ovine, caprine and equine IgG subclasses to defined types of immunoglobulin receptors in Gram-positive cocci. *Comparative Immunology, Microbiology and Infectious Diseases*. 1981 Jan 1; 4(3-4):317-28.

Potter MI. Immunoglobulin-producing tumors and myeloma proteins of mice. *Physiological Reviews*. 1972 Jul; 52(3):631-719.

Rabbitts TH, Forster A, Dunnick W, Bentley DL. The role of gene deletion in the immunoglobulin heavy chain switch. *Nature.* 1980 Jan; 283(5745):351.

Richardson JS, Richardson DC, Thomas KA, Silverton EW, Davies DR. Similarity of three-dimensional structure between the immunoglobulin domain and the copper, zinc superoxide dismutase subunit. *Journal of Molecular Biology*. 1976 Apr 5; 102(2):221-35.

Rovigatti U, Mirro J, Kitchingman G, Dahl G, Ochs J, Murphy S, Stass S. Heavy chain immunoglobulin gene rearrangement in acute nonlymphocytic leukemia. *Blood.* 1984 May 1; 63(5):1023-7.

Schanfield MS, Gershowitz H. Studies on the Immunoglobulin Allotypes of Asiatic Populations. *Human Heredity*. 1971; 21(2):168-72.

Schroeder HW Jr, Cavacini L. Structure and function of immunoglobulins. *J Allergy Clin Immunol*. 2010 Feb;125(2 Suppl 2):S41-52. doi: 10.1016/j.jaci.2009.09.046. PubMed PMID: 20176268; PubMed Central PMCID: PMC3670108.

Shulman ST, Melish M, Inoue O, Kato H, Tomita S. Immunoglobulin allotypic markers in Kawasaki disease. *The Journal of Pediatrics*. 1993 Jan 1; 122(1):84-6.

Siden E, Alt FW, Shinefeld L, Sato V, Baltimore D. Synthesis of immunoglobulin mu chain gene products precedes synthesis of light chains during B-lymphocyte development. *Proceedings of the National Academy of Sciences.* 1981 Mar 1; 78(3):1823-7.

Smith G, Matthews JB, Smith AJ, Browne RM. Immunoglobulin-producing cells in human odontogenic cysts. *Journal of Oral Pathology & Medicine*. 1987 Jan; 16(1):45-8.

Staudt LM, Lenardo MJ. Immunoglobulin gene transcription. *Annual Review of Immunology*. 1991 Apr; 9(1):373-98.

Takahashi M, Takagi N, Yagi Y, Moore GE, Pressman D. Immunoglobulin production in cloned sublines of a human lymphocytoid cell line. *The Journal of Immunology*. 1969 Jun 1; 102(6):1388-93.

Terry WD, Hood LE, Steinberg AG. Genetics of immunoglobulin κ-chains: chemical analysis of normal human light chains of differing Inv types. *Proceedings of the National Academy of Sciences*. 1969 May 1; 63(1):71-7.

Tønder O, Thunold S. Receptors for immunoglobulin Fc in human malignant tissues. *Scandinavian Journal of Immunology*. 1973 May; 2(3):207-16.

Tonegawa S. complete immunoglobulin heavy-chain genes. *Nature.* 1980 Aug; 286:14.

Turesson I. Distribution of immunoglobulin-containing cells in human bone marrow and lymphoid tissues. *Acta Medica Scandinavica*. 1976 Jan 12; 199(1-6):293-304.

Vaerman JP, Heremans JF. Origin and molecular size of immunoglobulin-A in the mesenteric lymph of the dog. *Immunology.* 1970 Jan; 18(1):27.

van de Winkel JG, Anderson CL. Biology of human immunoglobulin G Fc receptors. *Journal of Leukocyte Biology*. 1991 May; 49(5):511-24.

van Loghem E, de Lange G. Immunoglobulin epitopes in primates. *Vox sanguinis.* 1979 Dec; 37(6):329-37.

Wakasugi K, Suzuki H, Imai A, Konishi S, Kishioka H. Immunoglobulin free light chain assay using latex agglutination. *International Journal of Clinical and Laboratory Research.* 1995 Dec 1; 25(4):211-5.

Yamada K, Ikeda I, Nakajima H, Shirahata S, Murakami H. Stimulation of proliferation and immunoglobulin production of human-human hybridoma by various types of caseins and their protease digests. *Cytotechnology.* 1991 Mar 1; 5(3):279-85.

Zegers BJ, Maertzdorf WJ, van Loghem E, Mul NA, Stoop JW, Van Der Laag J, Vossen JJ, Ballieux RE. Kappa-chain deficiency: an immunoglobulin disorder. *New England Journal of Medicine.* 1976 May 6; 294(19):1026-30.

In: Immunoglobulins
Editor: Manzoor Ahmad Mir
ISBN: 978-1-53616-903-4

Chapter 3

ORGANIZATION AND EXPRESSION OF IMMUNOGLOBULIN GENES

Manzoor A. Mir*, Umar Mehraj, Safura Nisar and Hina Qayoom
Department of Bioresources, School of Biological Sciences,
University of Kashmir, India

ABSTRACT

The immune system of vertebrates is remarkable as it generates antibodies that can respond to limitless array of antibodies. Each immunoglobulin secreted by the plasma cell has a unique amino acid sequence in its variable region. In the constant region there is only one invariant sequence among the limited numbers. The reason for the combination of variation and constancy in a single protein molecule lies in the organization of the immunoglobulin genes. In germ-line DNA, multiple gene segments encode portions of a single immunoglobulin heavy or light chain. These gene segments are carried in the germ cells but cannot be transcribed and translated into complete chains until they are rearranged into functional genes. During the maturation of B-cells in the bone marrow, random shuffling of these certain gene segments, by a dynamic genetic system capable of generating more than 10^6 combinations. Subsequent processes increase the diversity of the repertoire of antibody binding sites to a very large number that exceeds 10^6 by at least two or three orders of magnitude. The processes in the development of B cell are carefully regulated: the maturation of a progenitor B cell progresses through an ordered sequence of Ig-gene rearrangements, coupled with modifications to the gene that contribute to the diversity of the final product. By the end of this process, a mature, immunocompetent B cell will contain coding sequences for one functional heavy chain variable-region and one light-chain variable-region. Thus, a single B cell is antigenically committed to a single

* Corresponding Author: Dr Manzoor Ahmad Mir (Ph. D), Department of Bioresources, School of Biological Sciences, University of Kashmir, Srinagar J&K, India-190006, Contact:-+919622901319. Email: drmanzoor@kashmiruniversity.ac.in.

epitope. In this chapter we will discuss about the gene rearrangements of immunoglobulin genes and the molecular pathways and processes involved in it.

Keywords: B cell, maturation, Ig gene shuffling, gene rearrangements, antibody diversity. Lambda, kappa, heavy chain, splicing

OBJECTIVES

- Discuss the theories of antibody diversity.
- Describe the multigene organization of Ig genes.
- Discuss the light chain and heavy chain gene family.
- Discuss the gene rearrangements of variable region.
- Give an outline of the mechanism of variable region gene rearrangements.

INTRODUCTION

The immunoglobulin sequencing data results revealed a number of features about the structure of immunoglobulin that were difficult to reconcile with classic genetic models. Any viable model of the immunoglobulin genes had to account or explain known findings about the structure of antibodies viz., (i) the limitless diversity of antibody specificity (ii) the presence of Ig heavy and light chains of a variable region at the amino terminal end and at the carboxyl-terminal, the constant region and (iii) existence of isotypes with the same antigenic specificity. A number of models were put forward to explain the immunoglobulin gene rearrangements, some of the theories are discussed below.

GERM LINE THEORIES AND SOMATIC-VARIATION THEORIES

The germ line theories proposed that the genome contributed by the germ cells, egg and sperm, contains a large repertoire of immunoglobulin genes; thus, these theories invoked no special genetic mechanisms to account for antibody diversity. They argued that the immense survival value of the immune system justified the dedication of a significant fraction of the genome to the coding of antibodies. The somatic variation theories proposed the concept of mutations and recombination in the immunoglobulin genes which generate the large diversity of antibody specifications. This theory maintained that the immunoglobulin gene family is relatively small in the genome and the diversity is a result of recombination. Neither the germ line theories nor the somatic variation theories could explain the findings revealed about the structural diversity of immunoglobulins.

Two-Gene Model

In 1965 W. Dreyer and J. Bennett put forward the two gene model to explain the known findings about the structure of immunoglobulins. The two scientists proposed that two separate genes encode for the variable and constant regions of an immunoglobulin. One gene encodes for the variable regions while the second gene encodes for the constant regions. The two gene model suggested that two genes come together and form a continuous message or mRNA that can be translated into a single heavy or light chain Figure 3.1.

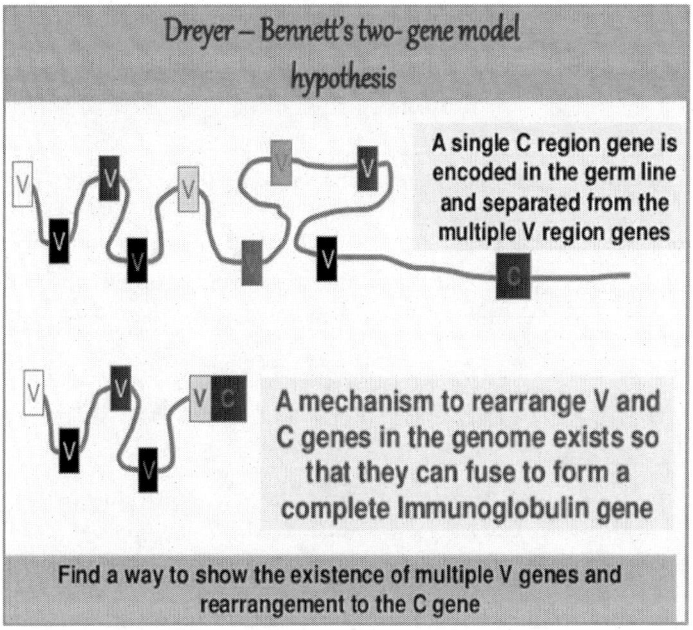

Figure 3.1. The two gene model of antibody diversity.

Multigene Organization of Immunoglobulin Genes

With the advancement of technology, cloning and sequencing of light and heavy chain of the immunoglobulin was accomplished. The sequencing suggested greater complexity in the development of antibody than predicted earlier. The reason responsible for the development of isotypic antibodies i.e., the kappa and lambda light and heavy chains are encoded by separate multigene families present of different chromosomes. In germ-line DNA, each of these multigene families contains several coding sequences, called gene segments, separated by noncoding regions. During B-cell maturation, these gene segments are rearranged and brought together to form functional immunoglobulin genes. Immunoglobulin (antibody) comprises of three genes encoding; kappa (κ) and lambda (λ) of light chain and the heavy chain of antibodies. Gene encoding antibodies present a peculiar fine structure. The functional gene is produced by joining together of several gene segments each of which may exist in more than one form. This provides a great diversity by creating a huge variety of functional genes from a relatively small number of different gene segments. Some of these gene segments may be split or interrupted. The different gene segments are separated from each other by spacer (noncoding) sequences, which include 9-12bp long signal sequences that are involved in the correct joining of different gene segments. The fine structure of antibody coding genes deviates from the κ one-gene-polypeptide concept since several gene segments join together to code for one polypeptide chain.

Table 3.1. Chromosomal locational of immunoglobulin genes in humans and mice

Gene	Human chromosome	Mouse chromosome
λ	22	16
κ	2	6
Heavy chain	14	12

The Features of Multi Gene Family

Each multigene family has distinct features. The κ and λ light-chain families contain V, J, and C gene segments; the rearranged VJ segments encode the variable region of the light chains. The heavy-chain family contains V, D, J, and C gene segments; the rearranged VDJ gene segments encode the variable region of the heavy chain. In each gene family, C gene segments encode the constant regions. Each V gene segment is preceded at its 5' end by a small exon that encodes a short signal or leader (L) peptide that guides the heavy or light chain through the endoplasmic reticulum. The signal peptide is cleaved from the nascent light and heavy chains before assembly of the finished immunoglobulin molecule. Thus, amino acids encoded by this leader sequence do not appear in the immunoglobulin molecule.

The Lambda (λ) Chain Multi Gene Family

In the mouse germ line, the λ multigene family contains three V_λ gene segments, four J λ gene segments, and four C λ gene segments. The J λ gene is a defective gene as it is incapable of encoding a protein and are called as pseudogene. The pseudogenes are indicated with the psi symbol (ψ). However, the interesting fact about the $J_\lambda4$'s constant region partner, the Cλ4 is a perfectly functional gene. The three function J_λ and V_λ encode for the variable region of the light chain. The three C_λ genes encodes the constant region of one of the three λ -chain subtypes each. In humans, there are 31 functional V_λ gene segments, 4 J_λ segments, and 7 C_λ segments. In additional to the functional gene segments, the human lambda complex contains many V_λ, J_λ, and C_λ pseudogenes Figure 3.2.

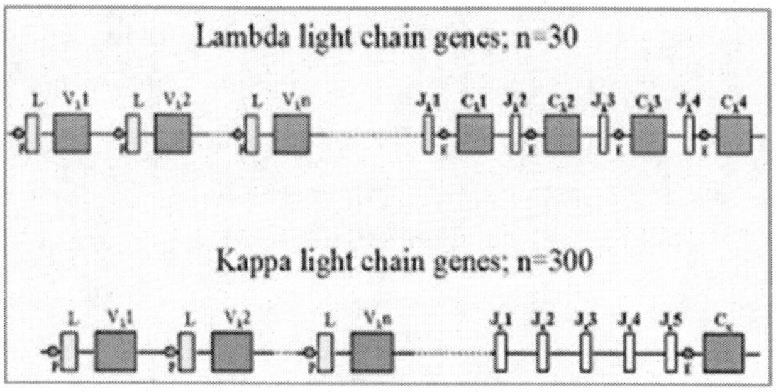

Figure 3.2. The light chain gene families.

The Kappa (κ) Chain Multigene Family

In the mouse germline the kappa (κ) chain multigene family contains approximately 85 Vκ gene segments and each segment with an upstream leader sequence. There are five Jκ segments and one among these is a pseudogene. In the kappa chain multigene family, there is only one Cκ segment. The Vκ and the Jκ encode for the variable region of the light chain while the Cκ encodes for the constant region. The Cκ being single, there there are no subtypes for the κ light chains. In humans the κ-chain multigene family has 40 V κ gene segments, 5 J κ segments and a single C κ segment for the constant region of the light chains.

The Heavy-Chain Multigene Family

The heavy chain multigene family organization is similar to, but more complex than, that of the κ and λ light-chain genes. For the heavy chain variable region an additional segment is present. The existence of this additional segment was first proposed by Leroy Hood and his team. Leroy and his colleagues compared the heavy-chain variable-region amino acid sequence with the VH and JH nucleotide sequences. Amino acids 1 to 94 were found to be encoded by the VH segment and JH was found to encode amino acids 98-113. The mRNA message for the amino acids 95 to 97 was neither found with the VH segment and nor with the JH segment. The additional nucleotide sequence was observed between the VH and JH gene segments by comparing the germ-line DNA sequence with a rearranged myeloma DNA. The nucleotide sequence corresponded to amino acids 95 to 97 of the heavy chain. This gene segment, which encoded amino acids within the third complementarity-determining region (CDR3), was designated D for diversity, because of its contribution to the generation of antibody diversity. In humans the heavy chain multigene family is situated on chromosome 14. It contains approximately 51 VH gene segments located upstream from a cluster of 27 functional DH gene segments. Each VH gene segment is preceded by a leader sequence a short distance upstream. Downstream from the DH gene segments are six functional JH gene segments, followed by a series of CH gene segments. Each CH gene segment encodes the constant region of an immunoglobulin heavy-chain isotype. The CH gene segments consist of coding exons and noncoding introns. Each exon encodes a separate domain of the heavy-chain constant region. A similar heavy-chain gene organization is found in the mouse Figure 3.3.

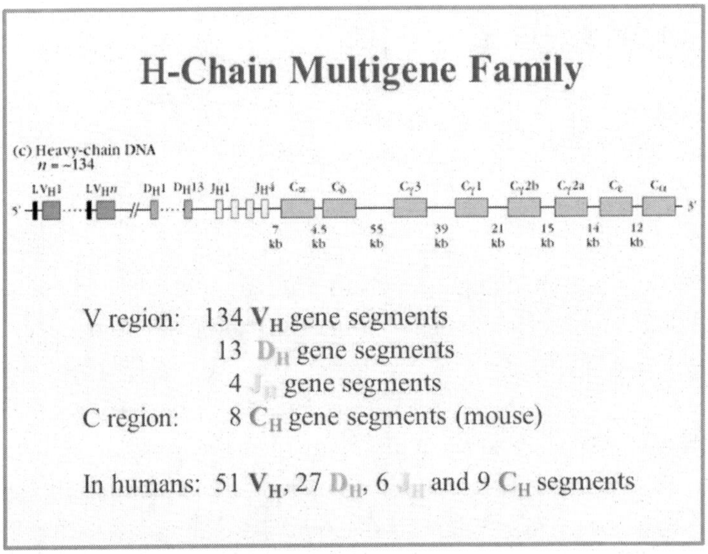

Figure 3.3. The Heavy chain multigene family of immunoglobulins.

VARIABLE-REGION REARRANGEMENTS

The variable region rearrangements happen in an orderly sequence in the bone marrow. Usually, the assortment of these gene segments occurs at B cell maturation. The heavy-chain variable-region genes rearrange first, then the light-chain variable-region genes. At the end of this process, each B cell contains a single functional variable region DNA sequence for its heavy chain and another for its light chain.

The gene rearrangements process of the variable-region produces mature, immunocompetent B cells and each such cell is committed to produce antibody with a binding site encoded by the particular sequence of its rearranged V genes. The steps in variable-region gene rearrangement occur in an ordered sequence, but they are random events that result in the random determination of B-cell specificity.

Light Chain DNA

Expression of both κ and λ light chains requires rearrangement of the variable-region V and J gene segments. In humans, among the four functional V λ genes, anyone can combine with any of the four functional J λ-C λ combinations. The VJ rearrangements in the mouse are slightly more complicated. DNA rearrangement can join the V λ 1 gene segment with either the J λ 1 or the J λ 3 gene segment, or the V λ 2 gene segment can be joined with the J λ 2 gene segment. In human or mouse κ light-chain DNA, any one of the V κ gene segments can be joined with any one of the functional J λ gene segments Figure 3.4.

Rearranged κ and λ genes contain the following regions in order from the 5' to 3' end: a short leader (L) exon, a noncoding sequence (intron), a joined VJ gene segment, a second intron, and the constant region. Upstream from each leader gene segment is a promoter sequence.

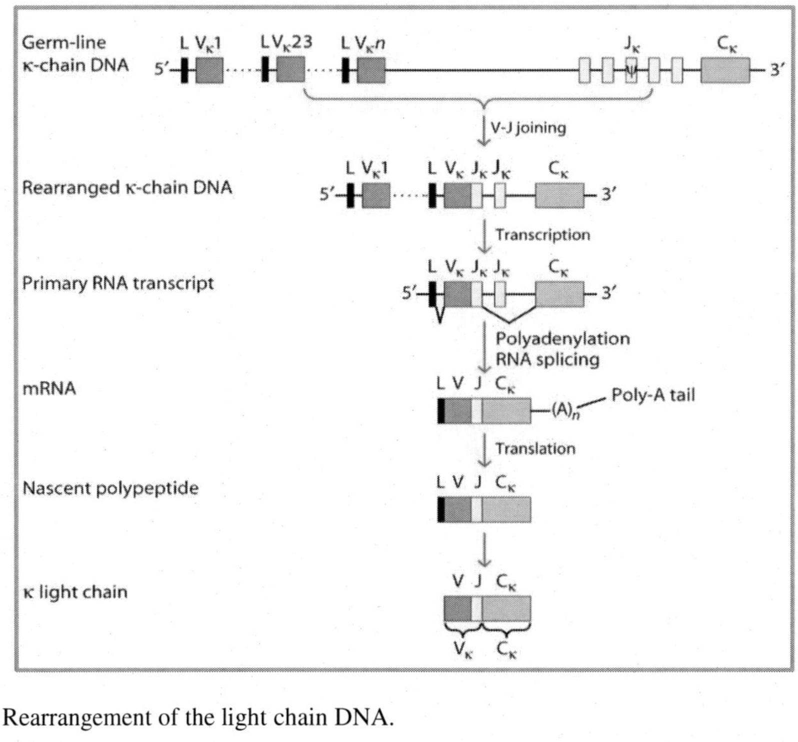

Figure 3.4. Rearrangement of the light chain DNA.

The rearranged light chain sequence is transcribed by RNA polymerase from the L exon through the C segment to the stop signal, generating a light-chain primary RNA transcript. The introns in the primary transcript are removed by RNA processing enzymes, and the resulting light-chain messenger RNA then exits from the nucleus. The light-chain mRNA binds to ribosomes and is translated into the light-chain protein. The leader sequence at the amino terminus pulls the growing polypeptide chain into the lumen of the rough endoplasmic reticulum and is then cleaved, so it is not present in the finished light-chain protein product.

The Rearrangements of Heavy-Chain DNA

Generation of a functional immunoglobulin heavy-chain gene requires two separate rearrangement events within the variable region. As illustrated in Figure 5.5, a DH gene segment first joins to a JH segment; the resulting DHJH segment then moves next to and joins a VH segment to generate a VHDHJH unit that encodes the entire variable region. In heavy-chain DNA, variable-region rearrangement produces a rearranged gene consisting of the following sequences, starting from the 5' end: a short L exon, an intron, a joined VDJ segment, another intron, and a series of C gene segments. As with the light-chain genes, a promoter sequence is located a short distance upstream from each heavy-chain leader sequence Figure 3.5.

Once heavy-chain gene rearrangement is accomplished, RNA polymerase can bind to the promoter sequence and transcribe the entire heavy-chain gene, including the introns. Differential polyadenylation and RNA splicing remove the introns and process the primary transcript to generate mRNA. The production of two different heavy-chain mRNAs allows a mature, immunocompetent B cell to express both IgM and IgD with identical antigenic specificity on its surface.

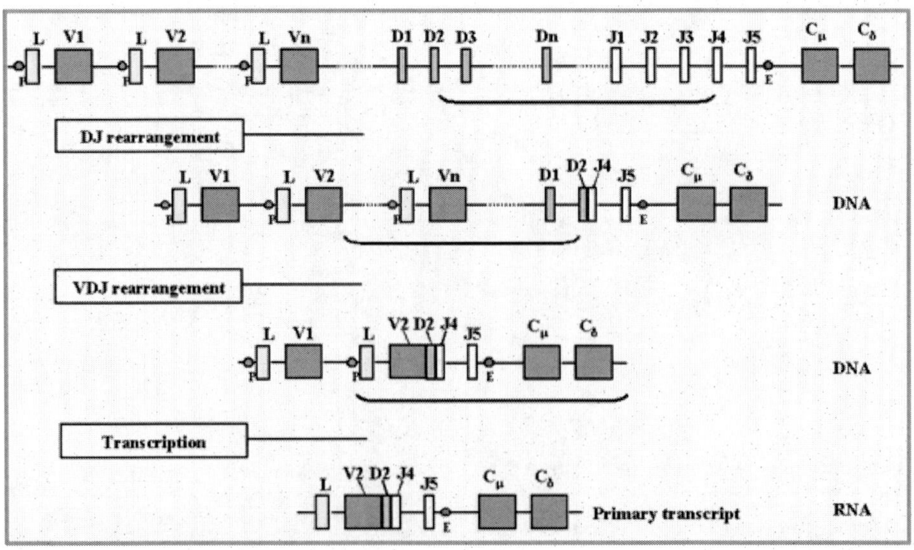

Figure 3.5. The gene rearrangement of heavy-chain DNA of immunoglobulin genes.

Mechanism of Variable Region Rearrangements

It is understood that rearrangement occurs between specific sites on the DNA called recombination signal sequences (RSSs). The signal sequences are composed of a conserved palindromic heptamer and a conserved AT- rich nonamer. These signal sequences are separated by non-conserved spacers of 12 or 23 base pairs called one-turn and two-turn respectively. They are within the lambda chain, k-chain. The processes of rearrangement in these regions are catalyzed by two recombination-activating genes: RAG-1 and RAG-2 and other enzymes and proteins. The segments joined due to signals generated RSSs that flank each V, D, and J segments. Only genes flank by 12 -bp that join to the genes flank by 23-bp spacer during the rearrangements and combinations to maintain VL-JL and VH-DH-JH joining.

FREQUENTLY ASKED QUESTIONS

Q1. Give an account of the germ line theories of antibody diversity.

Answer: The germ line theories proposed that the genome contributed by the germ cells, egg and sperm, contains a large repertoire of immunoglobulin genes; thus, these theories invoked no special genetic mechanisms to account for antibody diversity. They argued that the immense survival value of the immune system justified the dedication of a significant fraction of the genome to the coding of antibodies.

Q2. Give an account of the germ line theories of antibody diversity.

Answer: The somatic variation theories proposed the concept of mutations and recombination in the immunoglobulin genes which generate the large diversity of antibody specifications. This theory maintained that the immunoglobulin gene family is relatively small in the genome and the diversity is a result of recombination.

Q3. What do you understand by the two gene model?

Answer: In 1965 W. Dreyer and J. Bennett put forward the two gene model to explain the known findings about the structure of immunoglobulins. The two scientists proposed that two separate genes encode for the variable and constant regions of an immunoglobulin. One gene encodes for the variable regions while the second gene encodes for the constant regions. The two gene model suggested that two genes come together and form a continuous message or mRNA that can be translated into a single heavy or light chain.

Q4. What are the features of a multi gene family?

Answer: Each multigene family has distinct features. The κ and λ light-chain families contain V, J, and C gene segments; the rearranged VJ segments encode the variable region of the light chains. The heavy-chain family contains V, D, J, and C gene segments; the rearranged VDJ gene segments encode the variable region of the heavy chain. In each gene family, C gene segments encode the constant regions. Each V gene segment is preceded at its 5' end by a small exon that encodes a short signal or leader (L) peptide that guides the heavy or light chain through the endoplasmic reticulum. The signal peptide is cleaved from the nascent light and heavy chains before assembly of the finished immunoglobulin molecule. Thus, amino acids encoded by this leader sequence do not appear in the immunoglobulin molecule.

Q5. Give an account of the rearrangements of Heavy-Chain DNA?

Answer: Generation of a functional immunoglobulin heavy-chain gene requires two separate rearrangement events within the variable region. As illustrated in Figure 5.5, a DH gene segment first joins to a JH segment; the resulting DHJH segment then moves next to and joins a VH segment to generate a VHDHJH unit that encodes the entire variable region. In heavy-chain DNA, variable-region rearrangement produces a rearranged gene consisting of the following sequences, starting from the 5' end: a short L exon, an intron, a joined VDJ segment, another intron, and a series of C gene segments. As with the light-chain genes, a promoter sequence is located a short distance upstream from each heavy-chain leader sequence. Once heavy-chain gene rearrangement is accomplished, RNA polymerase can bind to the promoter sequence and transcribe the entire heavy-chain gene, including the introns. Differential polyadenylation and RNA splicing remove the introns and process the primary transcript to generate mRNA. The production of two different heavy-chain mRNAs allows a mature, immunocompetent B cell to express both IgM and IgD with identical antigenic specificity on its surface.

Q6. Briefly describe the rearrangements of light chain Ig DNA?

Answer: Expression of both κ and λ light chains requires rearrangement of the variable-region V and J gene segments. In humans, among the four functional V λ genes, anyone can combine with any of the four functional J λ-C λ combinations. The VJ rearrangements in the mouse are slightly more complicated. DNA rearrangement can join the V λ 1 gene segment with either the J λ 1 or the J λ 3 gene segment, or the V λ 2 gene segment can be joined with the J λ 2 gene segment. In human or mouse κ light-chain DNA, any one of the V κ gene segments can be joined with any one of the functional J λ gene segments.

Q7. Briefly discuss the mechanism of variable region rearrangements?

Answer: It is understood that rearrangement occurs between specific sites on the DNA called recombination signal sequences (RSSs). The signal sequences are composed of a conserved palindromic heptamer and a conserved AT- rich nonamer. These signal sequences are separated by non-conserved spacers of 12 or 23 base pairs called one-turn and two-turn respectively. They are within the lambda chain, k-chain. The processes of rearrangement in these regions are catalyzed by two recombination-activating genes: RAG-1 and RAG-2 and other enzymes and proteins. The segments joined due to signals generated RSSs that flank each V, D, and J segments. Only genes flank by 12 -bp that join to the genes flank by 23-bp spacer during the rearrangements and combinations to maintain VL-JL and VH-DH-JH joining.

Q8. Give an account of the kappa (κ) chain multigene family?

Answer: In the mouse germline the kappa (κ) chain multigene family contains approximately 85 Vκ gene segments and each segment with an upstream leader sequence. There are five Jκ segments and one among these is a pseudogene. In the kappa chain multigene family, there is only one Cκ segment. The Vκ and the Jκ encode for the variable region of the light chain while the Cκ encodes for the constant region. The Cκ being single, there there are no subtypes for the κ light chains. In humans the κ-chain multigene family has 40 Vκ gene segments, 5 Jκ segments and a single Cκ segment for the constant region of the light chains.

Q9. Give an account of the Lambda (λ) chain multi gene family?

Answer: In the mouse germ line, the λ multigene family contains three Vλ gene segments, four Jλ gene segments, and four Cλ gene segments. The Jλ gene is a defective gene as it is incapable of encoding a protein and are called as pseudogene. The pseudogenes are indicated with the psi symbol (ψ). However, the interesting fact about the Jλ4's constant region partner, the Cλ4 is a perfectly functional gene. The three function Jλ and Vλ encode for the variable region of the light chain. The three Cλ genes encodes the constant region of one of the three λ -chain subtypes each. In humans, there are 31 functional Vλ gene segments, 4 Jλ segments, and 7 Cλ segments. In additional to the functional gene segments, the human lambda complex contains many Vλ, Jλ, and Cλ pseudogenes.

Q10. Describe the heavy chain multigene family organization?

Answer: The heavy chain multigene family organization is similar to, but more complex than, that of the κ and λ light-chain genes. For the heavy chain variable region an additional segment is present. The existence of this additional segment was first proposed by Leroy Hood and his team. Leroy and his colleagues compared the heavy-chain variable-region amino acid sequence with the VH and JH nucleotide sequences. Amino acids 1 to 94 were found to be encoded by the VH segment and JH was found to encode amino acids 98-113. The mRNA message for the amino acids 95 to 97 was neither found with the VH segment and nor with the JH segment. The additional nucleotide sequence was observed between the VH and JH gene segments by comparing the germ-line DNA sequence with a rearranged myeloma DNA. The nucleotide sequence corresponded to amino acids 95 to 97 of the heavy chain. This gene segment, which encoded amino acids within the third complementarity-determining region (CDR3), was designated D for diversity, because of its contribution to the generation of antibody diversity. In humans the heavy chain multigene family is situated on chromosome 14. It contains approximately 51 VH gene segments located upstream from a cluster of 27 functional DH gene segments. Each VH gene segment is preceded by a leader sequence a short distance upstream. Downstream from the DH gene segments are six functional JH gene segments, followed by a series of CH gene segments. Each CH gene segment encodes the constant region of an immunoglobulin heavy-chain isotype. The CH gene segments consist of coding exons and noncoding introns. Each exon encodes a separate domain of the heavy-chain constant region. A similar heavy-chain gene organization is found in the mouse.

Multiple Choice Questions

Q1. The variable region of the heavy chain and the variable region of the light chain in a given antibody molecule:

A. are identical to each other in primary amino acid sequence.
B. are each encoded by single germ-line gene segments.
C. in combination with one another, define the specificity of the antibody for an antigen.
D. contain the hypervariable and constant region portions of an antibody molecule.

Q2. Which of the following statements about immunoglobulin is true?

A. The ratio of IgA to IgG is higher in the intestinal lumen than in serum.
B. Both IgA and IgE can cross the placenta.
C. Immunoglobulin heavy chain class switching requires the function of the Recombinase Activating Genes (RAGs).
D. The structure of IgA is the same in both saliva and serum.

Q3. During the rearrangement of immunoglobulin genes

A. An unsuccessful rearrangement on one chromosome may be followed by the successful rearrangement on the other chromosome.
B. A successful rearrangement of one chromosome does not prevent the subsequent rearrangement on the other chromosome.
C. Two non-functional rearrangements of the heavy chain gene is usually followed by a switch to a different isotype.
D. Every nucleotide at recombination points of the heavy chain are encoded in germline sequence the one of genetic elements.

Q4. Immunoglobulin heavy chain class switching from IgM to IgA:

A. Occurs frequently during T-independent antibody responses.
B. Requires the function of Recombination Activating Genes (RAGs).
C. Involves genetic recombination at the immunoglobulin heavy chain locus.
D. Is driven by interaction between a B cell and an antigen-presenting cell.

Q5. Which of the following are antigen-independent steps in B cell differentiation?

A. Expression of surface IgM.
B. Expression of surface IgG.
C. Affinity maturation.
D. Secretion of large amounts of soluble Ig molecules.

Q6. Which of the following is only contained in heavy chains and not in light chains?

A. Leader (L)
B. Joining (J)
C. Diversity (D)
D. Variable (V)
E. Constant (C)

Q7. The two types of light chains are:

A. Alpha (a) and beta (b)
B. Sigma (s) and gamma (g)
C. Epsilon (e) and delta (d)
D. Lambda (l) and kappa (k)

Q8. During the final step of synthesis for heavy and light immunoglobulin chains, which of the following peptides is cleaved?

A. Leader (L)
B. Joining (J)
C. Diversity (D)
D. Variable (V)
E. Constant (C)

Q9. "Isotype switching" of immunoglobulin classes by B cells involves:

A. Simultaneous insertion of VH region adjacent to each CH region.
B. Placement of a VH region adjacent to a different CH region.
C. Activation of homologous genes on chromosome 6 of both alleles.
D. Switching of light-chain types (kappa and lambda).
E. Alternative splicing of RNA to produce various isotypes.

Q10. In somatic recombination, the____ rearranges first. The____ region of the antibody molecule is generated via somatic recombination as it binds to the antigen and contains both a constant region and a variable region.

A. Heavy chain; Fab
B. Light chain; Fab
C. Heavy chain; Fc
D. Light chain; Fc

Answer Key

1(C), 2(A), 3(A), 4(C), 5(A), 6(C), 7(D), 8(A), 9(B), 10(A)

References

Butler J. E. Immunoglobulin gene organization and the mechanism of repertoire development. *Scandinavian Journal of Immunology*. 1997 Apr;45(4):455-62.

Chen, J., Y. Shinkai, F. Young, and F. W. Alt. 1994. Probing immune functions in RAG-deficient mice. *Curr. Opin. Immunol.* 6:313.

Cook, G. P., and I. M. Tomlinson. 1995. The human immunoglobulin VH repertoire. *Immunol. Today* 16:237.

Dreyer, W. J., and J. C. Bennett. 1965. The molecular basis of antibody formation. *Proc. Natl. Acad. Sci. U.S.A.* 54:864.

Enhancer complexes located downstream of both human immunoglobulin C alpha genes. *J. Exp. Med.* 186:845.

Fugmann, S. D., I. L. Lee, P. E. Shockett, I. J. Villey, and D. G. Schatz. 2000. The RAG proteins and V(D)J recombination: Complexes, ends and transposition. *Annu. Rev. Immunol.* 18:495.

Gavilondo, J. V., and J. W. Larrick. 2000. Antibody engineering at the millennium. *Biotechniques* 29:128.

Hayden, M. S., L. K. Gilliand, and J. A. Ledbetter. 1997. Antibody engineering. *Curr. Opin. Immunol.* 9:201.

Hesslein, D. G., and D. G. Schatz. 2001. Factors and forces controlling V(D)J recombination. *Adv. Immunol.* 78:169.

Honjo T. Immunoglobulin genes. *Annual Review of Immunology*. 1983 Apr;1(1):499-528.

Hozumi, N., and S. Tonegawa. 1976. Evidence for somatic rearrangement of immunoglobulin genes coding for variable and constant regions. *Proc. Natl. Acad. Sci. U.S.A.* 73:3628.

Ichihara Y., Matsuoka H., Kurosawa Y. Organization of human immunoglobulin heavy chain diversity gene loci. *The EMBO journal*. 1988 Dec 1;7(13):4141-50.

Kuby J. *Immunology*, Text book.

Lydyard P, Whelan A, Fanger M. *Bios Instant Notes in Immunology*. Taylor & Francis; 2004 Mar 1.

Maloney, D. G., et al. 1997. IDEC-C2B8 (Rituximab) anti-CD20 monoclonal antibody therapy in patients with relapsed lowgrade non-Hodgkin's lymphoma. *Blood* 90:2188.

Manis, J. P., M. Tian, and F. W. Alt. 2002. Mechanism and control of class-switch recombination. Trends Immunol. 23:31. Matsuda, F., K. Ishii, P. Bourvagnet, Ki Kuma, H. Hayashida, T. Miyata, and T. Honjo. 1998. The complete nucleotide sequence of the human immunoglobulin heavy chain variable region locus. *J. Exp. Med.* 188:2151.

Max, E. E. 1998. Immunoglobulins: Molecular Genetics. In *Fundamental Immunology*, 4th ed., W. E. Paul, ed. Lippincott-Raven, Philadelphia. Mills, F. C., N. Harindranath, M. Mitchell, and E. E. Max. 1997.

Oettinger, M. A., et al. 1990. RAG-1 and RAG-2, adjacent genes that synergistically activate V(D)J recombination. *Science* 248:1517.

Shamblott M. J., Litman G. W. Genomic organization and sequences of immunoglobulin light chain genes in a primitive vertebrate suggest coevolution of immunoglobulin gene organization. *The EMBO Journal*. 1989 Dec 1;8(12):3733-9.

Shimizu A., Takahashi N., Yaoita Y., Honjo T. Organization of the constant-region gene family of the mouse immunoglobulin heavy chain. *Cell.* 1982 Mar 1;28(3):499-506.

Tonegawa, S. 1983. Somatic generation of antibody diversity. *Nature* 302:575.

Van Gent, D. C., et al. 1995. Initiation of V(D)J recombination in a cell-free system. *Cell* 81:925.

Winter, G., and C. Milstein. 1990. Man-made antibodies. *Nature* 349:293.

In: Immunoglobulins
Editor: Manzoor Ahmad Mir
ISBN: 978-1-53616-903-4

Chapter 4

ANTIGEN-ANTIBODY INTERACTION

Manzoor A. Mir*, Umar Mehraj, Safura Nisar, and Hina Qayoom
Department of Bioresources, School of Biological Sciences,
University of Kashmir, Srinagar, J&K, India

ABSTRACT

Ag-Ab interaction is highly specific and occurs in a similar way as a bimolecular association of an enzyme-substrate. The binding between antigens and immune components involves weak non-covalent interactions. The binding forces are relatively weak and *reversible* and consist mainly of van der Waals forces, electrostatic forces and hydrophobic forces. The combined strength of all noncovalent interactions between a single antigen-binding site of an antibody and a single epitope of the antigen is called *affinity* of that antibody for the concerned *epitope*. A low affinity antibody binds the concerned antigen weakly and it tends to dissociate readily from the Ag-Ab complex. Many antigens are complex antigens in that they have multiple, repeating epitopes. Such antigens usually bind an antibody molecule having multiple binding sites at more than one binding site. Further, interaction between such an antibody molecule and a complex antigen molecule at one binding site will increase the probability of reaction between them at a second binding site. The cumulative strength of the multiple interactions between a multivalent antibody and antigen is called *avidity*. When an antibody elicited by one antigen (more precisely, an immunogen) reacts with an unrelated antigen, it is called *cross-reaction*. Cross reaction occurs when two unrelated antigens have an identical or a very similar epitope. When the epitopes are similar (but not identical) the affinity of the antibody for the cross-reacting epitope is usually less than that for the original epitope, which elicited this antibody. In blood the antigens bound to antibodies forming an antigen-antibody complex known as immune complex which is then taken to the cellular systems where it can be broken or deactivated. The reactions between antigen and antibody occur in three

* Corresponding Author: Dr Manzoor Ahmad Mir (Ph. D), Department of Bioresources, School of Biological Sciences, University of Kashmir, Srinagar J&K, India-190006, Contact:-+919622901319. Email: drmanzoor@kashmiruniversity.ac.in.

stages. Stage1: The reaction comprises the formation of Ag-Ab complex. Stage2: Generate visible events like precipitation, agglutination, etc. Stage3: Involves destructions of antigen or its neutralization.

Keywords: antigen, antibody, affinity, avidity, specificity, cross reaction, agglutination, neutralization, precipitation

OBJECTIVES

- Brief idea about antigen and antibody.
- Understanding the concept of Antigen antibody interaction.
- Describe the properties of antigen antibody interaction.
- Discuss various types of antigen antibody interaction.
- Explain effects of antigen antibody interaction.

ANTIGEN

Antigens are unfamiliar substances which are recognized by the receptors on lymphocytes, which then evoke the immune response. Antigens which are able to induce immune response are known as immunogens (Figure 4.1). Every immunogen is an antigen but not every antigen is an immunogen. There are some low molecular weight substances which are not capable of inducing immune response when present alone they are known as haptens. They when conjugated with a high molecular weight substance known as carrier molecule to form carrier-hapten conjugate they then induce immune response against both the hapten and the carrier. The immune response is shown against all known bio-chemical families of compounds, like carbohydrates, proteins, nucleic acids and lipids. Immune response is also generated against drugs, food additives. Small synthetic peptides can also be induced only when these are combined with carriers. On the surface of antigens are regions called, "antigen determinants/epitopes" that fit and bind to receptor molecules of complementary structures on the surface of the lymphocytes. The binding of the lymphocytes' receptors to the antigens' surface molecules stimulate the lymphocytes to multiply and to initiate immune response which include the production of antibody, the activation of cytotoxic cells or both against the antigens. Complex antigens present as "antigenic mosaic" can evoke immune responses from a variety of specific lymphocytes. Some antigenic determinants are better than other in evoking an immune response, apparently because a greater number of responsive lymphocytes are present. It is possible for two or more different substances to have an epitope in common. In these cases, immune components induced by one antigen are able to react with all other antigens carrying the same epitope. Such antigens are known as cross-reacting antigens.

There are certain factors that influence immunogenicity of an antigen

- Foreignness
- Molecular weight
- Chemical composition.
- Genotype of the recipient animal.
- Dosage and route of administration.

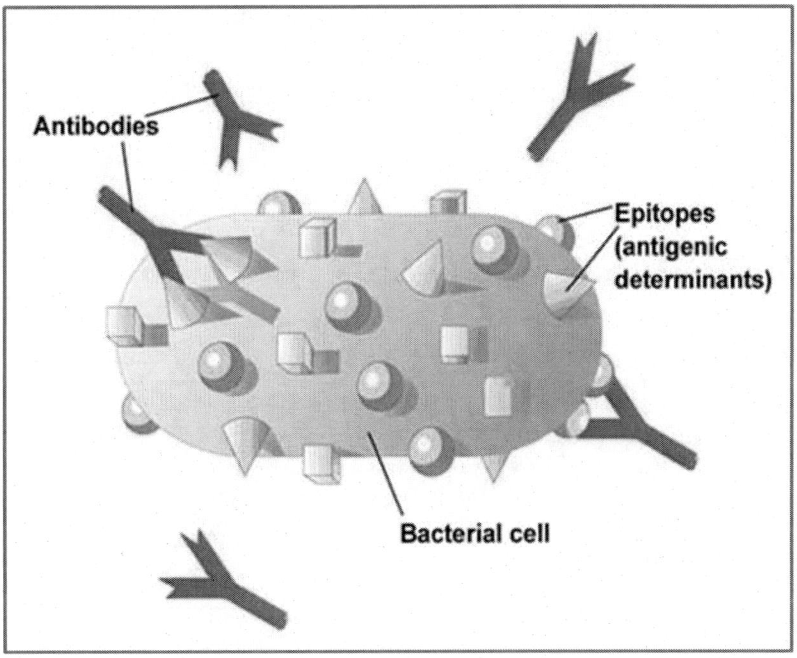

Figure 4.1. Illustrating binding of specific antibodies to Specific antigenic epitopes (image source: https://microbenotes.com/introduction-to-antigen/).

Foreignness

This is the most important feature of an immunogen and the effective immunogen must be recognized as non-self by the biological system. Degree of immunogenicity depends on degree of foreignness. The adaptive immune system recognizes and eliminates only foreign agents. Self-antigens are not recognized and thus individuals are tolerant to their self-molecules even though these same molecules have capacity to act as immunogens in other individuals of same species. For example, if bovine serum albumin is injected in cow, rabbit and chicken, the order of immunogenicity will be least for cow and most for chicken.

Molecular Weight

For an antigen to show immunogenicity it must have a certain molecular weight. There exists a co-relation between the molecular weight and immunogenicity (Figure 4.2). Small compounds that have a molecular weight of less than 1000 Da (e.g., aspirin, progesterone etc.) are not immunogenic and those molecules which have a molecular weight between 1000 and 6000Da (adrenocorticotropic hormone ACTH) may or may not be immunogenic whereas those with a molecular weight greater than 6000 Da (e.g.; albumin, tetanus toxin etc.) are generally immunogenic.

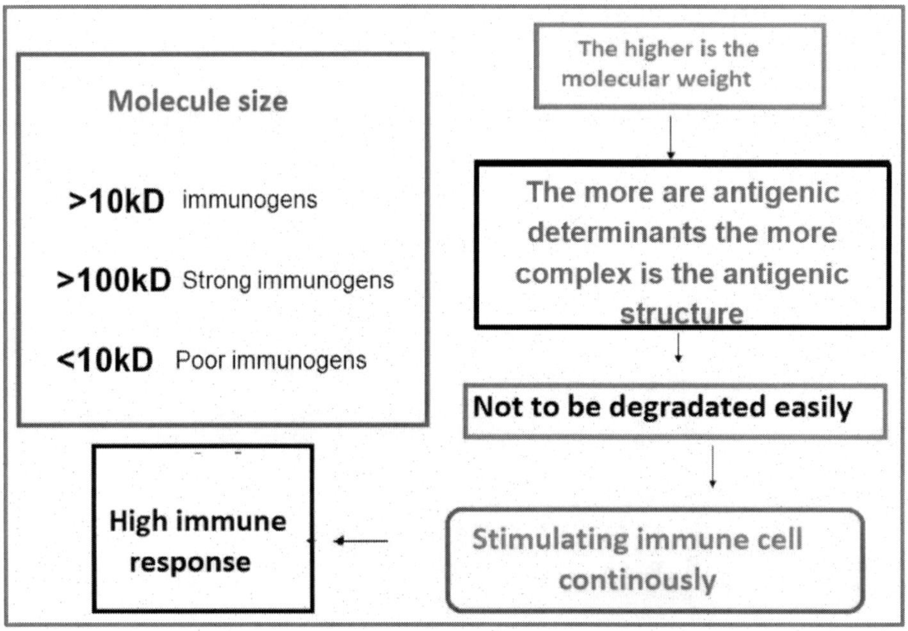

Figure 4.2. Illustrating the relation between the molecular size and immunogenicity of an antigen.

Chemical Composition

Chemical complexity contributes to immunogenicity. Co-polymers of different amino acids or sugars are usually more immunogenic than homopolymers of their constituents. Some high molecular weight compounds lack immunogenicity because these are not sufficiently chemically complex. For example, a simple molecule of homopolymer of amino acid lysine with a molecular weight of 50,000Da is not an immunogen despite having high molecular weight. Complexity of such molecules is increased by attaching various substances (such as dinitrophenol or other low molecular weight compound known as hapten) which by themselves are not immunogenic, the entire macromolecule becomes immunogenic. An increase in the chemical complexity of compound is accompanied by the increase in its immunogenicity (Figure 4.3).

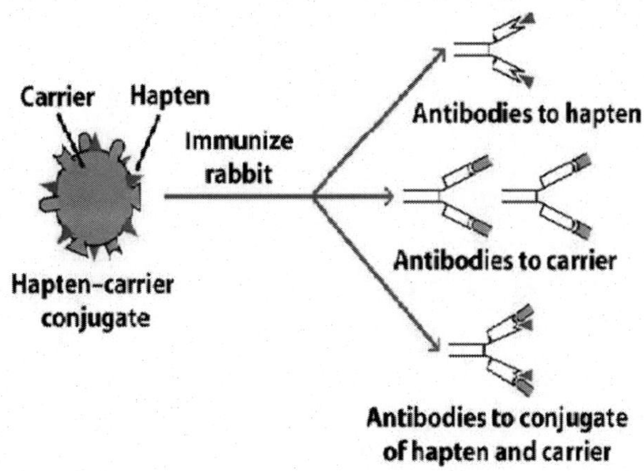

Injection with:	Antibodies formed:
Hapten (DNP)	None
Protein carrier (BSA)	Anti-BSA
Hapten-carrier conjugate (DNP-BSA)	Anti-DNP (major) Anti-BSA (minor) Anti-DNP/BSA (minor)

Figure 4.3. Diagram showing that an increase in chemical complexity is accompanied by the increase in its immunogenicity (adapted from Kuby et al. 7th Ed.).

Genotype of the Recipient Animal

The genetic constitution of an animal which is being immunized influences the type of immune response that the animal shows as well as the degree of response. Genetic control of immune responsiveness is largely confined to gene within the major histocompatibility complex (MHC).

Dosage and Route of Administration

An appropriate dose is necessary to mount good immune response. Insufficient dose will not stimulate immune response and excessively high dose will induce tolerance. Administration route strongly determines which immune organ and cell population will be involved in the response. Immunogens can be administered through a number of common routes. Intravenous (into a vein), intra dermal (into the skin), subcutaneous (beneath the skin), peritoneal (in the peritoneal cavity). Immunogens administered through the most common route evoke the strongest immune response.

ANTIBODIES

Antibodies are the globular proteins that belong to immunoglobulin family. These are synthesized by B-cells in billions of forms. Each antibody has different amino acid sequence and a different antigen binding site. Antibodies appear in circulation after infection and have ability to specifically react with epitopes of antigen introduced in the organism. There are five major classes of immunoglobulins: IgG, IgA, IgD, IgM and IgE. They have similar basic structures and contain four polypeptide chains which are fused together by disulfide bonds forming a symmetrical molecular structure. Immunoglobulin's are the most abundant protein components in the blood, consisting about 20% of the total protein components in the blood plasma. These antibodies are present on B-cell membrane and also secreted by the plasma cells. Antibodies belong to a class of proteins called globulins and are multifunctional glycoprotein's produced in direct response against antigens by vertebrates and essential for the deterrence and resolution of infection by various pathogens. They initiate their biological effect by binding non-covalently to antigen without modifying the covalent structure of the antigen. Antibody molecules are referred to as multifunctional molecules as they perform more than one function. One part of the immunoglobulin is involved in the binding of the antigen while the other part of the molecule may be involved in binding to receptors on phagocytes, the activation of complement pathways and the activation of the NK cells of the immune system. Antibodies are Y shaped molecules each containing four polypeptide chains out of which two are identical heavy chains (H) and two are identical light chains (L) (Fig 4.4). These chains are held together by several disulphide bonds. The amino terminal ends of the polypeptide chain show considerable variations in amino acid composition. These are the regions which are known as variable (V) regions. Each light chain contains about 220 amino acids residue and heavy chain contain about 440 amino acids residue. Antibody is a dimer of H-L chains. Each heavy and light chain are connected with each other by disulphide bridges and other covalent linkages.

Both the chains have a variable domain at N-terminal ends and a constant domain at C-terminal ends. Light chains have a constant domain (CL) of about 110 amino acids and a variable domain (VL) of the same size whereas the heavy chains has a variable (VH) of about 330 or 440 amino acids long and constant domain (CH) is about 110 amino acids long (Figure 4.5).

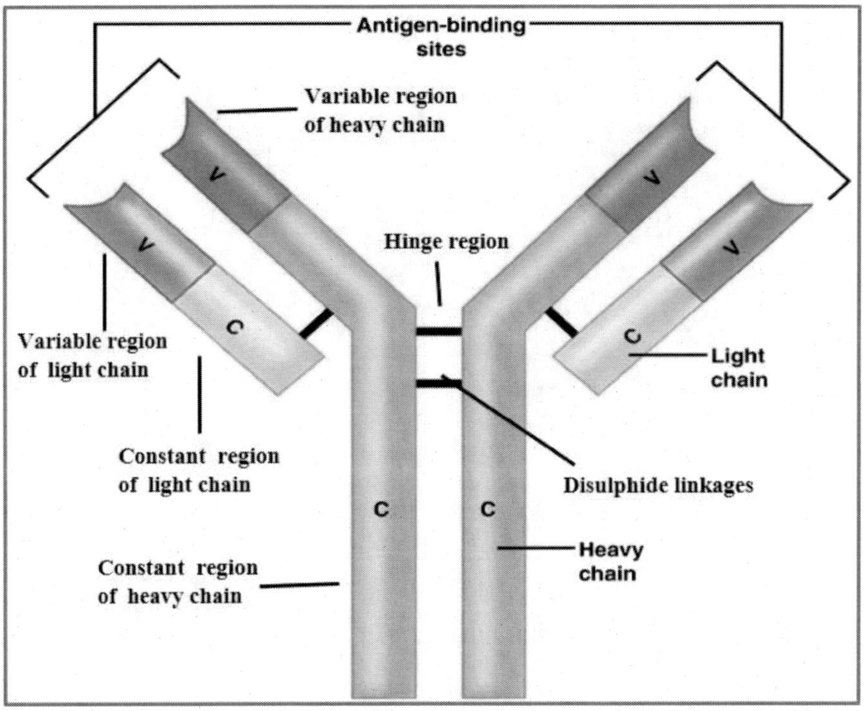

Figure 4.4. Representing the structure of an antibody.

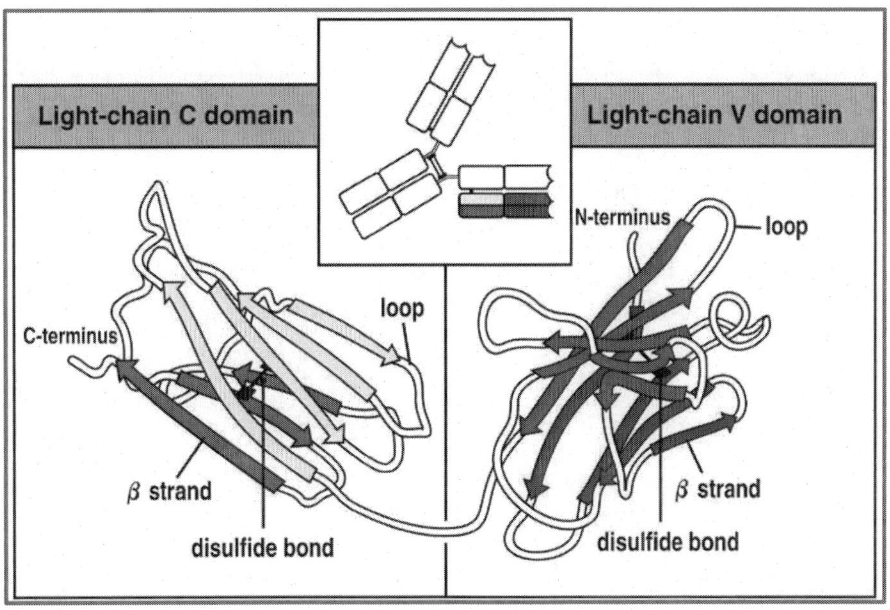

Figure 4.5. Showing the arrangement of polypeptides and the disulphide bridge in an antibody. (Image source: NCBI).

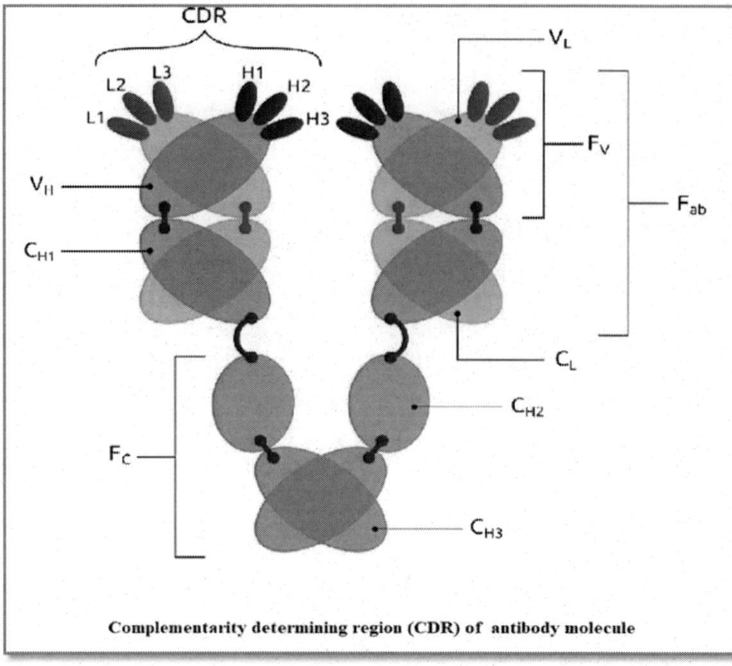

Figure 4.6. Showing the arrangement of complementarity determining region (CDR) responsible for specificity of an antibody molecule. (Image Source: NCBI).

Each light chain consists of one variable domain (VL) and one constant domain (CL). The heavy chains have a variable domain and three constant domains (CH1, CH2&CH3). Antigen binding sites are located on the variable regions of light and heavy chains. Immunoglobulins have a hinge region which is composed of short segments of amino acids predominantly made of cysteine and proline residues. It is the area of heavy chains between the first and second C region domains and is held together by disulphide bonds. The flexible hinge region is found in IgG, IgA and IgD. It is absent in IgM and IgE. It allows the distance between the two antigen bin ding sites to vary. The structure of immunoglobulins includes several features which are necessary for their participation in the immune response. These features are

a) Specificity
b) Biologic activity

Specificity is assigned to a defined region of antibody molecule containing the complementarity determining region (CDR). CDR limits the antibody to combine to only with those substances that contain a particular antigenic structure (Figure 4.6).

Complementary Determining Regions (CDRs)

Complementarity-determining regions (CDRs) are part of the variable chains in immunoglobulins (antibodies) and T cell receptors, where these molecules bind to their specific antigen. A set of CDRs constitutes a paratope which binds with the epitope of the antigen during antibody-antigen interaction. The region within the variable region that generates the antigen binding site are called the complementarity determining regions or CDRs. Being the most variable parts of the immunoglobulin molecules, CDRs are crucial to the diversity of antigen specificities generated by lymphocytes. There are three CDRs (CDR1, CDR2 and CDR3), arranged non-consecutively, on the amino acid sequence of a variable domain of an antigen receptor. Since the antigen receptors are typically composed of two variable domains (on two different polypeptide chains, heavy and light chain), there are six CDRs for each antigen receptor that can collectively come into contact with the antigen. A single antibody molecule has two antigen receptors and therefore contains twelve CDRs total. There are three CDR loops per variable domain in antibodies. Sixty CDRs can be found on a pentameric IgM molecule.

General Features of Immunoglobulins

Antigen Binding

Immunoglobulins bind particularly to one or a few closely related antigens. Each immunoglobulin actually binds to a particular antigen determinant. Antigen binding by the antibodies is the primary function of antibodies and can result in the protection of host.

Effector Functions

Usually the binding of an antibody to an antigen has no direct biological effect. Somewhat, the notable biological effects are a consequence of a secondary "effector functions" of antibodies. The immunoglobulins mediate a variety of functions. To carry out a particular effector function requires antibody bind to its antigen. The effector functions include fixation of complement and binding to various cell types.

Antibodies form basis of humoral immunity by searching out and neutralizing antigens or marking them for elimination.

Antigen-Antibody Interaction

Antigen-antibody interaction is a specific chemical interaction between antibodies formed by B-cells of white blood cells and antigens during immune reactions. It is the fundamental reaction in the body by which the body is protected from the complex extraneous molecules, such as pathogens and their chemical toxins. Antigen antibody complex is formed in blood where a highly attracted antigen bind to the antibody (Figure 4.7). The immune complex thus formed is transported to cellular system where it can be destroyed or deactivated. In body these reactions form the basis of antibody mediated immunity in infectious diseases, or of tissue injury in some types of hypersensitivity and auto immune diseases.

In the laboratory, antigen antibody interactions help in the diagnosis of infections in epidemiological surveys, in the identification of infectious agents and of noninfectious agents such as enzymes. These reactions can be used for detection and quantitation of either antigens or antibodies. Antigen antibody reactions in vitro are known as serological reactions. The reaction between antigen and antibody is a bimolecular association and it does not lead to an irreversible chemical alteration in either the antibody or the antigen (Figure 4.8).

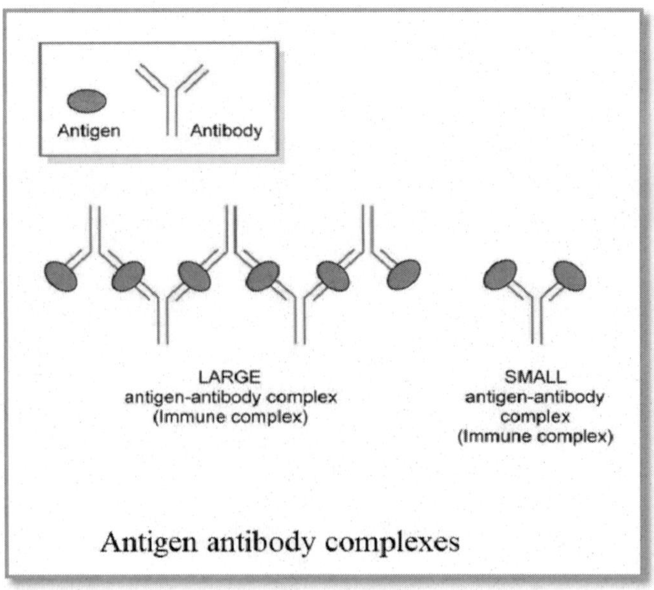

Figure 4.7. Showing the formation of large and small antigen-antibody (immune) complexes.

Both the antigen and antibodies are part of the human body's immune system. Antigens are usually large, complex foreign substance that cause the production of antibodies. The body recognizes these foreign antigens as invaders and moves to destroy them with lymphocytes or white blood cells, which secrete antibodies. The body produces antibodies only for the purpose of destroying antigens.

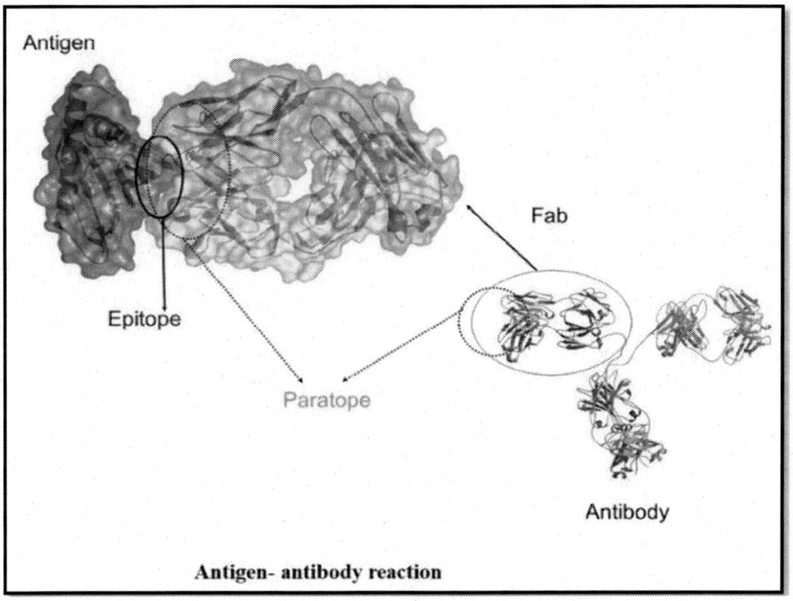

Figure 4.8. Representing the reaction between the antigen-antibody.

The association between antigen and antibody involves various non-covalent interactions between the antigenic determinant(epitope) of the antigen and a variable region (VL/VH) domain of the antibody molecule. Antigen antibody interactions are spontaneously reversible under the conditions of temperature, pressure, pH, and ionic strength that generally prevail in living organisms. Several types of non-covalent bonds have been shown to contribute to antigen antibody binding these include Vander wails force, hydrogen bonds, ionic bonds and hydrophobic interactions. Individually the strength of these bonds is in the range of one to few Kcal/mol versus 50-100Kcal/mol for covalent bonds.

The reactions between antigens and antibodies occur in three stages. The primary stage is the initial interaction between the two without any visible effects. No covalent bonds are involved in the interactions therefore the binding forces are relatively weak. They consist mainly of Vander Waals forces, electrostatic forces and hydrophobic forces all of which require is a very close fit between an epitope and the antibody. The primary reaction can be detected by estimating free and bound antigen or antibody separately in the reaction mixture by a number of physical and chemical methods including the use of markers such as fluorescent dye, ferritin, or radioactive isotopes.

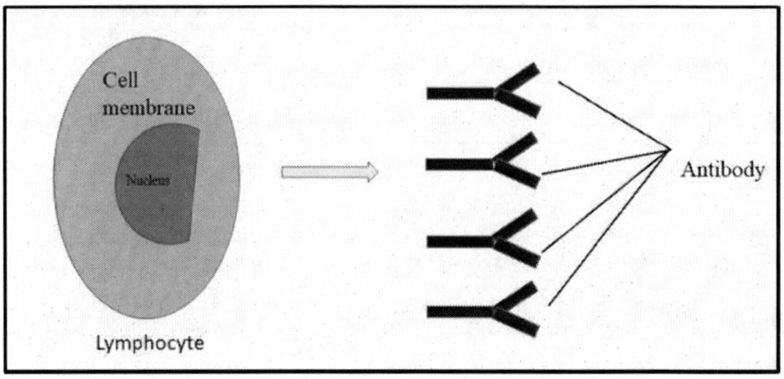

Figure 4.9. Representing the action of release of antibodies from the B-lymphocyte.

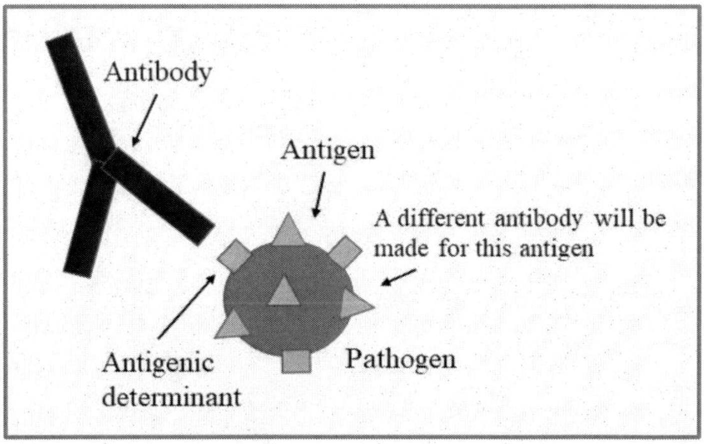

Figure 4.10. Illustrating the specificity of binding between antigen and antibody.

Because of the low energy levels of the energy involved in the interaction between antigen and antibody, antigen antibody complexes are readily dissociated by low or high pH or by high salt concentration, which interfere with the hydrogen bonding of water molecule.

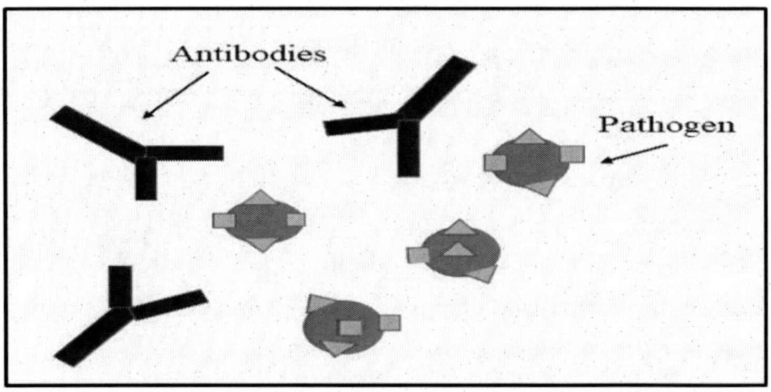

Figure 4.11. Antibodies attach to antigens; (This prevents the pathogen from entering a host cell).

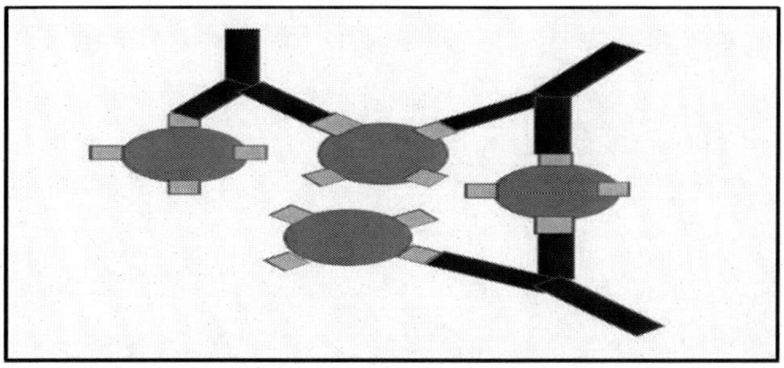

Figure 4.12. Antibodies cause pathogens to clump together.

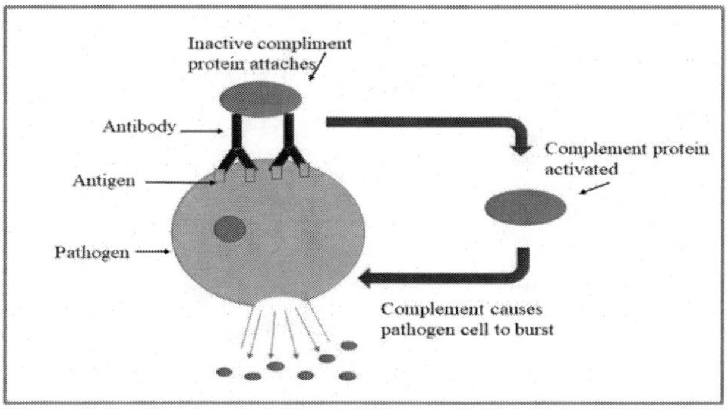

Figure 4.13. Antibody activates compliment protein.

In most of the cases but not all primary stage is followed by secondary stage leading to the clearly apparent leading events such as precipitation, lysis of cell, agglutination, killing of live antigens, neutralization of toxins and other biologically active agents, immobilization of motile organisms, phagocytosis, etc. (Figure 4.9-4.13).

GENERAL FEATURES OF ANTIGEN-ANTIBODY REACTIONS

Antigen –antibody interaction has following general characteristics:

- The reaction is specific, the specificity however is not absolute and 'cross-reactions' may occur due to antigenic similarity and relatedness.
- When an antigenic determinant present in a large molecular or on a 'carrier' particle reacts with its antibody, whole molecule or particle are agglutinated.
- There is no change in antigen or antibody during the reaction.
- The combination occurs at the surface; hence the surface antigens are immunologically relevant. Antibodies to the surface antigens of infectious agents are protective.
- The association is firm but reversible. The firmness of the union is influenced by the affinity and avidity of the reaction.

- Both the antigen and antibody participate in the formation of agglutinates and precipitates.
- Antigens and antibodies can bond in varying proportions, unlike chemicals with fixed valences.

FORCES THAT BIND ANTIGEN TO ANTIBODY

The forces that bind antigen to antibody become larger as intercellular distances decrease.

These forces are, in essence, no different from the 'non-specific' interactions which occur between any two unrelated proteins (Figure 4.14). These intermolecular forces are:

- Electrostatic forces (attraction between oppositely charged ionic groups)
- Hydrogen bonding (reversible hydrogen bridges between hydrophilic groups)
- Hydrophobic bonding (similar to the manner in which oil droplets in water merge to form a single large drop, side chains of valine, leucine and phenylalanine tend to associate in an aqueous environment. This may account for over 50% of the strength in an antigen antibody bond).
- Van der Waals forces: interaction between the electrons in the external orbits of the two different macro molecules.

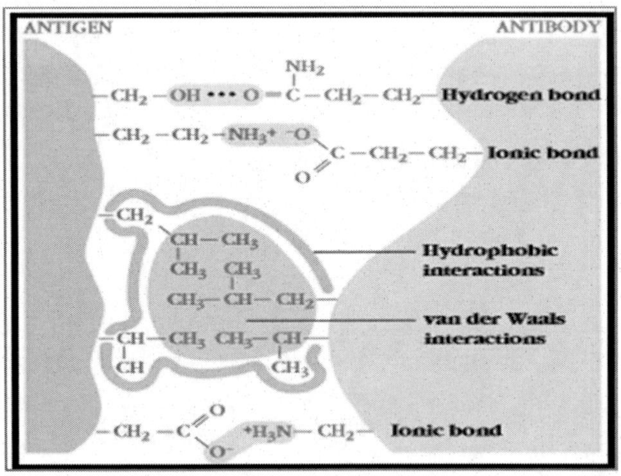

Figure 4.14. The interaction between an antibody and an antigen depends on four types of noncovalent forces: (1) hydrogen bonds, in which a hydrogen atom is shared between two electronegative atoms; (2) ionic bonds between oppositely charged residues; (3) hydrophobic interactions, in which water forces hydrophobic groups together; and (4) van der Waals interactions between the outer electron clouds of two or more atoms. In an aqueous environment, noncovalent interactions are extremely weak and depend upon close complementarity of the shapes of antibody and antigen (adapted from kuby immunology, 6th edition).

PROPERTIES OF ANTIGEN- ANTIBODY INTERACTIONS

A strong antigen-antibody interaction depends on a very close fit between the antigen and antibody which requires high degree of specificity. Properties of antigen antibody can be defined with the help of following points. They are:

- Affinity
- Avidity
- Specificity
- Cross reactivity

Affinity

Affinity is the sum of the attractive and repulsive forces operating between the antigenic determinants and combining site of antibody. There exists a weak association between low-affinity antibodies and antigen, which dissociates easily whereas high affinity antibodies bind antigen more tightly and remain bound longer. A strong antigen antibody interaction depends on a very close fit between antigen and antibody (Figure 4.15). The combined strength of the non-covalent interactions between a single antigen binding site on an antibody and a single epitope is the affinity of the antibody for that epitope.

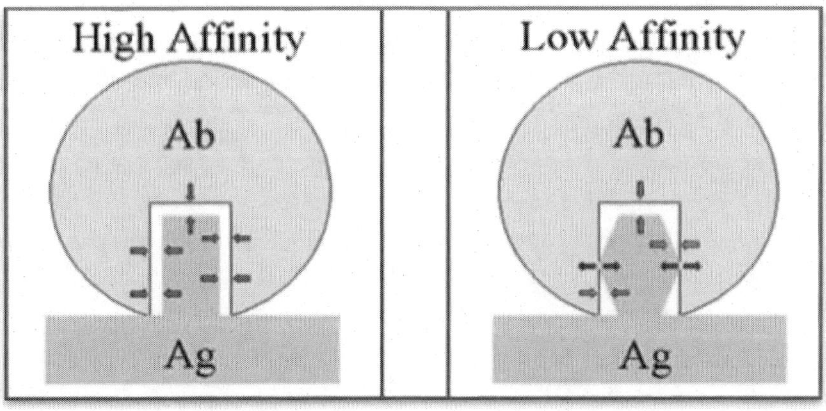

Figure 4.15. Representing Affinity between Antigen-Antibody (Source: Immunology-chapter 7, Immunoglobulins-antigen-antibody reactions and selected tests, by Gene Mayer, Ph.D).

The binding of an antibody (Ab) to its antigen (Ag) is reversible, so the binding reaction can be expressed as:

$$Ag + Ab \leftrightarrow Ag\text{-}Ab, \text{ Or}$$

Ka [Ag][Ab]= kd [Ag-Ab]

Where [Ag] and [Ab] are the concentrations of unbound Ag and Ab, [Ag-Ab] is the concentration of bound antigen and antibodies, ka and kd are dissociation and association constants.

At equilibrium, the equilibrium constant K=ka/kd=[AgAb]/[Ag][Ab]. If all binding sites of an Ab have an equilibrium constant, then

r/c= nk- rk, or, k=r/(n-r)c

where r is moles of ligands bound/mole of antibody, c=concentration of unbound(free) ligand in the moles/litre, and n is the valency of antibody (number of ligand binding sites). k values can vary.

Avidity

Avidity is used to designate the strength of binding of a multivalent antigen with an antibody possessing at least two combining sites or an antiserum containing antibodies of differing specificities. The strength of Ag-Ab reaction is enhanced when several different antibodies bind simultaneously to different epitopes on the antigen molecule, cross-linking antigen molecules very tightly (Figure 4.16). Thus, a more stable bonding between antigen and antibody will be established due to "bonous effect" of multiple antigen- antibody bonds; the increased stability of overall antigen-antibody reaction corresponds to an increased avidity. Avidity is more than the sum of the individual affinities. Affinity describes the strength of interaction between antibody and antigen at single antigenic sites. Avidity is a more informative measure of overall stability or strength of antigen-antibody complex, which is controlled by three major factors; antibody-epitope affinity; the valence of both the antigen and antibody; and structural arrangements of the interacting parts.

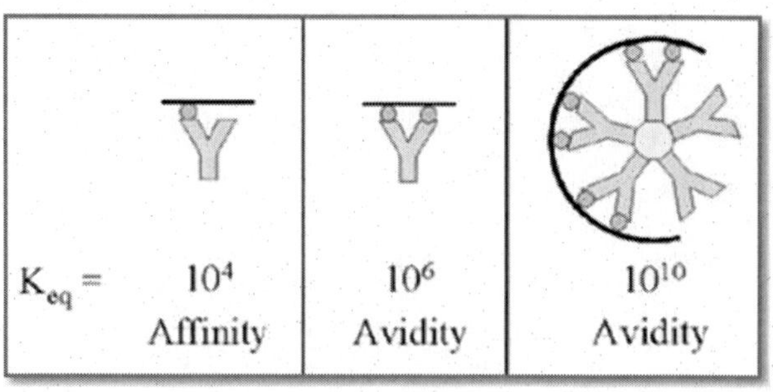

Figure 4.16. Representing Difference in affinity and avidity of antigen-antibody binding (Source: Immunology-chapter 7, Immunoglobulins-antigen-antibody reactions and selected tests, by Gene Mayer, Ph.D).

Specificity

Specificity of an antigen antibody interaction refers to the ability of an individual antibody combining sites to react with only one antigenic determinant or the ability of a population of antibody molecules to react with only one antigen. One antibody can react with only its own antigen because the antigen antibody interactions are highly specific in nature. Specificity of an antibody molecule refers to its ability to discriminate between the epitope that stimulated its production and other determinants. High -affinity antibodies are more specific than low-affinity antibodies. A strong antigen antibody interaction depends on a very close fit between the antigen and antibody which requires high degree of specificity.

Cross-Reactivity

The most notable feature of antigen antibody interactions is its high specificity and affinity. But in some cases, antibodies induced by one antigen can cross react with an unrelated antigen, which is known as cross reactivity. Such cross reactions occur if two different antigens share an identical or very similar epitope.in the case where epitopes are similar, the antibody's affinity for cross reacting epitope is usually less than that for the original epitope (Figure 4.17).

Cross reactivity is often observed among polysaccharide antigens that contain similar oligosaccharide residues. The ABO blood group antigens are the glycoproteins expressed on red blood cells. Fine differences in the terminal residues of the sugars attached to these surface proteins distinguish the A and B blood group antigens. An individual lacking one or both of these antigens will have serum antibodies to the missing antigens. Therefore, anti-A antibodies are found in the serum of group O and B individuals and anti-B antibodies are found in the serum of group O and A individual. Group AB individuals are believed not to have anti-A nor anti-B antibodies because they express both the antigens on red blood cells (Table 4.1).

Table 4.1. ABO blood types

Blood type	Antigens on RBCs	Serum antibodies
A	A	Anti-B
B	B	Anti-A
AB	A and B	Neither
O	Neither	Anti-A and anti-B

The antibodies are induced not by the exposure to the red blood cells but by the exposure to the cross-reacting microbial agents present on common intestinal bacteria.

These microbial antigens induce the formation of antibodies individuals lacking the similar blood-group antigens lacking on their red blood cells. The blood-group antibodies, although induced by microbial antigens, will cross react with similar oligosaccharides on foreign red blood cells, contributing the basis for blood typing testes and accounting for necessity of compatible blood types during blood transfusions.

A number of viruses and bacteria have antigenic determinants that are identical or similar to the normal host cell components, in some cases these microbial agents induce antibodies which cross react with the host cells and cause tissue damaging immune reaction.

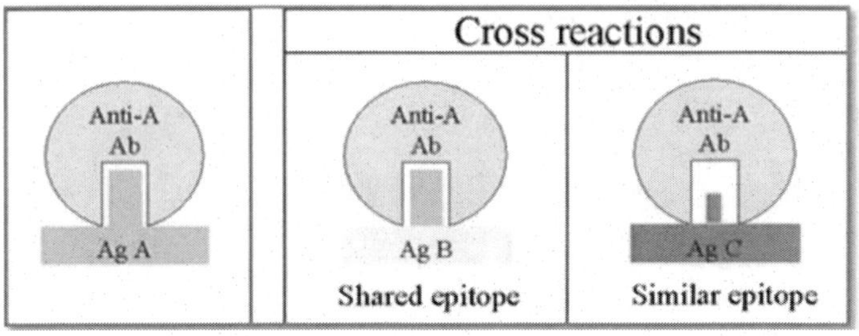

Figure 4.17. Showing Cross reactive interaction between antigen and antibody. (Source: Immunology-chapter 7, Immunoglobulins-antigen-antibody reactions and selected tests, by Gene Mayer, Ph.D).

Factors Affecting Antigen-Antibody Interaction

The study of antigen –antibody reactions in vitro is called serology. Serological reactions are the basis for all diagnostic immunology tests. Antigen-antibody reactions depend on the number of factors.

Affinity-the higher the affinity of the antibody for antigen, the more stable will be the interaction.

Avidity-reactions between multivalent antigens and multivalent antibodies are more stable.

Antigen to antibody ratio-the ratio between the antigen and antibody influences the detection of antigen- antibody complexes because the size of the complexes formed is related to the concentration of the antigen and antibody.

Physical form of antigen-the physical form of the antigen influences the reaction with the antibody. When the antigen is particulate, the reaction of the antibody with the antigen can be detected by agglutination (clumping) of the antigen. If the antigen is soluble, one generally looks for the precipitation of the antigen after the production of large insoluble antigen-antibody complexes.

TYPES OF ANTIGEN-ANTIBODY REACTION

- Agglutination.
- Precipitation.
- Complement Fixation.
- Enzyme Linked Immunosorbent Assay.
- Radioimmunoassay.
- Western blotting

Agglutination

Agglutination is the aggregation of cells due to antibody binding. The word agglutination has been derived from a Latin word agglutinate means "to glue to." Agglutination is the interaction between di- or multivalent antibodies and multivalent, bacterial, or other cellular antigen resulting in visible clumping. The general term agglutinin is used to describe antibodies that agglutinate particulate antigens. Agglutination is a serological reaction. This reaction is highly specific and cross reaction occurs between antibodies and antigens. The cross-linked products are visible to naked eye because of larger size of the antigens (Figure 4.18). Excess of antibodies inhibit agglutination reactions; this inhibition is known as prozone effect. Antigens for this reaction are large, easily sediment particles.

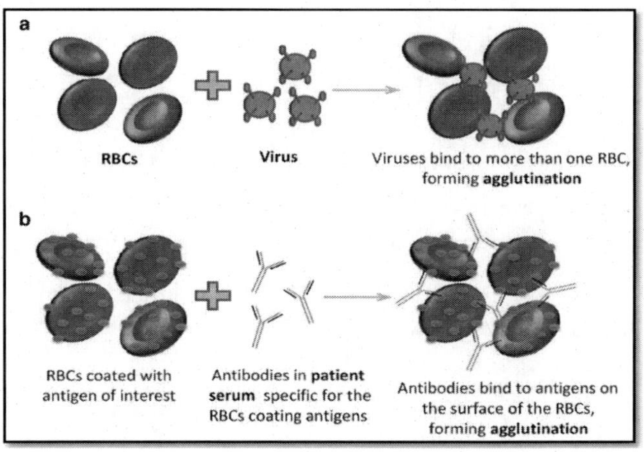

Figure 4.18. Showing the agglutination reaction between an antigen and antibody. (Image Source: Alhabbab R.Y. (2018) https://doi.org/10.1007/978-3-319-77694-1_3).

Prozone Effect

Antibodies when present at high concentrations, antibody binding sites are present in great numbers and epitopes are present in less number. Agglutination may not occur at high ratio of antibody to antigen because every epitope of one antigen will bind only to a single antibody molecule, preventing cross linking of different antigens. Prozone effects are observed by performing the assays at a variety of antigens or antibody concentration. Higher levels of agglutination can be seen at optimum dilution of antibody concentration. When using polyclonal antibodies incomplete antibodies (class IgG) also causes prozone effect.

TYPES OF AGGLUTINATION REACTIONS

Agglutinations reactions can be broadly divided into three groups:

- Active/Direct agglutination
- Passive agglutination
- Hem agglutination

Active/Direct Agglutination

Agglutination reactions where the antigens are found naturally on a particle are known as direct agglutination. Cells displaying surface antigens can clump or agglutinate when mixed with their antisera. Direct agglutinations provide a simple and fast method for identifying cells if known antibodies are available (Figure 4.19). The agglutination reaction can also be used as a semi quantitative assay. If a patient is suspected of having typhoid fever, the patient's serum is mixed with a culture of *Salmonella typhii*. If agglutination reaction occurs, shown as clumping of the bacteria, then the patient is said to have an infection of *S. typhii* at present or had it in past because certain antibodies can persist in patient's blood for years after the patient has recovered from the infection.

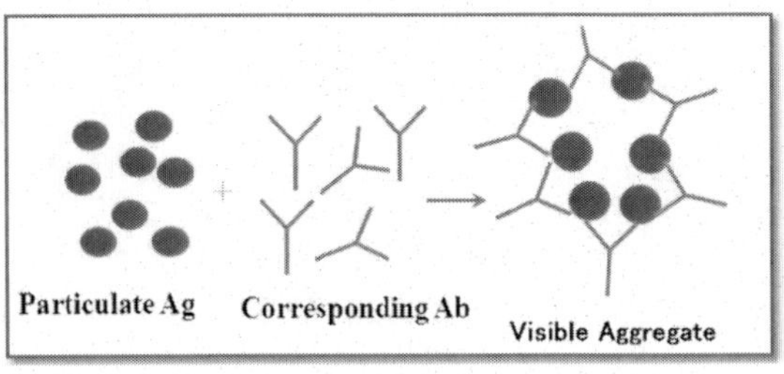

Figure 4.19. Showing the active agglutination reaction between antigen and antibody.

To determine whether a patient is suffering from typhoid fever, the amount or titer of the antibody will be determined at the onset of illness. A given number of cells are added to a series of tubes, each containing the same volume of antiserum, at different dilutions. Over time, clumping is observed and relative strength of an antiserum is expressed as the reciprocal of the highest dilution that causes agglutination.

Passive Agglutination

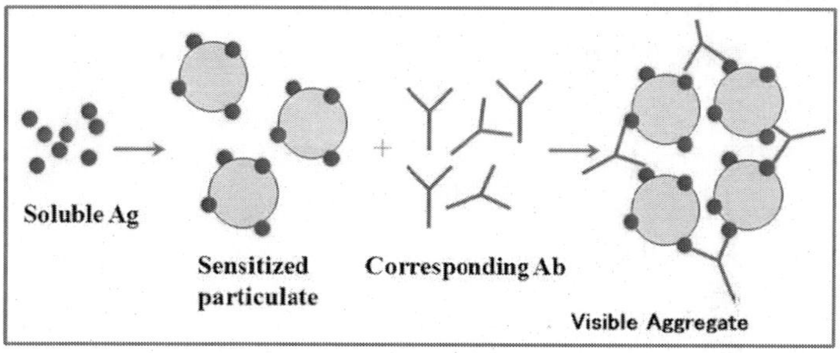

Figure 4.20. Showing the passive agglutination reaction between antigen and antibody.

It is defined as the agglutination of particles that have been coated with soluble antigen, by antiserum specific for the adsorbed antigen. Passive hem agglutination is a kind of passive agglutination in which red blood cells, usually modified by mild treatment with tannic acid or chromium chloride, are used to adsorb soluble antigen onto their surface, and which then agglutinate in the presence of antiserum specific for the adsorbed antigen (Figure 4.20). Serum containing antibodies is serially diluted into micro titer plate wells and the antigen-coated erythrocytes are then added to each well. Agglutination is assayed by the size of the characteristic spread pattern of agglutinated erythrocytes on the bottom of the well. Passive agglutination can be performed with tanned erythrocytes or synthetic particles, such as latex beads. The use of synthetic beads offers the advantages of consistency, uniformity and stability. The first step in this test is linking together of the latex particles by the antibody molecules that specifically attach to the antigenic determinants on the surface of particle. There is a formation of large lattice through these large clumps and are visible to naked eye within minutes.

Hemagglutination

RBCs are used as carrier particles in hem agglutination tests. RBCs of sheep, human, chick etc. are commonly used in the test. When RBCs are coated with antigen detect antibodies in the serum, the test is called indirect hem agglutination (IHA) test. Hemagglutination uses red blood cells as the biological carriers of bacterial antigens, and purified polysaccharides for determining the presence of corresponding antibodies in a specimen. When antibodies are attached to the RBCs to detect microbial antigen it is known as reverse passive hem agglutination (RPHA).

Precipitation Reaction

Precipitation reaction is the reaction, in which a soluble antibody reacts with a soluble antigen to give an insoluble product or the precipitate. Soluble antibodies that aggregate soluble antigens are called precipitins. Soluble antigen that induces the formation of a specific precipitin is called a precipitinogen (Figure 4.21).

When the antigens, which must have at least two epitopes per molecule are cross-linked by the bivalent antibodies, a lattice formation occurs which ultimately develops into a visible precipitate. For the lattice to be formed, the bivalent antibody will bind to epitopes on two different antigens. A second antibody molecule combines with the second epitope on one of the antigen molecules and a third epitope on another antigen molecule, so that the complex is formed. The complex continues to grow and when it is sufficiently large, it becomes insoluble and can be visible as a precipitate. Thus, formation of the visible precipitate takes sufficiently longer time than formation of the soluble antigen-antibody complexes. Precipitation reaction can occur using poly clonal antibodies or mixture of monoclonal antibodies. Polyclonal antibodies can form large aggregates that precipitates out of the solution. If the antigen is monovalent or a single monoclonal antibody is used, the antibody can link only two molecules of antigen and no precipitate is formed.

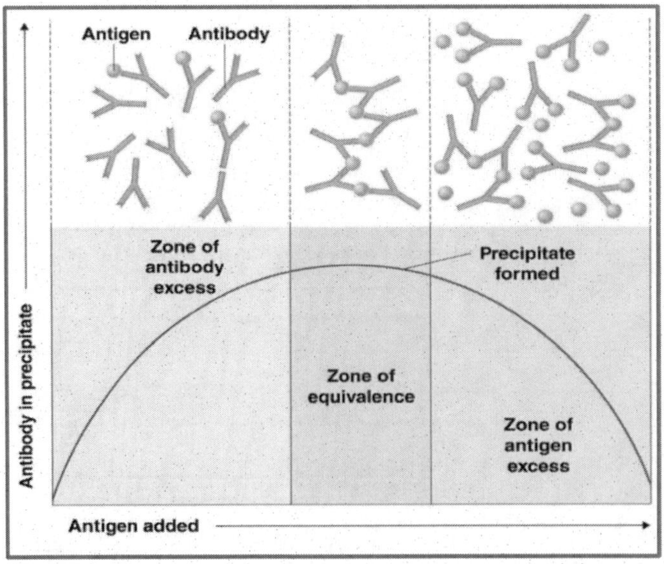

Figure 4.21. Graph showing the formation of specific precipitate induced by the specific antigen. (Image Source: NCBI).

Types of Precipitation Reactions

Precipitation reactions occur both in solution and in gel phase, where antigen-antibody forms a precipitate.

- Fluid phase precipitation
- Gel phase precipitation

Fluid Phase Precipitation

It is a double diffusion method, where in a capillary tube an antigen solution is layered over an antibody solution. Both antigen and antibody will diffuse towards each other and at an interface when the antigen is recognized by antibody there occurs formation of precipitate. This method is used to identify unknown antigen or unknown antibody.

Gel Phase Precipitation

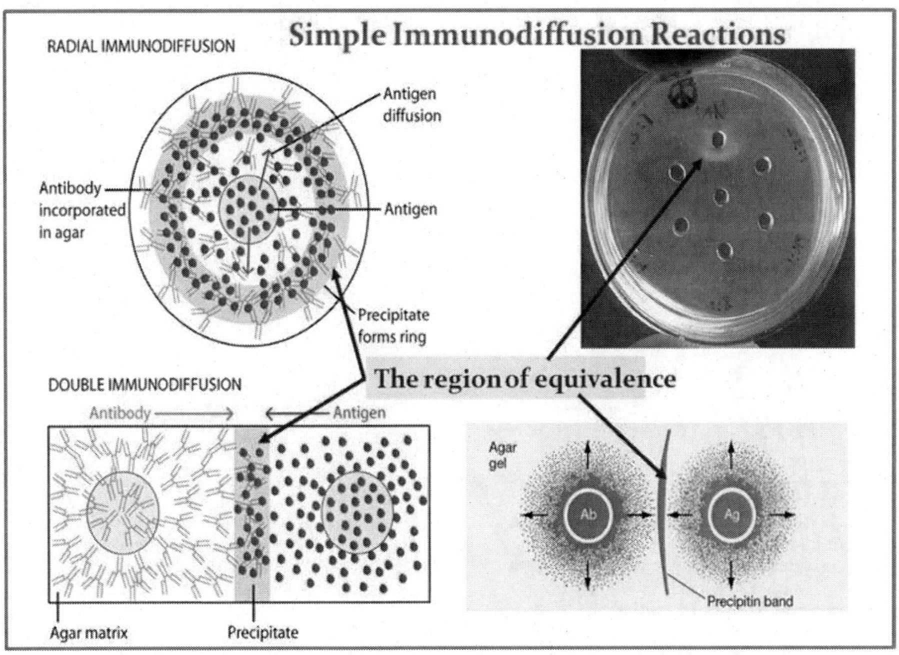

Figure 4.22. Diagrammatic representation of radial and double immunodiffusion. (Precipitation reactions in gels yield visible precipitin; no visible precipitate forms in regions of Ab or Ag excess). (Image source: Slidshare).

In this reaction gel reaction gel is used as a semisolid medium. The gel contains pores that allow the movement of molecules (Figure 4.22). Agarose gel is used which allows soluble antigen and/or antibody to diffuse through the pores until the antigen and antibody reach the optimal concentration for lattice formation. Smaller molecules move through gel faster than larger molecules.

Complement Fixation

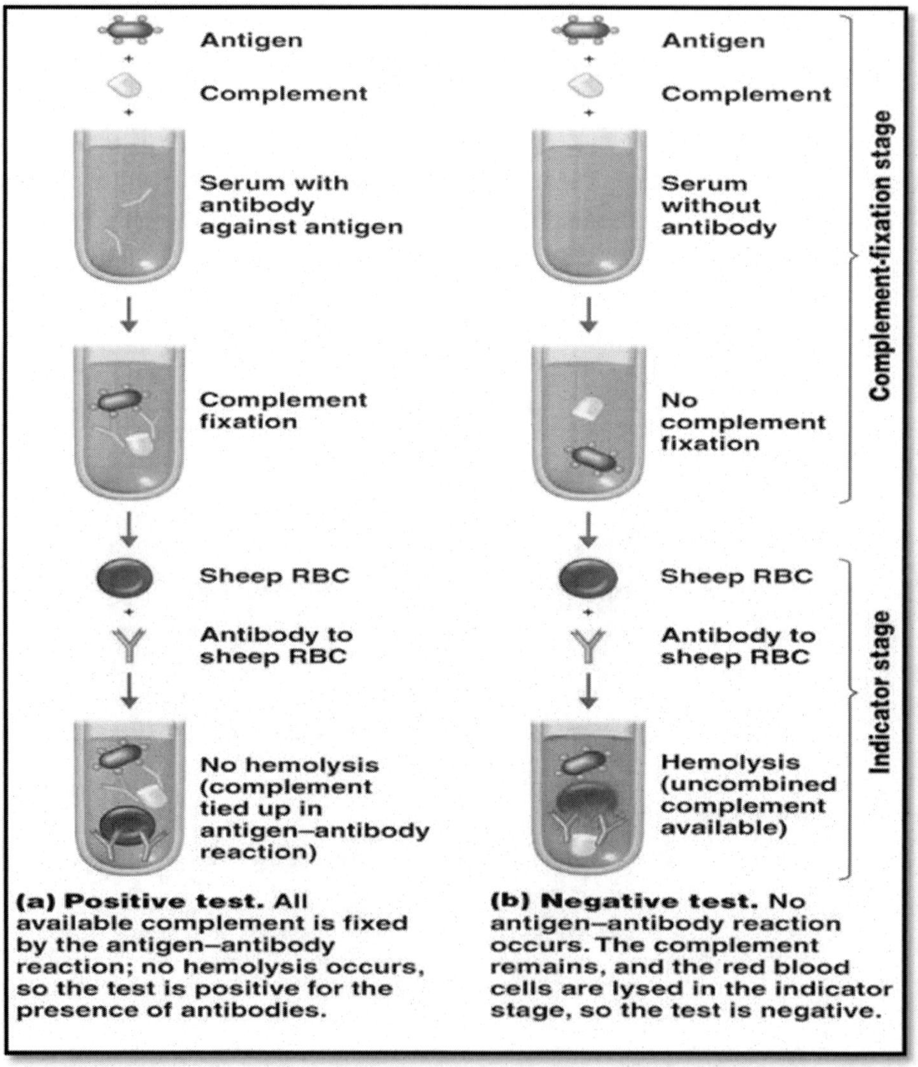

Figure 4.23. Illustrating steps in complement fixation test (Source: https://laboratoryinfo.com/complement-fixation-test/).

It can be used detect the presence of either specific antibody or specific antigen in a patient's serum. Complement is the activity of blood serum that completes the action of antibody. The complement system is a system of serum proteins that react with antigen-antibody complexes. Antibody binding with antigen is not sufficient to remove the antigen from body. Hence antigen –antibody complex initiates the activation and binding of complement system to it. Binding of complement to Ag-Ab complex is known as complement fixation (Figure 4.23). The activated complement enzymes continue further immune reactions that finally removes the foreign agent.

Inactive enzymes of the complement get activated in a cascade manner in presence of Ag-Ab complex, and bind to the pathogen from membrane associated complex.

Complement fixation assay can be used to check presence of specific antibody or specific antigen in an individual's serum. It is widely used to diagnose infections particularly microbial infections and rheumatoid diseases that are not easily detected by culture methods.

Enzyme Linked Immunosorbent Assay (ELISA)

The ELISA is a fundamental tool of clinical immunology, which has been used as a diagnostic tool in medicine and plant pathology, as well as a quality-control check in various industries. ELISA, or Enzyme-linked Immunosorbent assay that uses a solid-phase enzyme immunoassay (EIA) to detect the presence of a substance, usually an antigen, or antibody in a liquid sample or wet sample (Figure 4.24). It depends on an enzyme-substrate reaction that generates a colored reaction product. An enzyme conjugated with an antibody reacts with a colorless substrate to generate a colored reaction product. A number of enzymes i.e., alkaline phosphatase, horseradish peroxidase, and β-galactosidase, have been employed for ELISA.

A number of variations of ELISA have been developed, allowing qualitative detection or quantitative measurement of either antigen or antibody. Four types of ELISA methods are there, but generally three methods are employed for detection of either antigen or antibody. Direct ELISA is the simplest type of ELISA. Antibody can be determined with an indirect ELISA whereas antigen can be determined with a sandwich ELISA or competitive ELISA. Each assay can be used qualitatively or quantitatively by comparison with standard curves prepared with known concentrations of antibody or antigen. ELISA and RIA will be discussed in detail in a separate chapter.

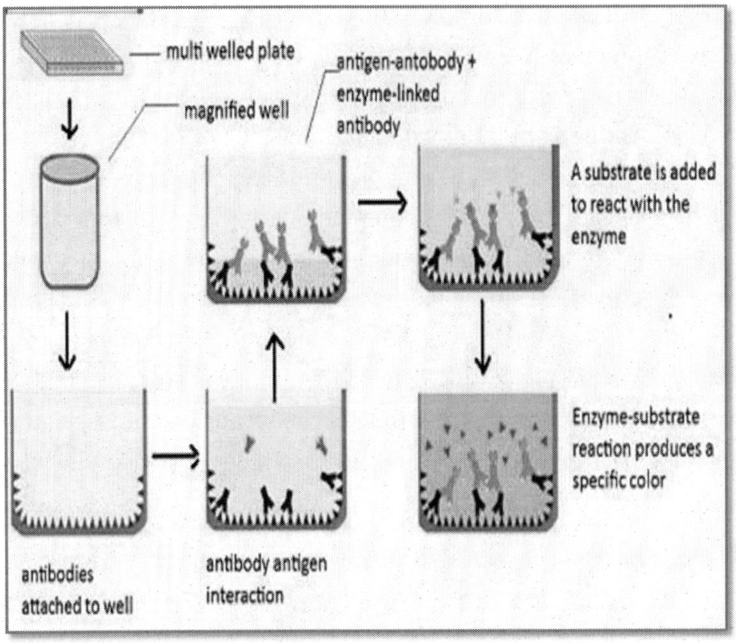

Figure 4.24. Showing general procedure in ELISA. (Source: http://www.aipmtbio.co.in).

Direct ELISA

The direct ELISA is a method for detecting and measuring antigen concentration in a sample. In direct ELISA, the presence of a particular antigen in a sample, is detected by using a capture monoclonal antibody. The procedure for direct ELISA includes (Figure 4.24).

- The wells of a micro titer plate are coated with a sample containing the target antigen.
- The antigen is fixed to the surface to make it immobile.
- The plate wells are then coated with a blocking buffer.
- In a separate reaction, an enzyme is linked to an antibody.
- The enzyme-antibody compound is added to the wells to adsorb to the antigen. The plate is washed to remove any surplus enzyme-antibody conjugate.
- Then a substrate is applied for the enzyme, and which is converted by the enzyme to evoke a chromogenic or fluorescent signal. Therefore, substrate detects the presence of the enzyme and the antigen.
- The amount of colored product is measured by a specialized spectrophotometric plate reader.

The advantage of direct ELISA is that it is relatively quick because of the use of only one antibody but direct ELISA requires the labeling of every primary antibody, which can be time-consuming and more expensive than in indirect methods. Certain antibodies may be unsuitable for direct labeling. Direct methods do not allow for signal amplification in contrast to methods that use a secondary antibody. Direct ELISA can be used to test specific antibody-to-antigen reactions, and helps to eliminate cross-reactivity between other antibodies.

Indirect ELISA

Indirect ELISA helps detecting antibody. By this method serum antibody against human immunodeficiency virus (HIV), the causative agent of AIDS is detected.

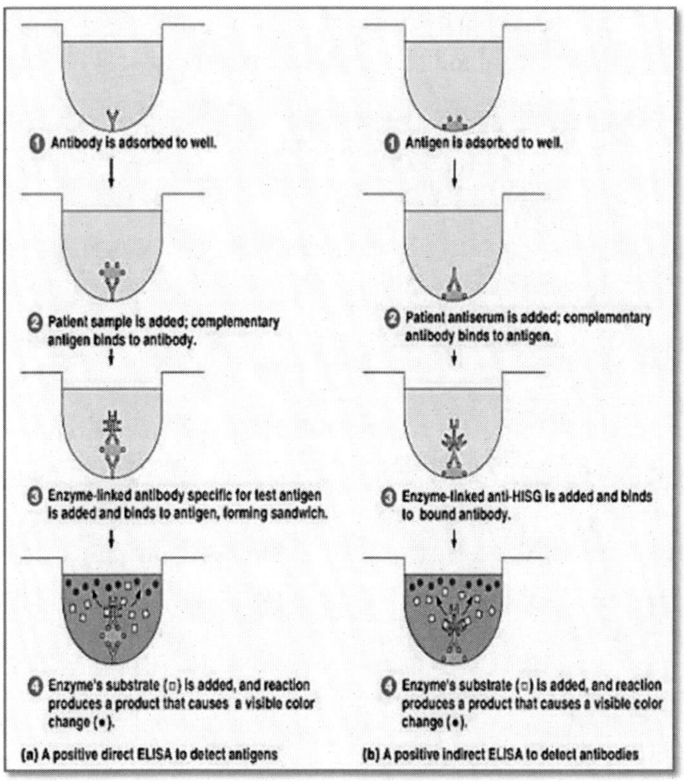

Figure 4.25. Illustrating steps of Direct and Indirect ELISA. (Source: Google scholar).

The procedure of indirect ELISA includes (Figure 4.25).

A buffered solution of the antigen to be tested for is added to each well of a micro titer plate. Coating is done through passive adsorption of the antigen to the assay micro plate. This process occurs though hydrophobic interactions between the micro titer plate and nonpolar protein residues. After incubation any excess antigen is removed by washing steps by flooding and emptying the wells with deionized water. A solution of nonreacting protein, such as bovine serum albumin or casein, is added to block any plastic surface in the well that remains uncoated by the antigen.

Serum or some other sample containing primary antibody is added to the antigen-coated micro titer well. The antibody specific to the test antigen, binds the coated antigen on incubation. Excess antibody or any unbound antibodies are removed by washing and is followed by addition of blocking solution.

Any free secondary antibody is washed away, and a substrate for the enzyme is then added. Often, this substrate changes color upon reaction with the enzyme, which shows the secondary antibody has bound to primary antibody. The higher the concentration of the primary antibody present in the serum, the stronger the color change. The amount of colored reaction product that forms is measured by specialized spectrophotometric plate readers.

The presence of antibody bound to the antigen is detected by adding an enzyme-conjugated secondary anti-isotype antibody, which binds to the primary antibody. Serum antibodies to HIV can be detected by in directed ELISA within six weeks of infection. In this assay, recombinant envelope and core proteins of HIV are adsorbed to solid-phase antigens to micro titer wells. Individuals infected with HIV will produce serum antibodies to epitopes on these viral proteins.

Sandwich ELISA

In this technique, the antibody is immobilized on a micro titer well. A sample containing unknown amounts of antigen is allowed to react with immobilized antibody. After the well is washed, a second enzyme-linked antibody specific for a different epitope on antigen is added and allowed to react with the bound antigen. Any extra free secondary antibody is removed by washing, substrate is added and colored amount is measured (Figure 4.26).

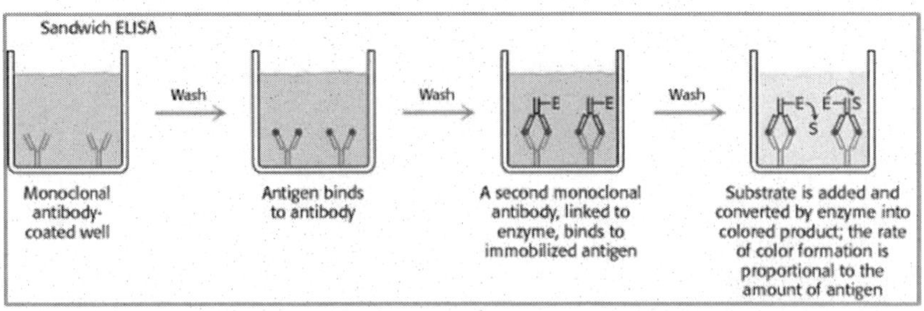

Figure 4.26. Illustrating the steps and compounds involved in sandwich ELISA.

Sandwich ELISA have proven useful for measurement of soluble cytokine concentrations in the tissue culture supernatants as well as in serum and body fluids.

Competitive ELISA

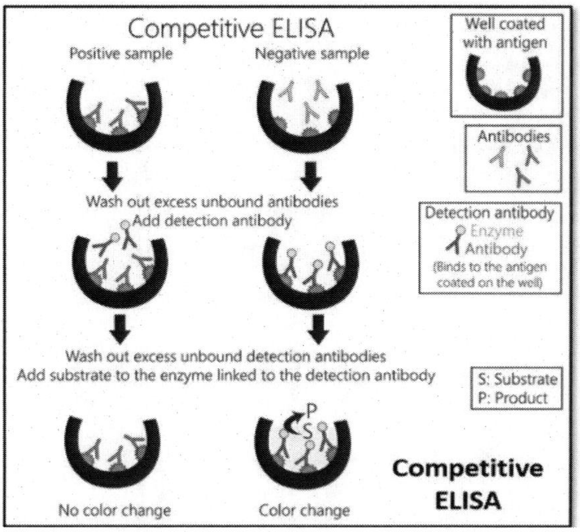

Figure 4.27. Showing the procedure involved in competitive ELISA and the comparison between the specific and unspecific sample. (Image Source: Apcam).

In this technique, antibody is first incubated in solution with a sample containing antigen. The antigen-antibody mixture is added to an antigen coated micro titer well. The more antigen presents in the initial solution phase sample; the less free antibody will be available to bind to the antigen coated well (Figure 4.27). The competitive ELISA provides extremely sensitive variation for measuring amount of antigen.

Radioimmunoassay (RIA)

Radioimmunoassay is a very sensitive in vitro assay technique used for separation of a protein from a mixture using the specificity of antibody-antigen binding and quantitation using radioactivity. The technique was first developed in 1960 by two endocrinologists, S.A. Berson and Rosalyn Yalow, to determine levels of insulin-anti-insulin complexes in diabetics. Although the RIA technique is extremely sensitive and extremely specific, requiring specialized equipment, it remains the least expensive method to perform such tests. It requires special precautions and licensing, since radioactive substances are used.

The principle of RIA involves competitive binding of radiolabeled antigen and unlabeled antigen to a high-affinity antibody. The method of RIA starts with making a known quantity of antigen radioactive by labeling it with gamma-radioactive isotopes of iodine attached to tyrosine. The labeled antigen is then mixed with antibody at a concentration that saturates the antigen-binding sites of the antibody. Then test samples of unlabeled antigen of unknown concentration are added in progressively larger amounts. The test sample may be a complex mixture, such as serum or body fluids, that contains the unlabeled antigen. This causes the unlabeled antigen to compete with the radio-labeled antigen for antibody binding sites (Figure 4.28).

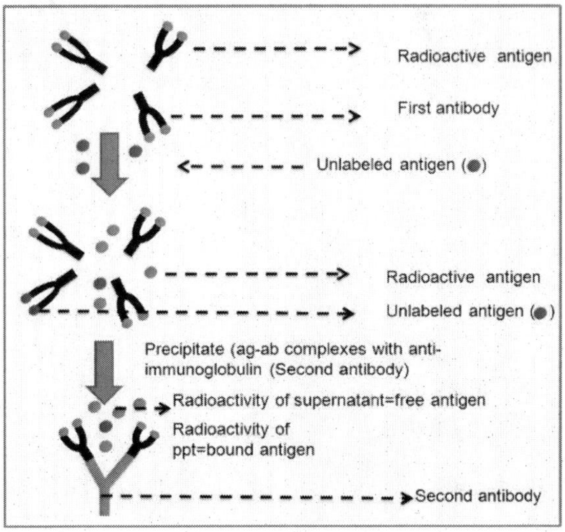

Figure 4.28. Showing the general procedure involved in Radioimmunoassay (RIA).

As the concentration of unlabeled antigen increases, more labeled antigen will be displaced from the binding sites. Then the bound antigens are separated by the unbound ones by precipitating the antigen-antibody complex with a secondary anti-isotype antiserum. For example, if the antigen-antibody complex contains rabbit IgG antibody, then goat anti-rabbit IgG will bind to the rabbit IgG and precipitate the complex. After separation, the radioactivity of the free antigen remaining in the supernatant is measured using a gamma counter.

Western Blotting

Western blotting, also known as immunoblotting, is a well-established and widely used technique for the detection and analysis of proteins. The method is based on building an antibody: protein complex via specific binding of antibodies to proteins immobilized on a membrane and detecting the bound antibody with one of several detection methods. The detection method was first described in 1979 and has since become one of the most commonly used methods in life science research.

Principle

- Western blotting technique is used for identification of particular protein from the mixture of protein.
- In this method labelled antibody against particular protein is used identify the desired protein, so it is a specific test. Western blotting is also known as immunoblotting because it uses antibodies to detect the protein.

Procedure/Steps

The following are the procedure of western blotting technique (Figure 4.29).

- Extraction of protein
- Gel electrophoresis: SDS PAGE
- Blotting: electrical or capillary blotting
- Blocking: BSA
- Treatment with primary antibody
- Treatment with secondary antibody (enzyme labelled anti Ab)
- Treatment with specific substrate; if enzyme is alkaline phosphatase, substrate is p-nitro phenyl phosphate which give color.

Step I: Extraction of Protein

Cell lysate is most common sample for western blotting. Protein is extracted from cell by mechanical or chemical lysis of cell. This step is also known as tissue preparation. To prevent denaturing of protein protease inhibitor is used. The concentration of protein is determined by spectroscopy. When sufficient amount of protein sample is obtained, it is diluted in loading buffer containing glycerol which helps to sink the sample in well. Tracking dye (bromothymol blue) is also added in sample to monitor the movement of proteins.

Step II: Gel Electrophoresis

The sample is loaded in well of SDS-PAGE Sodium dodecyl sulfate- poly-acrylamide gel electrophoresis. The proteins are separated on the basis of electric charge, isoelectric point, molecular weight, or combination of these all. The small size protein moves faster than large size protein. Protein are negatively charged, so they move toward positive (anode) pole as electric current is applied.

Step III: Blotting

The nitrocellulose membrane is placed on the gel. The separated protein from gel get transferred to nitrocellulose paper by capillary action. This type of blotting is time consuming and may take 1-2 days. For fast and more efficient transfer of desired protein from the gel to nitrocellulose paper electro-blotting can be used. In electro-blotting nitrocellulose membrane is sandwich between gel and cassette of filter paper and then electric current is passed through the gel causing transfer of protein to the membrane.

Step IV: Blocking

Blocking is very important step in western blotting. Antibodies are also protein so they are likely to bind the nitrocellulose paper. So before adding the primary antibody the membrane is non-specifically saturated or masked by using casein or Bovine serum albumin (BSA).

Step V: Treatment with Primary Antibody

The primary antibody (1° Ab) is specific to desired protein so it form Ag-Ab complex.

Step VI: Treatment with Secondary Antibody

The secondary antibody is enzyme labelled. For e.g., alkaline phosphatase or Horseradish peroxidase (HRP) is labelled with secondary antibody. Secondary antibody (2° Ab) is antibody against primary antibody (anti-antibody) so it can bind with Ag-Ab complex.

Step VII: Treatment with Suitable Substrate

To visualize the enzyme action, the reaction mixture is incubated with specific substrate. The enzyme convert the substrate to give visible colored product, so band of color can be visualized in the membrane. Western blotting is also a quantitative test to determine the amount of protein in sample.

Application

1) To determine the size and amount of protein in given sample.
2) Disease diagnosis: detects antibody against virus or bacteria in serum.
3) Western blotting technique is the confirmatory test for HIV. It detects anti- HIV antibody in patient's serum.
4) Useful to detect defective proteins. For e.g; Prions disease.
5) Definitive test for Creutzfeldt-Jacob disease, Lyme disease, Hepatitis B and Herpes

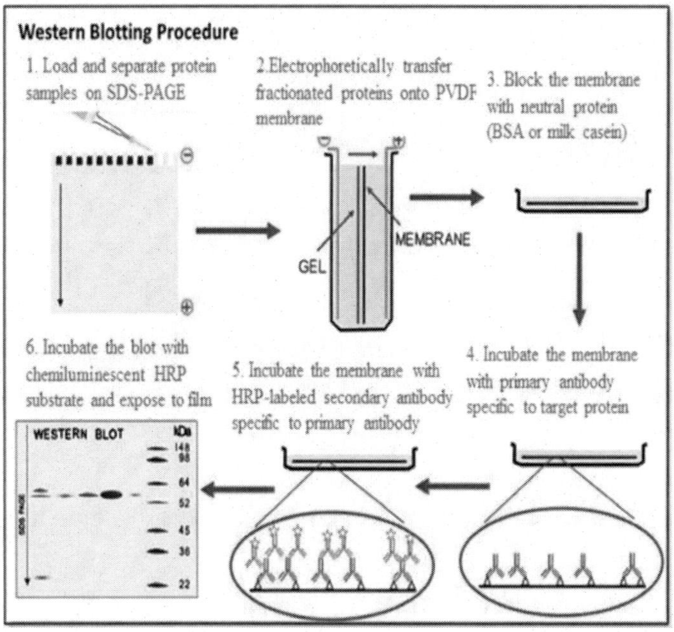

Figure 4.29. Illustrating the general procedure of western blotting technique. (Image source: NCBI).

EFFECTS OF ANTIGEN-ANTIBODY INTERACTIONS

The effects of antigen antibody interactions in the host are normally beneficial (Figure 4.30).

The beneficial effects of antigen-antibody interactions are:

- Initiation of immune responses

- Neutralization of toxins and viruses
- Activation of complement

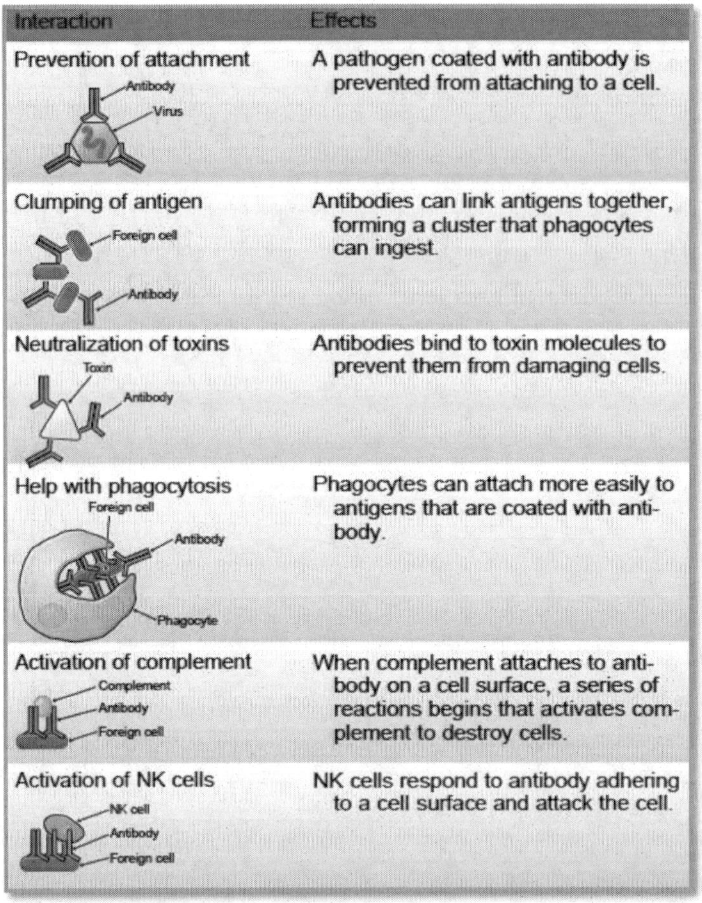

Interaction	Effects
Prevention of attachment	A pathogen coated with antibody is prevented from attaching to a cell.
Clumping of antigen	Antibodies can link antigens together, forming a cluster that phagocytes can ingest.
Neutralization of toxins	Antibodies bind to toxin molecules to prevent them from damaging cells.
Help with phagocytosis	Phagocytes can attach more easily to antigens that are coated with anti-body.
Activation of complement	When complement attaches to anti-body on a cell surface, a series of reactions begins that activates com-plement to destroy cells.
Activation of NK cells	NK cells respond to antibody adhering to a cell surface and attack the cell.

Figure 4.30. Representing antigen-antibody interactions and their effects. (source: http://encyclopedia.lubopitko-bg.com).

There are some antigen antibody interactions that can be damaging, these interactions occur in immune complex diseases.

Initiation of Immune Responses

Initiation of specific humoral immune response is the most important natural physiological process of the interaction of antigen with antibody. When antigen binds to antibody receptors on B lymphocytes, differentiation and replication follows, with the formation of antibody secreting plasma cells and specific population of memory cells.

Neutralization of Toxins and Viruses

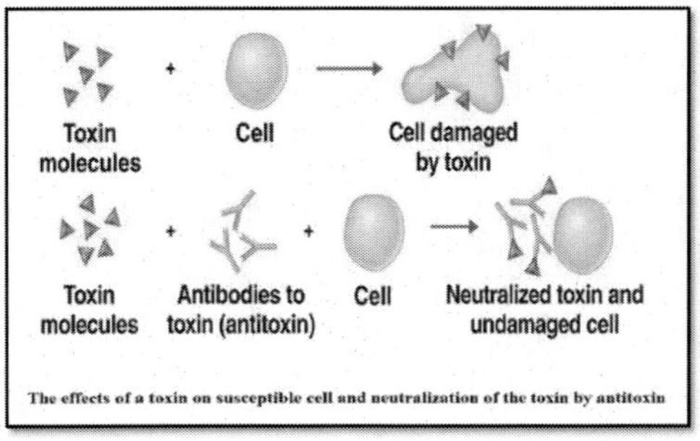

Figure 4.31. Showing the binding of antibody to toxin to neutralize its effects and further protecting the undamaged cell. (Image source: NCBI).

When antibodies raised against toxins are allowed to react with the toxin's antigenic determinants, they may block toxic sites. Alternatively, antibodies might bind to portions of toxins that bind to the target cells. In either case, binding of antibody to toxin neutralizes its toxicity (Figure 4.31). Antibodies that neutralize toxins are known as antitoxins. Antibodies engendered against viral antigens can complex with surface determinants on contact with free virus and prevent the virus from binding to the host cell receptors. Viral activity is thus neutralized.

Activation of Complement

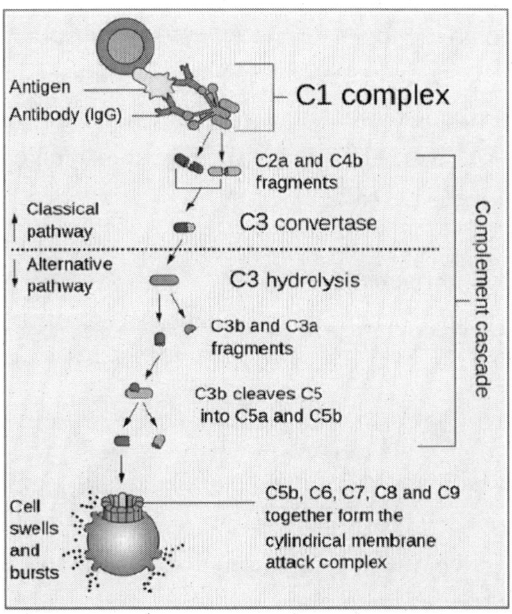

Figure 4.32. Representing the various events involved in complement activation, which thereby causes cell lysis thereby protecting other healthy cells. (Image source google scholar).

When an antigen and an antibody form a complex in the presence of complement, a number of events occur. If the events occur on the surface of cell or virus, the cells or virus may be lysed or become phagocytized. Fragments of complement can be released as are result. These fragments can induce dilation and increased permeability of blood vessels, as a part of inflammatory response. The importance of complement activation for protection against infectious disease has been dramatically demonstrated from the studies of disease susceptibility in those suffering complement deficiencies.

Frequently Asked Questions

Q1. What is a paratope?

Answer: The portion of antibody molecule that binds to the epitope is called as paratope. Paratope determine the specificity of immunological reactions.

Q2. What is antigenic specificity?

Answer: Antigenic specificity is the ability of the host cells to recognize an antigen specifically as a unique molecular entity and distinguish it from another with exquisite precision. Antigen specificity is due primarily to the side-chain conformations of the antigen. It is measurable and need not be linear or of a rate-limited step or equation.

Q3. Describe briefly about precipitation reaction?

Answer: It is a type of antigen–antibody reaction, in which the antigen occurs in a soluble form. When a soluble antigen reacts with its specific antibody, at an optimum temperature and P^H in the presence of electrolyte antigen antibody complex forms insoluble precipitate. This reaction is called precipitation reaction. A lattice is formed between the antigens and antibodies; in certain cases, it is visible as an insoluble precipitate. Antibodies that aggregate soluble antigens are called precipitins.

Q4. What are heterophile antigens?

Answer: A substance that occurs in unrelated species of animals, but has similar serologic properties among them they are also known as heterogenetic antigen.

Q5. What is avidity?

Answer: Avidity is a measure of the overall stability of the complex between antibodies and antigens and is governed by three factors, the intrinsic affinity of the antibody for the epitope, the valency of the antibody and antigen, and the geometric arrangement of the interacting components.

Q6. What is a cross reaction?

Answer: Antigen-antibody reactions are specific, but in some cases antibody elicted by one antigen can cross react with another antigen. This reaction is called as cross reaction and the antigens that produces cross-reaction are called cross reactive antigens.

Q7. What are the forces responsible for antigen antibody reactions?

Answer: The process that holds antigen-antibody together is called non-specific interactions. Intermolecular forces may be classified into four-

1. Electrostatic bonds.
2. Hydrogen bonds.
3. Hydrophobic interactions
4. Vanderwall interaction.

Q8. What is a monoclonal antibody?

Answer: A type of protein made in the laboratory that can bind to substances in the body, including cancer cells. There are many kinds of monoclonal antibodies. A monoclonal antibody is made so that it binds to only one substance. Monoclonal antibodies are being used to treat some types of cancer. They can be used alone or to carry drugs, toxins, or radioactive substances directly to cancer cells.

Q9. What are incomplete antibodies?

Answer: An incomplete form of antibody that may coat antigen, but which according to the lattice theory does not have a second receptor for attachment to another molecule of antigen; in the case of Rh_ erythrocytes, such an anti-Rh antibody may coat the cells but not cause them to agglutinate in saline; however, agglutination does occur when such coated cells are suspended in serum or other protein media such as albumin, therefore called serum agglutinin.

Q10. What are antigen-antibody interactions?

Answer: A reaction that occurs when an antigen combines with a corresponding antibody to produce an immune complex. A substance that induces the immune system to form a corresponding antibody is called an immunogen. All immunogens are also antigens because they react with corresponding antibodies; however, an antigen may not be able to induce the formation of an antibody and therefore may not be an immunogen. For instance, lipids and all low-molecular-weight substances are not immunogenic. However, many such substances, termed Haptens, can be attached to immunogens, called carriers, and the complex then acts as a new immunogen capable of eliciting antibody to the attached hapten.

Q11. What are immune complex?

Answer: Immune complexes are three-dimensional structures forming a lattice-type of sequence between antigens and antibodies with a ratio that may vary essentially according to the size of the antigen, the density of antigenic epitopes, and the isotype, subclass, poly/monoclonality and affinity of the antibody. The interaction between antigen and antibody does not result from chemical bonding bur rather is due to electrostatic forces. Therefore, the physicochemical properties of a given immune complex preparation are the result of a dynamic process between build-up and breakdown, contributing to constant variability of the characteristics of immune complexes *in vivo*. The immune complex structure is favoured by the Y-shaped configuration of the immunoglobulin molecules and by their abundance of water-repelling nonpolar groups known to favor auto aggregation of immunoglobulin M (IgM) or IgG in solution.

Q12. What is agglutination?

Answer: Agglutination is defined as the formation of clumps of cells or inert particles by specific antibodies to surface antigenic components (direct agglutination) or to antigenic components adsorbed or chemically coupled to red cells or inert particles (passive hemagglutination and passive agglutination, respectively). Erythrocytes are also agglutinated by non-antibody substances such as plant proteins, viruses, salts of heavy metals, inorganic colloidal acids and bases, and basic proteins (protamine, histones).

Q13. What is ELISA?

Answer: ELISA (enzyme-linked Immunosorbent assay) is a plate-based assay technique designed for detecting and quantifying substances such as peptides, proteins, antibodies and hormones. Other names, such as enzyme immunoassay (EIA), are also used to describe the same technology. In an that is linked to an enzyme. Detection is accomplished by assessing the conjugated enzyme activity via incubation with a substrate to produce a measurable product. The most crucial element of the detection strategy is a highly specific antibody-antigen interaction.

Q14. What are adjuvants?

Answer: Adjuvants are the components used to increase the potential any antigens to generate the immune response. They are commonly used to boost the immune response when an antigen has low immunogenicity or when only small amounts of an antigen are available.

Q15. What are the applications of precipitation reaction?

Answer: The application of the precipitation reaction are listed below:

1. Detection of unknown Ab to diagnose infection e.g., VDRL test for syphilis.
2. Standardization of toxins and antitoxins.

3. Identification of Bacteria e.g., Lancified grouping of streptococci.

4. Identification of bacterial component e.g., Ascoli's thermo precipitin test for *Bacillus anthracis*

MULTIPLE CHOICE QUESTIONS

Q1. Agglutination reaction is more sensitive than precipitation reaction for the detection of

A. antigens
B. antibodies
C. complement
D. antigen-antibody complexes

Q2. In which of the following case a large lattice is formed?

A. antibody is in excess
B. antigens and antibodies are in optimal proportion
C. antigen is in excess.
D. none of these

Q3. The ability of the immune system to recognize self-antigens versus non-self antigen is an example of:

A. specific immunity
B. tolerance
C. cell mediated immunity
D. antigenic immunity

Q4. Monoclonal antibodies recognize a single:

A. antigen
B. bacterium
C. Epitope
D. B cells

Q5. The specificity of an antibody is due to

A. Its valency
B. The heavy chains
C. The Fc portion of the molecule
D. The variable portion of the heavy and light chain

Q6. Which of the following substances will not stimulate an immune response unless they are bound to a larger molecule?

A. Antigen
B. Virus
C. Hapten
D. Antibody

Q7. The bonds involved in antigen-antibody interactions are

A. Weak hydrogen bonds and Vander walls forces
B. Strong covalent bonds
C. Strong di-sulphide bonds
D. All of these

Q8. Which of the following statement is true

A. All immunogens are antigens but all antigens are not immunogens
B. All immunogens are antigens and all antigens are immunogens
C. All immunogens are not antigens but all antigens are immunogens
D. All immunogens are proteins and all proteins are immunogens

Q9. Any agents that may stimulate the immune system and enhance the response without having any specific antigenic effect by itself.

A. Antigens
B. Allergens
C. Adjuvants
D. Carriers

Q10. Majority of antigens are

A. Proteins
B. Carbohydrates
C. Nucleic acids
D. Lipids

Q11. Which of the following is a hapten

A. Cyanide
B. Paracetamol
C. Penicillin
D. None of the above

Q12. Generally antibodies produced against a pathogen is

A. Monoclonal

B. Homogenous

C. Polyclonal

D. All of same specificity

Q13. What is the ELISA test intended to measure?

A. Antibody to HIV only

B. Antigen to HIV only

C. Presence of free, circulating virus in patient

D. Antibodies directed against HLA molecule

Q14. Agrose a porus matrix used in various molecular biology technique is made by repetitive units of

A. Maltose

B. agarobiose

C. Agar

D. None of these

Q15. The most effective Immunoglobulin (Ig) is:

A. IgM

B. IgG

C. IgA

D. IgE

Answer Key

1(A), 2(B), 3(B), 4(C), 5(D), 6(C), 7(A), 8(A), 9(C), 10(A), 11(C), 12(C), 13(A) 14(A), 15(A)

Assignments

Q1. Give general features of antigen antibody reactions.

Q2. Which chemical bonds are responsible for antigen-antibody interaction?

Q3. What are antibodies? What are different types of antibodies?

Q4. Explain in detail precipitation reaction? What are its types?

Q5. Give the principle of RIA. What are its limitations and uses?

REFERENCES

Aarvak T, Natvig JB. Cell-cell interactions in synovitis: antigen presenting cells and T cell interaction in rheumatoid arthritis. *Arthritis Research & Therapy.* 2000 Nov; 3(1):13.

Absolom DR. The nature of the antigen-antibody bond and the factors affecting its association and dissociation. *CRC Critical Reviews in Immunology.* 1986; 6(1):1-46.

Adamczyk M, Moore JA, Yu Z. Application of surface plasmon resonance toward studies of low-molecular-weight antigen–antibody binding interactions. *Methods.* 2000 Mar 1; 20(3):319-28.

Arnon R, Van Regenmortel MH. Structural basis of antigenic specificity and design of new vaccines. *The FASEB Journal.* 1992 Nov; 6(14):3265-74.

Atassi MZ. Antigenic structures of proteins: their determination has revealed important aspects of immune recognition and generated strategies for synthetic mimicking of protein binding sites. *European Journal of Biochemistry.* 1984 Nov; 145(1):1-20.

Batista FD, Harwood NE. The who, how and where of antigen presentation to B cells. *Nature Reviews Immunology.* 2009 Jan; 9(1):15.

Braden BC, Poljak RJ. Structural features of the reactions between antibodies and protein antigens. *The FASEB Journal.* 1995 Jan; 9(1):9-16.

Brownlee J. Antigen-antibody interaction. Melbourne, *Australia: Complex Intelligent Systems Laboratory*, Centre for Information Technology Research, Faculty of Information and Communication Technologies, Swinburne University of Technology. 2007.

Cheng PC, Dykstra ML, Mitchell RN, Pierce SK. A role for lipid rafts in B cell antigen receptor signaling and antigen targeting. *Journal of Experimental Medicine.* 1999 Dec 6; 190(11):1549-60.

Cohen BE, Rosenthal AS, Paul WE. Antigen-Macrophage Interaction: II. Relative Roles of Cytophilic Antibody and Other Membrane Sites. *The Journal of Immunology.* 1973 Sep 1;111(3):820-8.

Colman PM. Structure of antibody-antigen complexes: implications for immune recognition. In *Advances in Immunology* 1988 Jan 1 (Vol. 43, pp. 99-132). Academic Press.

Davies DR, Cohen GH. *Interactions of Protein Antigens with Antibodies*. Proceedings of the National Academy of Sciences. 1996 Jan 9;93(1):7-12.

Davies DR, Padlan EA, Sheriff S. Antibody-antigen complexes. *Annual Review of Biochemistry.* 1990 Jul; 59(1):439-73.

Davis MM, Bjorkman PJ. T-cell antigen receptor genes and T-cell recognition. *Nature.* 1988 Aug; 334(6181):395.

Dyson HJ, Wright PE. Antigenic peptides. *The FASEB Journal.* 1995 Jan; 9(1):37-42.

Glorioso JC, Smith JW. Immune interactions with cells infected with herpes simplex virus: antibodies to radioiodinated surface antigens. *The Journal of Immunology.* 1977 Jan 1; 118(1):114-21.

Guyre PM, Graziano RF, Goldstein J, Wallace PK, Morganelli PM, Wardwell K, Howell AL. Increased potency of Fc-receptor-targeted antigens. *Cancer Immunology, Immunotherapy.* 1997 Nov 1; 45(3-4):146-8.

Harper M, Lema F, Boulot G, Poljak RJ. Antigen specificity and cross-reactivity of monoclonal anti-lysozyme antibodies. *Molecular Immunology.* 1987 Feb 1; 24(2):97-108.

Homenko A, Kapilevich B, Kornstein R, Firer MA. Effects of 100 GHz radiation on alkaline phosphatase activity and antigen–antibody interaction. Bioelectromagnetics: *Journal of the Bioelectromagnetics Society,* The Society for Physical Regulation in Biology and Medicine, The European Bioelectromagnetics Association. 2009 Apr; 30(3):167-75.

Jefferis R, Deverill I. The antigen antibody reaction. In *Principles and Practice of Immunoassay* 1991 (pp. 1-18). Palgrave Macmillan, London.

Jin L, Fendly BM, Wells JA. High resolution functional analysis of antibody-antigen interactions. *Journal of Molecular Biology.* 1992 Aug 5; 226(3):851-65.

Jones S, Thornton JM. Principles of protein-protein interactions. *Proceedings of the National Academy of Sciences.* 1996 Jan 9; 93(1):13-20.

Kumagai I, Tsumoto K. *Antigen–Antibody Binding.* eLS. 2001 May 30.

Lamm ME. Interaction of antigens and antibodies at mucosal surfaces. *Annual Reviews in Microbiology.* 1997 Oct; 51(1):311-40.

MacCallum RM, Martin AC, Thornton JM. Antibody-antigen interactions: contact analysis and binding site topography. *Journal of Molecular Biology.* 1996 Oct 11; 262(5):732-45.

Mariuzza RA, Phillips SE, Poljak RJ. The structural basis of antigen-antibody recognition. *Annual Review of Biophysics and Biophysical Chemistry.* 1987 Jun; 16(1):139-59.

Metzger H. Early molecular events in antigen-antibody cell activation. *Annual Review of Pharmacology and Toxicology.* 1979 Apr; 19(1):427-45.

Metzger H. Effect of antigen binding on the properties of antibody. In *Advances in Immunology* 1974 Jan 1 (Vol. 18, pp. 169-207). Academic Press.

Metzger H. The effect of antigen on antibodies: recent studies. In *Contemporary Topics in Molecular Immunology* 1978 (pp. 119-152). Springer, Boston, MA.

Mian IS, Bradwell AR, Olson AJ. Structure, function and properties of antibody binding sites. *Journal of Molecular Biology.* 1991 Jan 5; 217(1):133-51.

Mond JJ, Lees A, Snapper CM. T cell-independent antigens type 2. *Annual Review of Immunology.* 1995 Apr; 13(1):655-92.

Najjar VA. Some aspects of antibody-antigen reactions and theoretical considerations of the immunologic response. *Physiological Reviews.* 1963 Apr 1;43(2):243-62.

Neri D, Montigiani S, Kirkham PM. Biophysical methods for the determination of antibody-antigen affinities. *Trends in Biotechnology.* 1996 Dec 1; 14(12):465-70.

Oda M, Uchiyama S, Robinson CV, Fukui K, Kobayashi Y, Azuma T. Regional and segmental flexibility of antibodies in interaction with antigens of different size. *The FEBS Journal*. 2006 Apr; 273(7):1476-87.

Pauling L, Campbell DH, Pressman D. The nature of the forces between antigen and antibody and of the precipitation reaction. *Physiological Reviews.* 1943 Jul 1; 23(3):203-19.

Platzer B, Stout MM, Fiebiger E. Antigen cross-presentation of immune complexes. *Frontiers in Immunology.* 2014 Apr 1; 5:140.

Reverberi R, Reverberi L. Factors affecting the antigen-antibody reaction. *Blood Transfusion*. 2007 Oct; 5(4):227.

Rodríguez-Pinto D. B cells as antigen presenting cells. *Cellular Immunology.* 2005 Dec 1; 238(2):67-75.

Rose ML. Endothelial cells as antigen-presenting cells: role in human transplant rejection. *Cellular and Molecular Life Sciences CMLS.* 1998 Sep 1; 54(9):965-78.

Sakurabayashi I. Dynamics of antigen-antibody interaction--an introduction. Nihon rinsho. *Japanese Journal of Clinical Medicin*e. 1995 Sep; 53(9):2126-33.

Schirrmacher V, Rubin B, Pross H. Cytotoxic Immune Cells with Specificity for Defined Soluble Antigens: V. Interaction of Antibody with the Cytotoxic Effector Cells in Immune or Non-Immune Mouse Spleen Cells. *The Journal of Immunology.* 1974 Jun 1; 112(6):2219-26.

Sela-Culang I, Kunik V, Ofran Y. The structural basis of antibody-antigen recognition. *Frontiers in Immunology.* 2013 Oct 8; 4:302.

Sutton BJ. Immunoglobulin structure and function: the interaction between antibody and antigen. *Current Opinion in Immunology.* 1989 Oct 1; 2(1):106-13.

Tainer JA, Deal CD, Geysen HM, Roberts VA, Getzoff ED. Defining antibody-antigen recognition: towards engineered antibodies and epitopes. *International Reviews of Immunology*. 1991 Jan 1;7(2):165-94.

Tarbell KV, Yamazaki S, Steinman RM. The interactions of dendritic cells with antigen-specific, regulatory T cells that suppress autoimmunity. In *Seminars in Immunology* 2006 Apr 1 (Vol. 18, No. 2, pp. 93-102). Academic Press.

Townsend A, Bodmer H. Antigen recognition by class I-restricted T lymphocytes. *Annual Review of Immunology*. 1989 Apr;7(1):601-24.

Vascotto F, Le Roux D, Lankar D, Faure-Andre G, Vargas P, Guermonprez P, Lennon-Dumenil AM. Antigen presentation by B lymphocytes: how receptor signaling directs membrane trafficking. *Current Opinion in Immunology*. 2007 Feb 1;19(1):93-8.

Webster DM, Henry AH, Rees AR. Antibody-antigen interactions. *Current Opinion in Structural Biology*. 1994 Jan 1; 4(1):123-9.

Wilson IA, Niman HL, Houghten RA, Cherenson AR, Connolly ML, Lerner RA. The structure of an antigenic determinant in a protein. *Cell.* 1984 Jul 1; 37(3):767-78.

Wilson IA, Stanfield RL, Rini JM, Areval JH, Schulze-Gahmen U, Fremont DH, Stura EA. *Structural Aspects of Antibodies and Antibody-Antigen Complexes.* Wiley, Chichester, UK; 1991.

Wilson IA, Stanfield RL. Antibody-antigen interactions. *Current Opinion in structural Biology*. 1993 Feb 1; 3(1):113-8.

Wilson IA, Stanfield RL. Antibody-antigen interactions: new structures and new conformational changes. *Current Opinion in Structural Biology.* 1994 Jan 1; 4(6):857-67.

In: Immunoglobulins
Editor: Manzoor Ahmad Mir
ISBN: 978-1-53616-903-4

Chapter 5

ANTIGEN DEPENDENT CELL MEDIATED CYTOTOXICITY (ADCC)

Manzoor A. Mir*[*], *Umar Mehraj, Safura Nisar, and Hina Qayoom
Department of Bioresources, School of Biological Sciences,
University of Kashmir, India

ABSTRACT

Antibody-dependent cell-mediated cytotoxicity (ADCC) was initially described by Moller in 1967 as the ability of natural killer (NK) cells – at the time identified only as lymphocytes – to destroy antibody-coated target cells. A nonphagocytic mechanism, whereby most leucocytes can kill target cells in the absence of complement and without major histocompatibility complex (MHC) restriction thus became apparent. The specificity of ADCC is conferred by target-bound antibody and can be induced by antibody concentrations well below those needed to activate complement-mediated lysis. Fc receptors (FcR) on the plasma membrane of the effector cell are necessary but not sufficient for ADCC activity. Once antibody has bound to the target cell, its interaction with either activating or inhibitory FcR on the cytotoxic effector cell appears to dictate the precise signaling pathways that are stimulated. When target-bound antibodies cross-link activating FcR, a cytotoxic pathway is initiated that results in the release of mediators such as perforin, tumor necrosis factor (TNFa), and reactive oxygen intermediates (ROI) – which induce target cell death. Conversely, ligation of inhibitory FcR prevents effector cell activation, resulting in survival of the target cell.

Keywords: cytotoxicity, effector cells, granzymes, perforins, FcR receptors, apoptosis, fas-l pathway, ROS/ROI pathway

---

[*] Corresponding Author: Dr Manzoor Ahmad Mir (Ph. D), Department of Bioresources, School of Biological Sciences, University of Kashmir, Srinagar J&K, India-190006, Contact:-+919622901319. Email: drmanzoor@kashmiruniversity.ac.in.

Objectives

- To be able to understand the brief history of ADCC.
- To be able to understand about the concept of ADCC.
- To be able to study the process involved in ADCC
- To understand the various mechanisms of ADCC.

Introduction

Several independent laboratory observations took place in the 1960s, which indicated that cells could be killed by other cells, yet the mechanisms of killing were unknown. Several hypotheses were formulated and various experiments were conducted paving the way for the discovery and characterization of what we now know as cytotoxic T lymphocytes (CTLs) and natural killer (NK) cells. However, early experiments pointed to the hypothesis that immune serum was in some cases necessary for some types of effector cells to mediate killing of target cells. In 1965, Erna Möeller, a researcher working at the Karolinska Institute, showed that incubation of mouse tumor target cells with heat-inactivated anti-serum from rabbits immunized with these tumor cells, followed by incubation with lymphoid cells from unimmunized mice, resulted in cytotoxicity. Such cytotoxicity required contact or "serum-induced aggregation" between the tumor targets and the lymphoid effectors. Furthermore, various experiments were performed to identify the aggregating and cytotoxicity-inducing factor contained in serum. In 1970, MacLennan, Loewi, and Harding, researchers working at the Canadian Red Cross Memorial Hospital, reported that the serum factor required for this type of cell-mediated cytotoxicity was an immunoglobulin with the "chemical properties of IgG." Subsequent experiments confirmed this finding by showing that the antibody required belonged to the IgG class and that the mechanism of antibody-dependent killing of target cells by serum factors did not require the heat-labile components of complement, as heat inactivation of serum maintained the killing effect. The phenomenon acquired the name of "antibody-dependent lymphocyte-mediated cytotoxicity" and gradually became known as "antibody-dependent cell-mediated cytotoxicity," or ADCC. The three basic components of ADCC were recognized as being effector cells, antibodies, and target cells coated with antigen. "Targets" could refer to cells expressing tumor, viral, or bacterial antigens; therefore, ADCC became known as an immune mechanism that could be potentially protective against certain types of cancers as well as infectious diseases. Substantial progress has been made in characterizing the various effector cells and the receptors involved in this phenomenon, and ADCC can now be defined as the immune mechanism through which

Fc-receptor-bearing effector cells can kill target cells that have surface antigens complexed with antibody.

Concept of ADCC Action

There are various number of cells which have the cytotoxic potential that express membrane receptors for the Fc region of the antibody molecule. When an antibody is specifically bound to a target cell, these receptor-bearing cells can bind to the antibody Fc region, and thus to the target cells, and subsequently cause lysis of the target cell. Eventhough, these cytotoxic cells are nonspecific for antigen, the specificity of the antibody directs them to specific target cells. This type of cytotoxicity is referred to as antibody-dependent cell-mediated cytotoxicity (ADCC). Among the cells that can mediate ADCC are NK cells, macrophages, monocytes, neutrophils, and eosinophils. Antibody dependent cell-mediated killing of cells infected with the measles virus can be observed in vitro by adding anti-measles antibody together with macrophages to a culture of measles infected cells. Similarly, cell-mediated killing of helminths, such as schistosomes or blood flukes, can be observed in vitro by incubating larvae (schistosomules) with antibody to the schistosomules together with eosinophils. Target-cell killing by ADCC appears to involve a number of different cytotoxic mechanisms, but not complement mediated lysis. When macrophages, neutrophils, or eosinophils bind to a target cell by way of the Fc receptor, they become more active metabolically; as a result, the lytic enzymes in their cytoplasmic lysosomes or granules increase. Release of these lytic enzymes at the site of the Fc mediated contact may result in damage to the target cell. In addition, activated monocytes, macrophages, and NK cells have been shown to secrete tumor necrosis factor (TNF), which may have a cytotoxic effect on the bound target cell. Since both NK cells and eosinophils contain perforin in cytoplasmic granules, their target-cell killing also may involve perforin-mediated membrane damage similar to the mechanism described for CTL-mediated cytotoxicity.

Antibody dependent cell mediated cytotoxicity (ADCC) is a cooperative response involving

- Binding of antibodies produced by B-cells to antigens
- Recognition of the constant fragment (Fc) by effector cells
- Release of damaging substances by effector cells
- Destruction of the pathogenic organism.

ADCC is an interaction between innate and adaptive immunity that is important because it allows innate immune effectors to recognize pathogens that do not express pathogen associated molecular patterns (PAMPs) hide these patterns from innate immune detection otherwise subvert innate immune function. It composes a big part of the immune response to parasites.

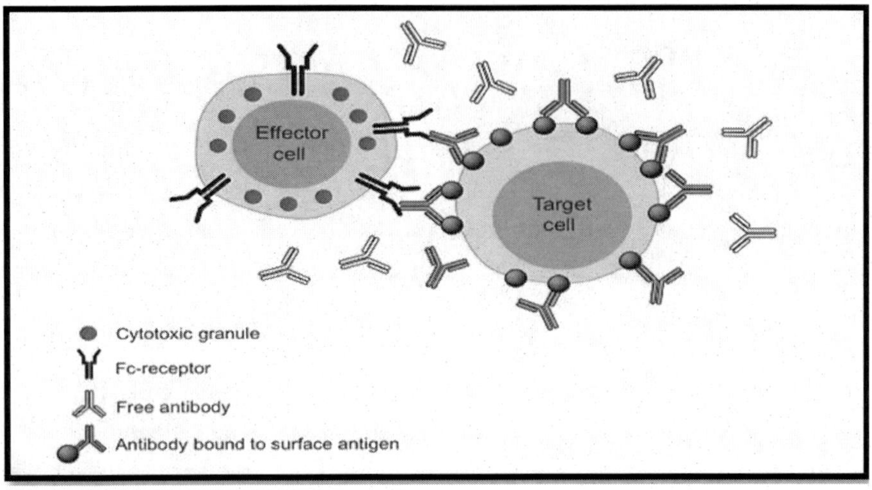

Source: Science direct.

Figure 5.1. Effector cells and targets. ADCC involves the interplay between a granular effector cell and a target cell expressing antigens on its surface. The granular effector cell must express Fc receptors on its surface for ADCC to occur.

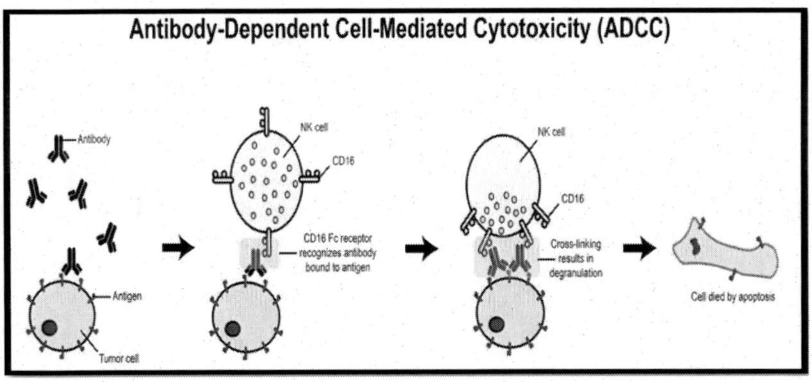

Source: https://step1.medbullets.com.

Figure 5.2. Overview of Antibody dependent cell-mediated cytotoxicity.

Mechanism of Antibody Dependent Cell Mediated Cytotoxicity

ADCC is the process by which antibodies coat a target cell and recruit effector cells to induce target cell death via non-phagocytic mechanisms (Figure 5.1). Antibodies can bind to their specific antigens on the target cell surface via; their antigen-binding fragment (Fab) portions and interact with effector cells via; their fragment crystallizable region (Fc) portions thereby acting as bridges that link the effector to a target. While several classes of human antibodies can mediate ADCC, including IgG, IgA, and IgE, IgG1 is the most used subclass for cancer therapeutic antibodies. In order for an effector cell to carry out ADCC it must express Fc receptors (FcR) that will bind the antibody. The known classes of FcR include FcγR, which bind IgG; FcαR, which bind IgA; and FcεR, which bind IgE. FcγR are the most important for tumor cell clearance by myeloid cells and are comprised of activating FcγRI (CD64), FcγRIIA (CD32A), FcγRIIIA (CD16A), and inhibitory FcγRIIB (CD32B) receptors. Once the FcγR binds antibody it triggers receptor cross-linking and downstream signal propagation. Activating FcγR signal via; their immunoreceptor tyrosine-based activation motifs while inhibitory FcγR signal via their immunoreceptor tyrosine based inhibitory motifs. Many effector cells also express other receptor types such as the inhibitory killer inhibitory receptors (e.g., KIR) and activating NKG2D receptors on natural killer (NK) cells. The delicate balance amongst the activating and inhibitory pathway signaling ultimately determines effector cell response. The myeloid cells capable of acting as ADCC effectors are NK cells, monocytes, macrophages, neutrophils, eosinophils, and dendritic cells. Once these effector cells have been activated, they mediate target cell death through three key mechanisms: cytotoxic granule release, Fas signaling, and elaboration of reactive oxygen species. The main and best characterized mechanism utilized in ADCC is the release of perforins and granzymes from effector cell granules. The effector cells such as NK cells, upon activation polarize and exocytose their granules in a calcium-dependent manner. Perforin and Granzyme B work in synergy to induce cell death. Perforins create pores in the cell membrane that facilitate granzyme B entry into the target cell, resulting in DNA fragmentation and apoptosis. Additionally, research has suggested that activated effector cells will upregulate Fas ligand expression in order to cause apoptosis in the target via, Fas signaling.

Fc Receptors Are Trigger Molecules for ADCC

The mammalian FcRs are a highly conserved family of glycoproteins of the immunoglobulin (Ig) superfamily, which are expressed predominantly on a majority of innate immune cells, including monocytes, macrophages, neutrophils, Dendritic cells (DCs), Basophils, PMNs, mast cells, and eosinophils; B cells; follicular DCs; and some other tissues (mucosal epithelium, endothelium, and placenta), where they bind free or antigen-bound antibodies via; the Fc portion to initiate diverse immunobiological processes, such as the functional activation of the cells, facilitation of antigen or cross-antigen presentation, and cytokine secretion. There are distinct FcRs for the major antibody isotypes, which are designated as FcγR, FcεR, FcαR, FcμR, and FcδR. Also, the neonatal FcR (FcRn) and the polymeric IgA receptor (pIgR) are respectively IgG- and IgA/IgM-binding and intracellular transport receptors that are expressed on mucosal epithelial cells lining the respiratory, gastrointestinal, and genital tracts. In addition, the FcRn is expressed on endothelial cells and the placenta. Although the precise biological functions of the recently identified FcR-like molecules (FcRLsss) are not fully known, their differential expression by B-lineage cells and overexpression in some B-cell malignancies have suggested that they may play a role in pathogenesis of B cell disease. FcR on the surface of cytotoxic effector cells are required for ADCC. Separate genes that are members of the immunoglobulin (Ig) supergene family encode different FcR: FcgR specifically bind IgG, FcaR bind IgA, FceR bind IgE and Fcm/aR bind both IgM and IgA. It is clear that members of the following FcR classes are all capable, when crosslinked, of activating ADCC (Figure 5.3) FcgRI (CD64), FcgRII (CD32), FcgRIII (CD16), FcaR (CD89) and FceRI. It has also been shown that engagement of at least one IgG-binding FcR, FcgRIIb, leads to inhibition of ADCC activation. The balance between activating and inhibitory FcR thus appears to be a crucial determinant of the magnitude of ADCC that is induced when the effector cell engages the antibody-coated target.

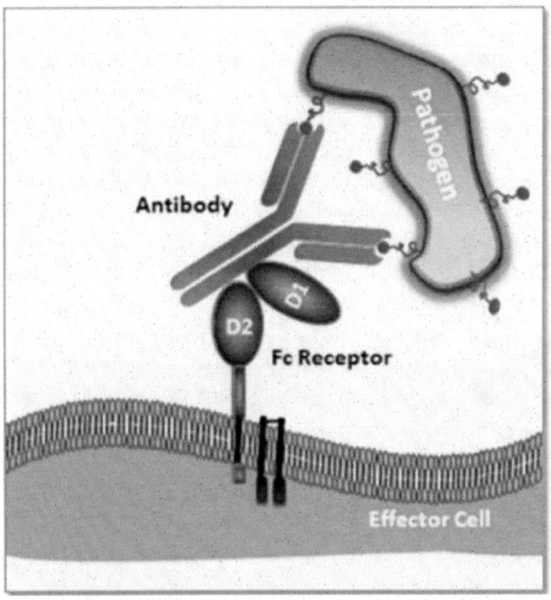

Source: Fc Receptor-Dependent Immunity, Antibody Fc, by Joseph U. Igietseme, Xiaoping Zhub and Carolyn M. Black.

Figure 5.3. Schematic illustration of Fc receptor interaction with an antibody-coated microbial pathogen.

THE FC RECEPTORS

Fc receptors are a series of cell surface proteins that recognizes with Fc region of antibody and that is the name of Fc receptor derived from. These Fc receptors are found on the membrane of certain immune cells, including B lymphocytes, natural killer cells, macrophages, neutrophils, and mast cells. Fc receptors binding to antibodies that are attached to infected cells or invading pathogens leads to the protective functions of the immune system.

Their activity stimulates phagocytic or cytotoxic cells to destroy microbes, or infected cells by antibody-mediated phagocytosis or antibody-dependent cell-mediated cytotoxicity as shown in the Figure 5.2.

Types of Fc Receptors

Generally speaking, there are several types of Fc receptors depending on the kind of antibody that they recognize. Among of these types, the following three types are major. One is Fc receptors recognizing Fc portion of IgG called Fc gamma receptor (FcγR). Other is Fc receptor recognizing C-terminal of IgA called Fc alpha receptor (FcαR). The last is Fc receptor recognizing C-terminal of IgE called Fc epsilon receptor (FcεR).

Gamma Receptor

Fc gamma receptor, also known as FcγR, is divided into three classes based on genetic homology: FcγR1 (CD64), FcγR2 (CD32) and FcγR3 (CD16). Each type is encoded by specific genes located in the same region of the long arm of chromosome 1. There are several different closely related genes for each FcgR which have different antibody affinities due to their different molecular structure, including FcγRI (CD64), FcγRIIa (CD32a), FcγRIIb (CD32b), FcγRIIc (CD32c), FcγRIIIa (CD16a), FcγRIIIb (CD16b) Figure 5.4.

ITAM: Immunoreceptor Tyrosine-Based Activation Motif; ITIM: Immunoreceptor Tyrosine-Based Inhibitory Motif

As shown in the Figure 5.4, FcγRI binds to IgG more strongly than all of other FcγRs (FcγRII or FcγRIII) do. FcγRI also has an extracellular portion composed of three immunoglobulin (Ig)-like domains, one more domain than others. This characteristic allows FcγRI to bind a signal IgG molecule (or monomer). But please note that all of FcγRs must bind multiple IgG molecules within an immune complex to be activated. Among of FcγRs, FcγRI, FcγRIIa, FcγRIIc and FcγRIIIa possess a classic cytoplasmic ITAM signaling domain, and can stimulate immune effector-cell activation. As a counterpoint to the activating FcgRs, FcgRIIb possesses a cytoplasmic ITIM. In the following Table 5.1, we compare the features of all FcγRs to each other.

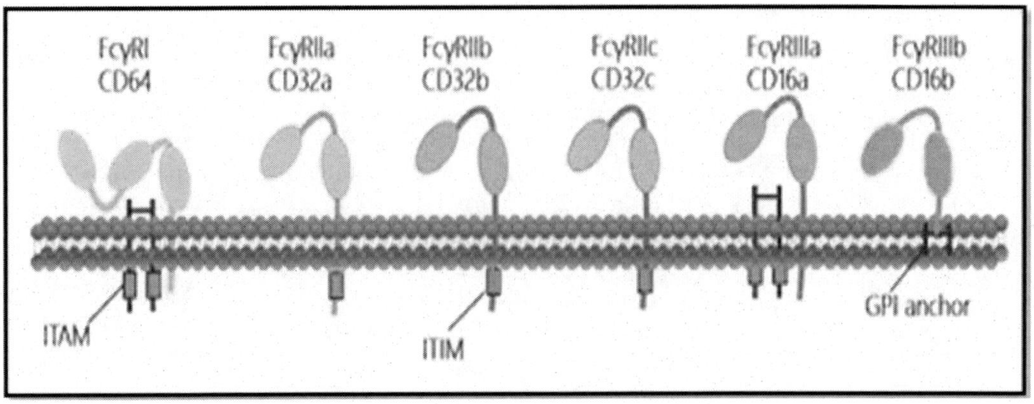

Figure 5.4. The structure of different type of Fc gamma receptors involved in the ADCC process.

Table 5.1. The features of Fc gamma receptors

	FcγRI	FcγRIIa	FcγRIIb	FcγRIIc	FcγRIIIa	FcγRIIIb
Function	Activating	Activating	Inhibitory	Activating	Activating	Activating
Expression	Monocyte, DC, macrophage, neutrophil and eosinophil	Monocyte, DC, macrophage, neutrophil	B cell, plasma cells, monocyte, DC, macrophage, neutrophil, mast cell and basophil	NK cell	NK cell Monocyte, DC, macrophages	Neutrophil, mast cell and eosinophil
Affinity	High	Low	Low	Low	Low	Low

Fc Alpha Receptor

Fc-alpha receptor, also known as FcαR, is the receptor specific for IgA which binds IgA with low affinity. Only one type Fc receptor belongs to the FcαR subgroup, which is called FcαRI (CD89). FcαRI is expressed on neutrophils, eosinophils, monocytes/ macrophages, dendritic cells, and Kupffer cells. FcαRI expression is constitutive and independent of the presence of IgA ligand because the receptor is expressed at similar levels on cells from patient deficient in IgA. The structure of Fc alpha receptor is shown in the Figure 5.5.

A single gene encoding FcαRI has been isolated, which is located within the leukocyte receptor cluster on chromosome 19. The FcαRI lacks canonical signal transduction domains but can associate with two FcRγ signaling chains that bear an activation motif (ITAM) in the cytoplasmic domain. FcαRI expressed alone mediates endocytosis and recycling of IgA.

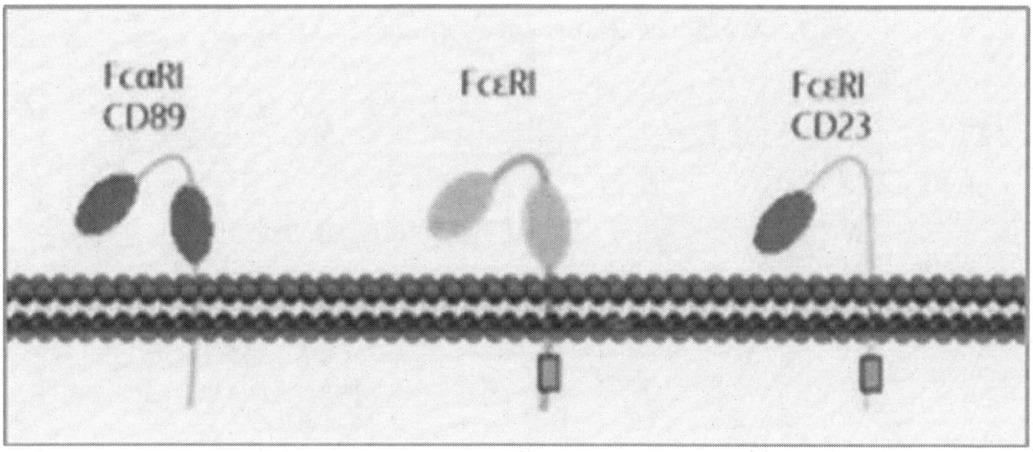

Figure 5.5. Showing the structure of Fc alpha receptor and Fc epsilon receptors.

Table 5.2. The features of Fc alpha receptor

Receptor name	principal antibody ligand	Affinity for ligand	Cell distribution	Effect following binding to antibody
FCαRI (CD89)	IgA	Low(Kd >10-6M)	Monocytes Macrophages Neutrophils Eosinophils	Phagocytosis Induction of microbe killing

Fc Epsilon Receptors

Fc epsilon receptor, also known as FcεR, is an Fc receptor that binding IgE. Currently, two types of FcεR are known. The high-affinity receptor FcεRI is capable of binding monomeric IgE (it has two Ig-like domains). The classical Fc epsilon receptor I is tetrameric. FcεRI is found on epidermal Langerhans cells, eosinophils, mast cells and basophils. Based on its cellular distribution, FcεRI plays a crucial role in controlling allergic responses. The low-affinity receptor FcεRII (CD23) interacts preferentially with complex IgE. FcεRII has multiple functions as a membrane-bound or soluble receptor; it controls B cell growth and differentiation and blocks IgE-binding of eosinophils, monocytes, and basophils. The structure of FcεRs are shown in the Figure 5.5.

Beside the above Fc receptors, there are several other Fc receptors, including Fcα/μR, and neonatal Fc receptor (FCGRT). Fcα/μR is an Fc receptor for IgA and IgM. FCGRT is kind of MHC class-I receptor and plays a role in monitoring IgG turnover.

Table 5.3. The features of Fc epsilon receptors

Receptor name	Principal antibody ligand	Affinity for ligand	Cell distribution	Effect following binding to antibody
FcεRII (CD23)	IgE	Low (Kd > 10^{-7} M)	B cells Eosinophils Langerhans cells	Possible adhesion molecule IgE transport across human intestinal epithelium Positive-feedback mechanism to enhance allergic sensitization (B cells)
FcαRI (CD89)	IgA	Low (Kd > 10^{-6} M)	Monocytes Macrophages Neutrophils Eosinophils	Phagocytosis Induction of microbe killing

Signaling Mechanism

Upon the Fc region of antibody binding to special Fc receptor, a series of signaling mechanisms are triggered in cell. Based on recent researches, the signaling mechanisms of Fc gamma receptor, Fc-alpha receptor and Fc epsilon receptor are more lucid than other Fc receptors. In this division, we briefly summarize the mechanism of Fc gamma receptor, Fc-alpha receptor and Fc epsilon receptor respectively.

Fc Gamma Receptor Signaling

Fc gamma receptors generate signals within their cells through an important activation motif, ITAM. All of the Fc gamma receptors (FcγR) belong to the immunoglobulin superfamily. Fc gamma receptor is essential participant in many immune system effector functions, such as phagocytosis of opsonized, release of inflammatory mediators and antibody-dependent cellular cytotoxicity. Here, we primary illustrate the mechanism of Fc gamma R mediated phagocytosis.

Phagocytosis plays a crucial role in host-defense mechanisms by the uptake and destruction of infectious pathogens. Specialized cell types take part in this process in higher organisms, including macrophages, neutrophils, and monocytes. Upon opsonization with antibodies (IgG), foreign extracellular materials are recognized by Fc gamma receptors.

Cross-linking of Fc gamma receptors initiates a variety of signals mediated by tyrosine phosphorylation of multiple proteins, which lead through the actin cytoskeleton rearrangements and membrane remodeling to the formation of phagosomes. Nascent phagosomes undergo a process of maturation that involves fusion with lysosomes. The acquisition of lysosomal proteases and release of reactive oxygen species are crucial for digestion of engulfed materials in phagosomes.

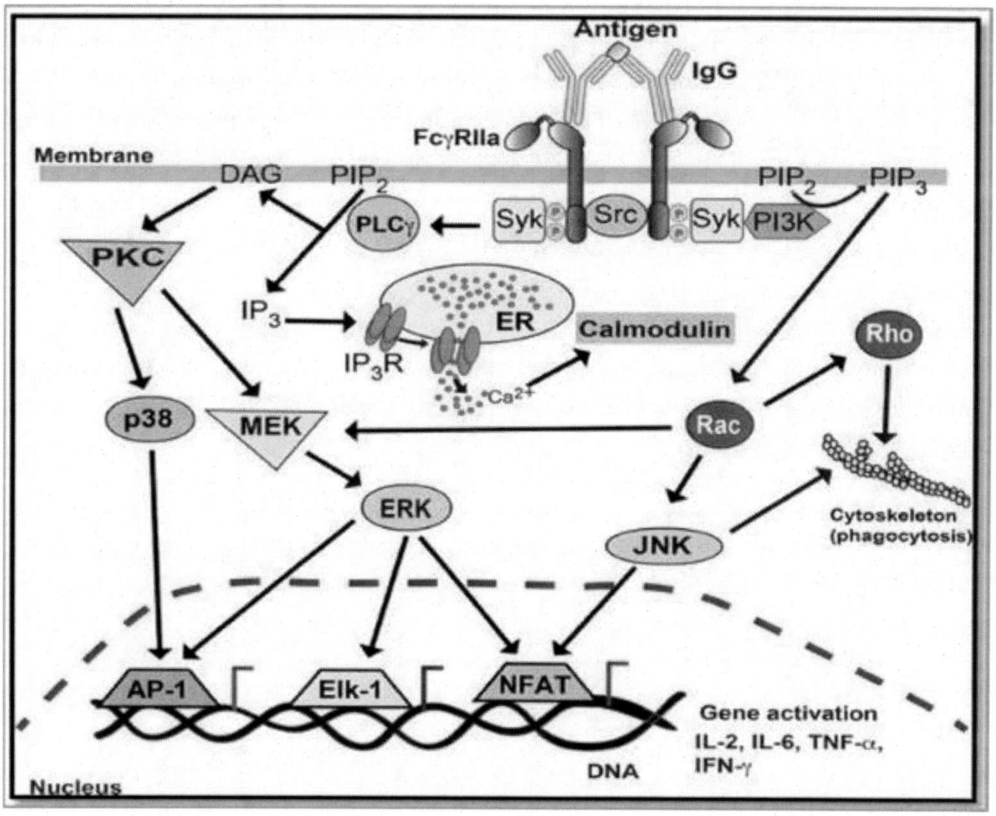

Figure 5.6. Signaling transduction pathway of the neutrophil FcγRIIa. Engagement of activating FcγRIIa by IgG-antigen immune complexes induces receptor cross-linking and phosphorylation of tyrosine residues in the immunoreceptor tyrosine-based activation motif domains (green oval) by Src family kinases. Phosphorylated tyrosines then become docking sites for Syk, which in turn phosphorylates multiple substrates leading to different signaling pathways that ultimately activate various cell responses. See text for details. P represents a phosphate group; Syk, spleen tyrosine kinase; PI3K, phosphatidylinositol-3 kinase; PIP2, phosphatidylinositol 4,5-bisphosphate; PIP3, phosphatidylinositol 3,4,5-trisphosphate; JNK, c-Jun N-terminal kinase; NFAT, nuclear factor of activated T cells; PLCγ, phospholipase Cγ; DAG, diacylglycerol; IP3, inositol triphosphate; IP3R, receptor for IP3; ER, endoplasmic reticulum; PKC, protein kinase C; MEK, MAPK/ERK kinase; ERK, extracellular signal-regulated kinase; p38, p38 MAP kinase; AP-1, activator protein 1; Elk-1, Ets LiKe gene1 (ETS) transcription factor; IL-2, interleukin-2; IL-6, interleukin-6; TNF-α, tumor necrosis factor α; and IFN-γ, interferon-γ.

Fc Alpha Receptor Signaling

FcαRI is a member of the multichain immune recognition receptor (MIRR) family. As mentioned in the section of Fc alpha receptor, the FcαRI lacks canonical signal transduction domains known as ITAM signaling motif. For this reason, the signaling transduction of FcαRI is dependent on association with the FcR γ–chain subunit, forming the trimer FcαRIα/γγ . The FcR γ-chain was initially described as a component of FcεRI and FcγRI, expressed on mast cells and monocytes, respectively. The FcαRIα-FcRγ interaction is strong and depends on oppositely charged residues in their transmembrane regions.

As the Figure 5.7 shows, FcRγ-ITAM are initially phosphorylated by the Src kinase lyn, which leads Recruitment and phosphorylation of Syk and Btk are modulated by cell stimulation with interferon-γ and/or phorbol ester, indicating that activation of these tyrosine kinases through FcαRI may depend on the level of cell priming at inflammatory sites.

FcαRI cross-linking triggers calcium release from intracellular stores in neutrophils and induction of NADPH oxidase activity that is sensitive to inhibition by PI 3-kinase inhibitors. FcαRI can also associate with Grb2, Shc, SHIP, and SLP-76, suggesting the formation of adaptor complexes to regulate signaling.

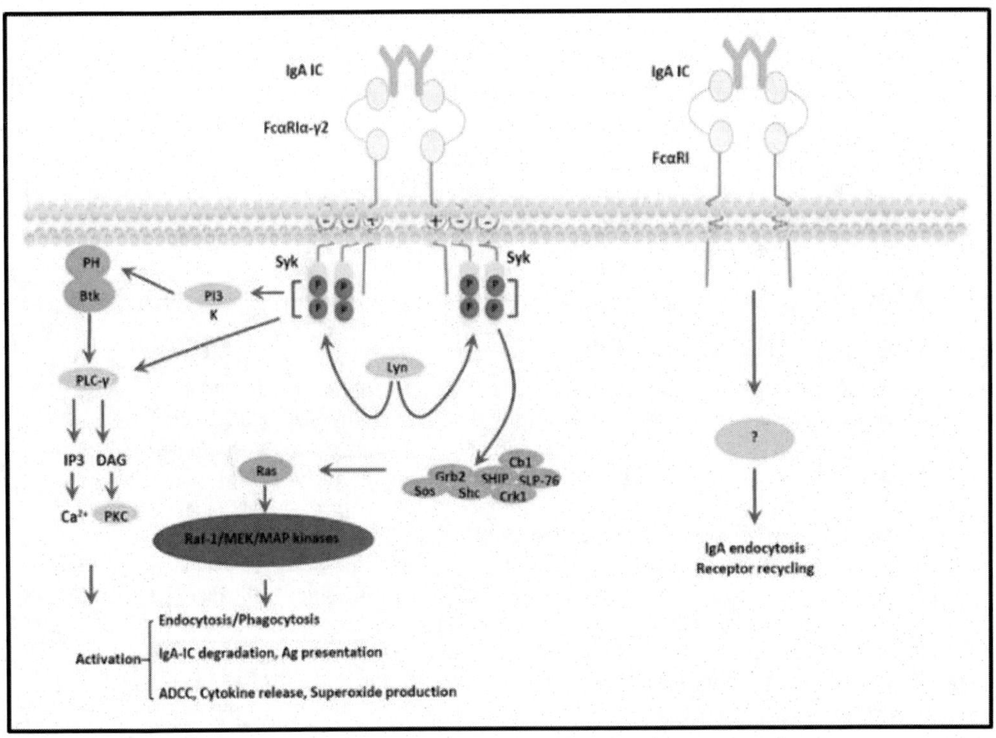

Figure 5.7. Signaling pathways triggered by the activation of FcαRI.

Fc Epsilon Receptor Signaling

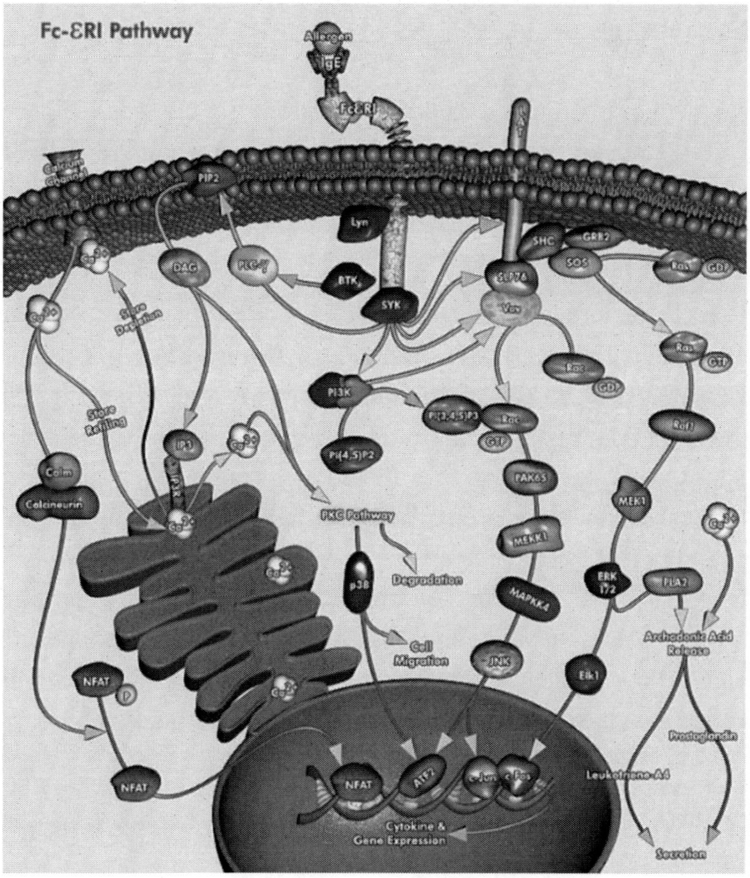

Figure 5.8. The Fc epsilon RI signaling pathway, and the molecular mechanisms involved in the signaling cascade.

Fc epsilon RI signaling pathways in mast cells are initiated via the interaction of antigen with IgE combined to the extracellular domain of the alpha chain of Fc epsilon RI. The activation pathways are regulated both positively and negatively by the interactions of numerous signaling molecules. Mast cells that are thus activated release preformed granules which contain biogenic amines (especially histamines) and proteoglycans (especially heparin). The activation of phospholipase A2 causes the release of membrane lipids followed by development of lipid mediators such as leukotrienes (LTC4, LTD4 and LTE4) and prostaglandins (especially PDG2). There is also secretion of cytokines, the most important of which are TNF-alpha, IL-4 and IL-5. These mediators and cytokines contribute to inflammatory responses.

CELLULAR EFFECTORS OF ADCC

Natural Killer Cells

NK cells are perhaps the most reliable effectors of ADCC. These cells can recognize pathogens which do not express MHC, and they act by releasing granzyme and perforin proteins which bore holes in the cellular membrane of pathogens, thereby killing them. NK cells appear not to express other classes of Fc receptors and, therefore, are not activated for ADCC by IgA, IgE or IgM. FcgRIIIa is present on both NK cells and macrophages as a transmembrane receptor that is associated with the FcR g-chain, an activating molecule involved in signal transduction. This association with the g-chain is essential both for surface expression of FcgRIIIa and for signal transduction following engagement of this receptor by IgG. In contrast, FcgRIIIb (CD16b) found on human neutrophils is expressed as a phosphatidyl inositol glycan-linked receptor that does not associate with the g-chain and does not activate tumor cell killing. Only FcgRIIIa, and not FcgRIIIb, is found in mice. Natural killer (NK) cells are capable of antibody-dependent cellular cytotoxicity or ADCC. NK cells have receptors on their surface for the Fc portion of certain subclasses of IgG. When the antibody IgG is made against epitopes on "foreign" membrane-bound cells, such as virus-infected cells and cancer cells, the Fab portions of the antibodies react with the "foreign" cell. The NK cells then bind to the Fc portion of the antibody.

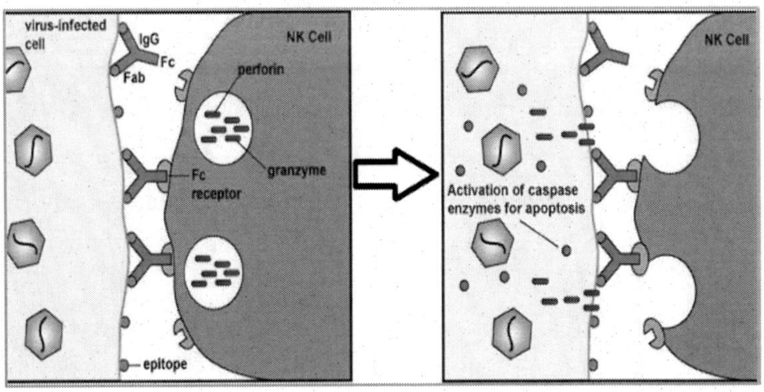

Figure 5.9. Destruction of Virus-Infected Cells by NK Cells through Antibody-Dependent Cellular Cytotoxicity (ADCC), (left) Step 1: The Fab portion of the antibody binds to epitopes on the "foreign" cell. The NK cell then binds to the Fc portion of the antibody. (right) Step 2: The NK cell is then able to contact the cell and release pore-forming proteins called perforins and proteolytic enzymes called granzymes. Granzymes pass through the pores and activate the enzymes that lead to apoptosis of the infected cell by means of destruction of its structural cytoskeleton proteins and by chromosomal degradation. As a result, the cell breaks into fragments that are subsequently removed by phagocytes. Perforins can also sometimes result in cell lysis.

The NK cell then releases pore-forming proteins called perforins, proteolytic enzymes called granzymes, and chemokines. Granzymes pass through the pores and activate the enzymes that lead to apoptosis of the infected cell by means of destruction of its structural cytoskeleton proteins and by chromosomal degradation (Figure 5.9 right panel and Figure 5.9 left panel). As a result, the cell breaks into fragments that are subsequently removed by phagocytes. Perforins can also sometimes result in cell lysis. (When NK cells are carrying out ADCC, they are sometimes also referred to as killer cells).

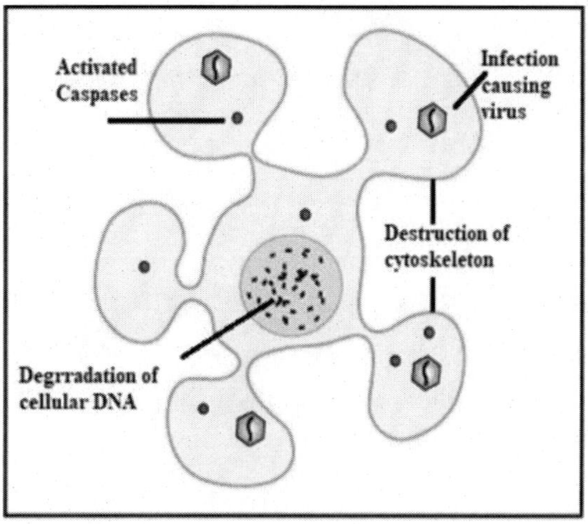

Figure 5.10. NK cells release pore-forming proteins called perforins and proteolytic enzymes called granzymes. Granzymes pass through the pores and activate the enzymes that lead to apoptosis, a programmed suicide of the infected cell. Apoptosis occurs when certain granzymes activate a group of protease enzymes called caspases that destroy the protein structural scaffolding of the cell, degrade the cell's nucleoprotein, and activate enzymes that degrade the cell's DNA. As a result, the infected cell breaks into membrane-bound fragments that are subsequently removed by phagocytes. If very large numbers of perforins are inserted into the plasma membrane of the infected cell, this can result in a weakening of the membrane and lead to cell lysis rather than apoptosis. An advantage to killing infected cells by apoptosis is that the cell's contents, including viable virus particles and mediators of inflammation, are not released as they are during cell lysis.

Monocytes, Macrophages and Dendritic Cells

Macrophages are involved in killing the target cell by releasing lytic and toxic enzymes at the target site. Monocytes and macrophages express all three classes of IgG receptors, FcgRI (CD64), FcgRII (CD32) and FcgRIII (CD16), as well as the receptor for IgA (CD89). Each has been shown to be capable of triggering phagocytosis and ADCC. Although only low levels of FcgRIIIa are expressed on freshly isolated monocytes, this receptor can be markedly increased when monocytes mature into macrophages or are cultured with activating cytokines. Cytokine activation has also been shown to induce expression of FceRI. It has been noted that activated macrophages have high FcgR IIIa/IIb ratios, favoring cell activation, whereas quiescent effectors have low IIIa/IIb ratios, providing a high threshold for cell activation. Human blood dendritic cells also express IgG FcR and are potent effectors in antibody-dependent cellular cytotoxicity (Figure 5.11).

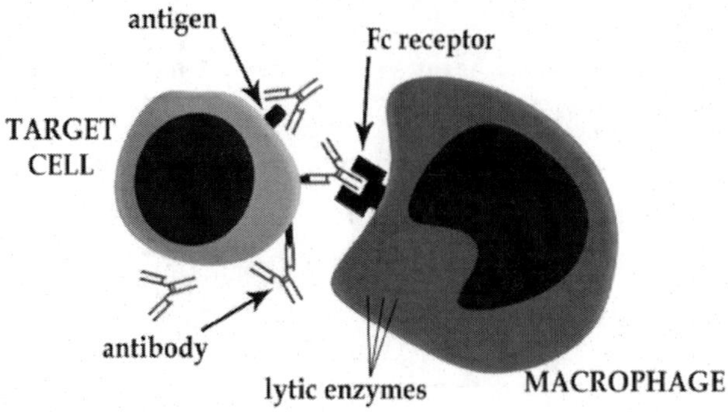

Figure 5.11. Antibody-dependent cellular cytotoxicity (ADCC). The Fc receptor on the macrophage is used to bind the constant Fc portion of antibodies to facilitate engulfment of cells coated with antibodies.

Neutrophils and Eosinophils

Neutrophils are involved in identifying pathogenic bacteria, and they act by releasing lytic enzymes and generating reactive oxygen species (ROS) which destroy the invading microbes. Eosinophils are involved in recognizing parasites, leading to their activation with the release of toxic substances which lead to the death of the pathogenic cells. Freshly isolated neutrophils do not lyse tumor targets through FcgR, but can be activated by granulocyte–macrophage colony-stimulating factor (GM-CSF) to mediate killing through FcgRIIa, and by interferon g (IFN-g) to mediate killing through FcgRI and FcgRIIa. The type of FcgRIII expressed by human neutrophils (FcgRIIIb) does not mediate cytotoxicity of tumor targets, even following cell activation by cytokines. On the other hand, FcgRIIIa on murine polymorphonuclear neutrophil (PMN) does induce ADCC of tumor cells. Cytolysis of red blood cells occurs via FcgRIIa, FcgRIIIa and FcgRIIIb even when resting human neutrophils are used as effectors. These observations suggest that erythroid and tumor targets are killed through different pathways, and that triggering of FcgRIa or FcgRIIa on neutrophils activates lytic mechanisms distinct from those initiated via FcgRIIIb. Of particular interest, neutrophils can efficiently kill human IgA-coated tumor cells and ADCC is highest when both IgA and IgG receptors are coordinately triggered. Furthermore, freshly isolated neutrophils carry out FcaR-dependent lysis without cytokine pre-activation. These findings suggest different pathways of cell activation via FcgR versus FcaR. Although freshly isolated eosinophils express FcgRII as well as low levels of FcgRIII and FceRI, they are not potent ADCC effectors unless activated with GM-CSF, interleukin 3 (IL-3) or IL-5. After cytokine activation, killing of tumor or erythrocyte targets is mediated through FcgRII but not FcgRIII. Thus, like neutrophils, eosinophils require cytokine activation to perform ADCC through FcgRII. Also, like neutrophils, freshly isolated eosinophils are able to mediate ADCC via FcaRI. Eosinophils have also been shown to mediate ADCC of parasites via FceR following activation by cytokines.

Table 5.4. Showing different cell types involved in ADCC their function and mechanism

ADCC Effector Cells		
Cell Type	Function	Mechanism
Macrophage	General destruction of recognized pathogens Presentation of ingested antigens	Release of lytic enzymes and toxic substances
Natural killer cells	Detection of abnormal cells without MHC expression Anti-tumor cell immunity	Release of granzymes and perforin proteins
Neutrophils	Antibacterial immunity	Release of lytic enzymes Generation of reactive oxygen species
Eosinophils	Antiparasitic immunity	Release of major basic protein and other toxic substances

MECHANISMS OF CYTOTOXICITY

It is now well established that antibodies act as a bridge between FcR on the effector cell and the target antigen on the cell that is to be killed. Crosslinking of receptors is required to activate the cytotoxic event. Even for high affinity receptors, binding of monomeric Ig is insufficient to transmit an activation signal. The nature of the biological response and the subsequent cytotoxicity initiated by FcR crosslinking depends on the integration of a variety of signals that are dependent on the class and subclass of both the antibody and the FcR involved, on activation of other signaling molecules, on the effector cell type, and on the nature of the target cell. Cell surface molecules other than FcR also may play a role by promoting adhesion between effector and target cells and/or by providing synergistic signals. Crosslinking of activating FcR initiates a signaling cascade that culminates in release of cytoplasmic granules and the synthesis of surface activation molecules including CD69 and CD25. It often begins with the stimulation of immunoreceptor tyrosine-based activation motif (ITAM) sequences ITAMs have been identified within the cytoplasmic domain of the so-called alpha or immunoglobulin binding chains of some FcR (e.g., FcgRIIa), and also within associated signaling molecules (e.g., g-chain in the case of FcgRIa, FcgRIIIa, FcaRI and FceRI). Granule release is dependent on activation of the mitogen-activated protein kinase (MAPK) family member ERK (extracellular signal-related kinase), and includes intermediate interactions involving the p56lck and many other protein tyrosine kinases. Many studies have focused on dissecting the mechanisms of target cell destruction mediated by cytolytic T cells (CTL) and NK effector cells. Although the manner in which T cells and NK cells recognize their targets is different, these cells share similar cytotoxic mechanisms. For one such mechanism it is proposed that, upon recognition of target, the contents of intracellular granules within the effector cell are released by a calcium-dependent exocytotic process. Perforin, or cytolysin, is one component that is released from granules. Perforin inserts and polymerizes within the target cell membrane, in a manner that is analogous to the membrane insertion of the terminal components of the complement cascade, resulting in the formation of a pore in the target cell (Figure 5.12). Calcium is required for the insertion and assembly of perforin in the target cell membrane. A critical role for perforin in host immune defense has been confirmed in perforin-deficient mice. Other components of granules released by exocytosis have also been implicated in the mechanism of CTL and NK cell-mediated cytotoxicity. One such granule component, granzyme B, may be involved in deoxyribonucleic acid (DNA) fragmentation observed in target cells after engaging a cytotoxic effector cell. However, studies using granzyme B deficient mice suggest a less important role for granzyme B in cytotoxicity. Another mechanism of target cell destruction is the engagement of Fas by Fas ligand (FasL), molecules expressed, respectively, on the surface of some target and effector cells. This interaction initiates a cascade of signals in the target cell resulting in apoptosis of the target. Crosslinking of FcgRIIIa on NK cells induces the expression of Fas L, enabling them to kill Fas+ targets. Therefore, FasL/Fas interaction may be an important mechanism of ADCC mediated by NK cells. Myeloid cells are able to destroy antibody-opsonized targets by phagocytosis as well as by extracellular lysis. Targets ingested phagocytically are destroyed in intracellular phagolysosomes of the effector cell. The mechanism(s) by which myeloid cells mediate extracellular lysis of antibody-opsonized target cells has been controversial and may be different from ADCC mediated by NK cells. In one in vitro system, the mechanism of ADCC mediated by myeloid cells has been shown to be distinct from that mediated by NK cells based on divalent cation requirements.

ADCC mediated by NK cells via FcgRIII was Ca2+ dependent and Mg2+ independent. In contrast, ADCC mediated by monocytes or by IFN-g-activated PMN via FcgRIa or FcgRIIa, or by peritoneal macrophages via FcgRIa, FcgRIIa or FcgRIIIa was Mg2+ dependent and Ca2+ independent. Thus, even though macrophages and NK cells both mediate ADCC via FcgRIIIa, the mechanism by which these two types of effector cells mediate their killing is distinct. Monocytes, macrophages and neutrophils release ROI when FcR are crosslinked, and ROI have been implicated as the mediators of target cell death. However, other mechanisms must also exist since neutrophils isolated from patients with chronic granulomatous disease CGD, whose cells lack the ability to form ROI, were able to mediate ADCC as well as neutrophils isolated from normal donors. Finally, it has been shown that when antibody binds to an apoptotic trigger molecule on the target cell, FcR crosslinking of the target-bound immunoglobulin can induce apoptotic cell death, apparently without a requirement for active participation of the effector cell such as release of toxic mediators by the effector cell. In conclusion multiple pathways may be involved in how an effector cell mediates ADCC of an antibody-opsonized target cell. These include perforin, cytolytic enzymes, reactive oxygen intermediates and apoptotic signaling both with and independent of Fas/FasL interactions. Several mechanisms may combine to mediate cytotoxicity depending on a variety of factors. The relative importance of each in vivo and ex vivo will continue to be dissected using knockout mice and/or effector cells from individuals deficient in specific effector mechanisms.

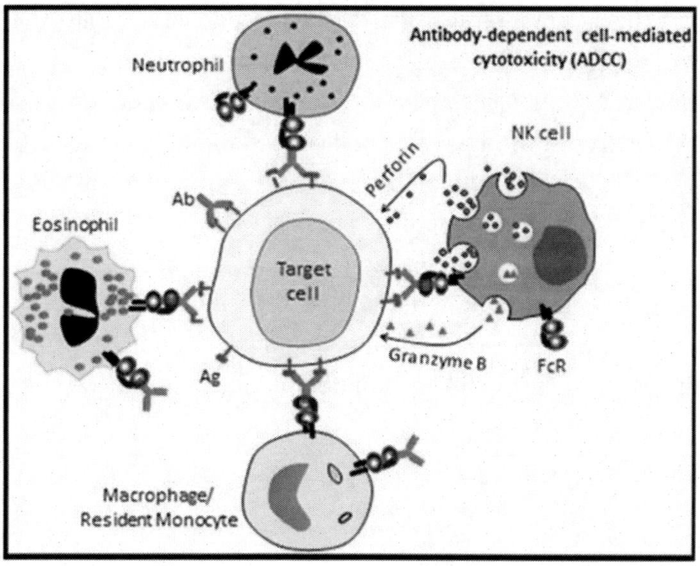

Source: Science direct.

Figure 5.12. Showing mechanism of ADCC of different cells.

MECHANISMS OF ADCC

Recognition of the Target Cell and Cross-Linking of the Fc Receptor on the Effector Cell

An obvious prerequisite for ADCC to occur is the interaction of antibody bound to the target cell with Fc receptors on the effector cell. This interaction is both regulated and facilitated by conformational changes that occur in the antibody molecule only after it has bound to its cognate antigen (Figure 5.1). After binding to surface antigens on the target cell, conformational changes occur in the Fc region of the antibody, which result in its increased affinity for a single Fc receptor on the effector cell. Glycosylation of the Fc region also plays an important role in modulating the affinity of antibody for Fc receptors; in particular, antibodies that are heavily fucosylated (during posttranslational modifications within the B cell) have decreased affinity for FcγRIIIA (CD16), whereas removal of fucose enhances their affinity for FcγRIIIA and their ability to mediate ADCC. Recent evidence indicates that binding between IgG and FcγRIIIA involves tight carbohydrate–carbohydrate interactions that are weakened or obliterated when IgG is fucosylated.

Compared to small, soluble antigens, the relatively larger size of tumors or virally infected cells coated with several antibody molecules on their surface can facilitate physical rearrangements and interactions between Fc receptors present on effector cells (Figure 5.13A). These interactions are often referred to as Fc-receptor ligation, agglutination, aggregation, or cross-linking. The main model to study "ADCC-like" signal transduction pathways relies on the assumption that the first step in generating an ADCC response is the ligation or cross-linking of Fc receptors on the surface of the effector cell as facilitated by a large, "particulate" antigen such as a viral-infected cell coated with surface antigen-specific antibody (Figure 5.13a). Experimentally, to simulate particulate antigen-induced Fc-receptor crosslinking, many researchers incubate NK cells with FcγRIIIA-specific antibodies, followed by incubation with a secondary antibody (Figure 5.13b). Another method of simulating antigen induced Fc-receptor cross-linking is by "reverse ADCC," an experimental setup in which the polarity of the bridging antibody is reversed (Figure 5.13c). Using these two CD16-crosslinking simulation strategies, ADCC-like signal transduction pathways have been dissected in both human and murine NK cells.

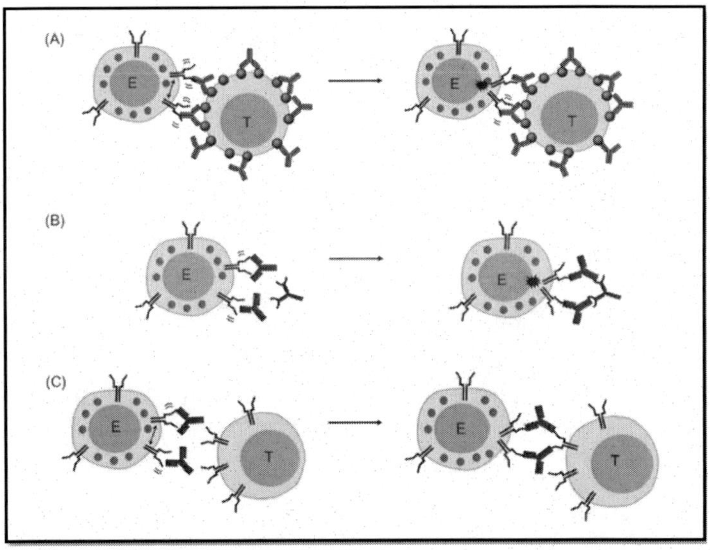

Figure 5.13. Antibody and FcgR ligation. ADCC involves cross-linking Fc receptors on the surface of effector cells. (A) The particulate nature of antibody complexed to antigens expressed on the surface of a target cell facilitates the physical rearrangement of Fc receptors, bringing them closer to each other and inducing their cross-linking (represented here by a black burst symbol). (B) ADCC in human NK cells can be simulated in vitro by incubating human NK effector cells with a mouse anti-human FcγRIIIA (mouse anti-human CD16) antibody (blue), followed by incubation with a secondary antibody (red), such as goat anti-mouse IgG. Addition of the secondary antibody will facilitate physical interactions between CD16 molecules, leading to their cross-linking. (C) ADCC in human NK cells can also be simulated in vitro via "reverse ADCC"— that is, by first incubating human NK effector cells with a mouse anti-human FcγRIIIA (mouse anti-human CD16) antibody (blue), followed by incubation with a mouse target cell expressing FcγRI receptors (gray). The particulate nature of the target cell will also facilitate physical interactions between CD16 molecules, leading to their cross-linking. Note the reversal in antibody polarity between (A) and (C). E, effector cell. T, target cell (Source: Google scholar).

Downstream Signals in the Effector Cell

For ADCC to occur, molecular signals must also be transduced when an FcγR-bearing effector cell recognizes an antibody-coated target cell. Much of what we know about ADCC signal transduction is based on experiments using FcγRIIIA-bearing NK cells as effector cells. Less is known about signaling in other effector cells expressing other FcγRs. In the current signaling model, the gamma (γ) subunit associated with the FcγRIIIA receptor plays a crucial role in signaling. It contains immunoreceptor tyrosine-based activation motifs (ITAMs), which are consensus sequences containing tyrosine residues that can be phosphorylated. ITAMs do not have intrinsic tyrosine kinase activity; instead, they become phosphorylated by cellular src kinases upon FcγRIIIA cross-linking. Phosphorylated ITAMs recruit the spleen tyrosine kinase (Syk) protein, which binds to the ITAMs via its SH2 domains and becomes activated (Figure 5.14). Recruitment and activation of Syk triggers three main pathways involved in ADCC: phospholipase C-gamma pathway (PLC-γ), phosphatidylinositol 3-kinase (PI-3K) pathway, and Vav/Rho-family G-proteins pathway. The PLC-γ pathway involves the Syk dependent phosphorylation of the PLC-γ isozymes. Activated PLC-γ cleaves membrane phosphatidylinositol 4,5-bisphosphate (PIP2) to generate inositol-1,4,5-trisphosphate (IP3) and diacylglycerol (DAG). IP3 triggers an increase in intracellular free calcium, which is required for cytotoxic granule mobilization and exocytosis. DAG activates protein kinase C (PKC), which is necessary for other NK cell functions not directly related to ADCC. The PLC-γ pathway is required for ADCC function and can be blocked by herbimycin A, an antibiotic that inhibits tyrosine kinase activity and ADCC. The PI-3K pathway involves the Syk dependent activation of PI-3K. PI-3K is a heterodimer composed of an 85-kDa adaptor protein (facilitates interactions with other tyrosine kinases) and a 110-kDa catalytic subunit. PI-3K phosphorylates membrane PIP2 into membrane phosphatidylinositol 3,4,5-triphosphate (PIP3). PIP3 mediates recruitment of pleckstrin homology (PH) domain-containing signaling proteins—that is, proteins (such as PLC-γ) that can be recruited to the membrane through binding to membrane-bound PIP3. Recruitment of PH domain-containing proteins to the membrane facilitates their interactions, thereby physically enhancing signal transduction pathways. The PI-3K pathway is also required for ADCC function and can be blocked by wortmannin, a fungal metabolite that inhibits ADCC and binds covalently to the p110 subunit of PI-3K. The PH domain-containing Vav protein and the Rho family guanine nucleotide-binding proteins (G proteins) are also involved in NK cell granule polarization required for ADCC function. Syk is involved in the phosphorylation of Vav, although the precise mechanism remains unclear. The cholera toxin A subunit (Ctx-A), a known inhibitor of G proteins, can inhibit ADCC activity; however, it is unclear where exactly this inhibition occurs, as Ctx-A may also have pleiotropic effects on IP3 and GTP-binding proteins. For the sake of simplicity, it is tempting to limit the description of ADCC signaling mechanisms to the three Syk-related pathways outlined in Figure 3; however, we must bear in mind four important concepts to avoid oversimplifications or dogmatic views of the downstream signals involved in ADCC:

1. Signaling cascades do not work in isolation, and there is a considerable degree of crosstalk between Fc receptors and integrins, surface receptors, and other intracellular proteins whose functions are still under investigation.
2. There is a considerable degree of redundancy involving certain steps along these three pathways.
3. ADCC signaling mechanisms involving FcγRI, FcγRII, and FcγRIIIb have yet to be characterized to the same extent as ADCC signaling involving FcγRIIIA.
4. Negative signals regulating each of the above-mentioned pathways exist.

With regard to the last point, we may cite the example of SH2-containing inositol phosphatase-1 (SHIP-1), a protein that modulates CD16-mediated ADCC in human NK cells. Whereas the three main ADCC signaling pathways described previously involve the phosphorylation of ITAMs present in the γ-subunit associated with FcγRIIIA, the SHIP-1 regulatory pathway engages the zeta (ζ) subunit associated with FcγRIIIA.

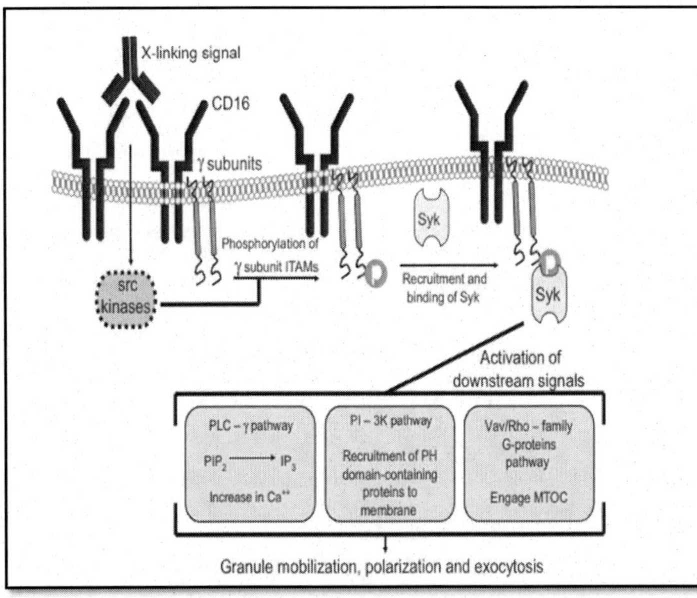

Figure 5.14. ADCC signaling pathways. ADCC mediated by NK cells involves three main signaling pathways. Following CD16 cross-linking, src kinases are activated and phosphorylate the ITAMs on the γ subunits associated with CD16. This is followed by recruitment of Syk and subsequent activation of the PLC-γ pathway, the PI-3K pathway and the Vav/Rho-family G-proteins pathway. The concerted action of these pathways leads to granule mobilization and exocytosis toward the target cell (Adapted from Antibody Fc book, by Victor Raúl Gómez Román, Joseph C. Murray, Louis M. Weiner).

Evidence supporting the role of these pathways in ADCC can be found in human immuno - deficiency virus (HIV) infection and certain congenital abnormalities; for example, following CD16 cross-linking, NK cells from HIV infected donors have significantly reduced levels of phosphorylated Syk in the cytoplasm compared to healthy controls. This reduction correlates with a reduced ability of NK cells to degranulate and a reduced level in the number of CD16 molecules expressed per NK cell among HIV-infected donors. Similarly, NK cells from patients with congenital hemophagocytic syndromes have been shown to be impaired in their ability to degranulate due to loss-of-function mutations in genes whose corresponding proteins are involved in vesicle trafficking and/or cytotoxic granule formation and release.

Mechanisms of Killing

CTLs and NK cells are able to kill target cells through three main pathways: the perforin/granzyme cell death pathway, the FAS ligand (FAS-L) pathway, and the reactive oxygen intermediates/reactive oxygen species (ROI/ROS) or oxidative burst pathway. Of these three pathways, the perforin/granzyme cell death pathway, also known as the granule exoctyosis pathway, has been the most widely investigated in the context of ADCC involving NK cells.

The Perforin/Granzyme Cell Death Pathway

Upon FcγRIIIA receptor cross-linking, the signaling pathways described earlier trigger the increase of intracellular calcium, the engagement of the microtubule-organizing center (MTOC) within the effector cell, and the polarization and release of cytotoxic granules containing perforin and granzyme. These granules are highly organized dual-function organelles containing mainly perforin, granzyme, and granulysin within a dense core surrounded by a pre-lysosomal compartment of acidic pH containing lysosomal hydrolases. The content of cytotoxic granules in NK cells has been reviewed elsewhere.

Killing of target cells via the perforin/granzyme cell death pathway essentially occurs in sequential stages involving the concerted action of perforin and granzyme. The most recent model for this pathway argues for the existence of a macromolecular complex containing perforin, granzyme, and other molecules being released by the effector cell. Granzyme B in this macromolecular complex interacts with the mannose 6-phosphate receptor (MPR) expressed on the surface of target cells, leading to MPR-mediated endocytosis of the macromolecular complex. Endosomes containing granzyme and perforin are formed within the target cell; endosome membrane disruption is achieved by perforin resulting in granzyme being released into the cytosol. Perforin by itself can mediate cell lysis and death but not apoptosis. For apoptosis to occur, granzyme B must either directly activate caspases that induce DNA fragmentation in the nucleus or cleave the Bid molecule, which, when truncated, can insert itself into mitochondria and release mitochondrial constituents that can in turn either potentiate caspase activation and nuclear DNA degradation or induce caspase independent apoptosis (Figure 5.15).

Experiments using human NK cells impaired in their ability to degranulate provide evidence that the perforin/granzyme cell death pathway is involved in ADCC. Such "degranulation impaired" NK cells are obtained from patients with congenital immune disorders known as hemophagocytic syndromes. For example, Griscelli syndrome type 2 (GS2) and familial hemophagocytic lymphohistiocytosis type 3 (FHL3) are two congenital conditions associated with loss of function of the genes encoding the Rab27a and Munc13-4 proteins. While Rab27a is a member of the Rab GTPase family of proteins that regulate vesicle trafficking, Munc13-4 is a member of a family of proteins that regulate membrane fusion, including the fusion of perforin-containing granules with the plasma membrane. Cells derived from patients with these hemophagocytic syndromes express normal levels of Fc receptors; however, upon FcgRIIIA cross-linking, cells from GS2 and FHL3 patients cannot degranulate.

In addition to these naturally occurring mutations affecting degranulation and ADCC, evidence for the involvement of the granule exocytosis pathway in ADCC also stems from experiments in which perforin/granzyme granule polarization and release are blocked chemically. This can be achieved by the addition of calcium ion inhibitors interfering with microtubule rearrangement and granule polarization or, as briefly mentioned earlier, by the addition of herbimycin A, wortmannin, or cholera toxin A subunit—compounds that affect the three main signaling pathways leading to calcium increase and granule polarization and release.

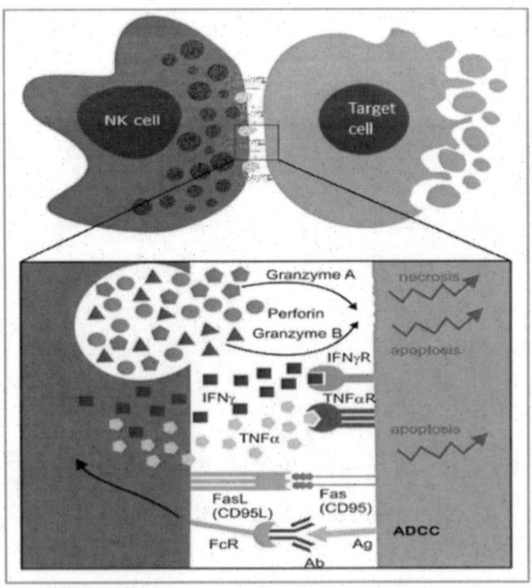

Figure 5.15. Immunological synapse and NK cell mediated cell death. Upon synapsing with target cells, NK cells release proteolytic enzymes, granzyme A and B, as well as perforins that induce necrotic death. The ligation of Fas Ligand (FasL) to Fas/CD95 receptor together with the binding of TNF-α to death receptors induce apoptosis of the tumor cell. NK cell FcγRIII (CD16) recognizes antibody Fc constant domains resulting in cytokine secretion and NK cell mediated antibody dependent cellular cytotoxicity (ADCC). The most abundantly secreted cytokines are TNF-α and IFN-γ.

Unlike previous models arguing for passive diffusion of cytotoxic granules into the target/tumor cell, the most recent model of the perforin/granzyme cell death pathway invokes a key role of novel receptors on the target cell that presumably mediate the uptake of the macromolecular complex containing perforin and granzyme. As mentioned above, the mannose 6-phosphate receptor (MPR) interacts with granzyme B and engages receptor-mediated endocytosis of the macromolecular complex. There is also evidence that platelet-activating factor (PAF), a lysolipid that promotes platelet aggregation, is co-released with perforin when NK cells degranulate. It has been postulated that this may potentiate lysis by allowing PAF released from the effector cell to form a molecular bridge with PAF receptors (PAF-Rs) present on target cells. Although the roles of MPR and PAF-R remain to be fully elucidated in the context of ADCC, it is conceivable that differences in MPR and PAF-R expression by target cells may translate into differences in target cell susceptibility to killing via ADCC and its associated perforin/granzyme pathway.

The Fas-L Signaling Pathway

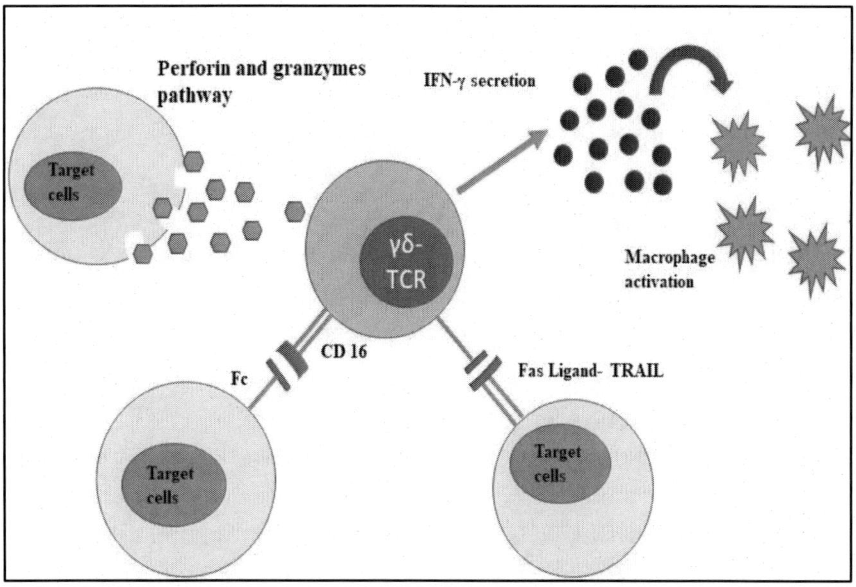

Figure 5.16. γδ T cells show strong and different anti-tumor properties in fact they can release a lot a direct-destructing factor like granzymes and perforin that induce the apoptosis of tumor cells. Furthermore, activated γδ TCR increases CD16 expression and this leads them to mediated antibody-dependent cellular cytotoxicity. On the other hand, these cells can upregulate Fas-L and so induce an apoptosis TRAIL-dependent apoptosis. Finally, γδ T cells can also destroy indirectly target (tumor) cells through regulation of macrophage recruitment via IFN-γ secretion.

Compared to the myriad of reports examining the role of the perforin/granzyme cell death pathway in ADCC, there is less but still substantial evidence to indicate that granule-independent mechanisms may also be involved in the killing of target cells. For example, in addition to inducing granule exocytosis, cross-linking of FcγRIIIA on NK cells induces transcriptional upregulation of FAS ligands (FAS-L), thereby enabling NK cells to become effectors that can kill target cells expressing FAS receptors. This may be relevant in the context of ADCC, as the killing by macrophages of antibody-coated mesanglial cells and vascular smooth muscle cells via ADCC can be inhibited by anti-FAS-L or anti-CD16 antagonistic antibodies.

The ROS/ROI Pathway

The role of the reactive oxygen species/reactive oxygen intermediates (ROS/ROI) pathway in ADCC remains controversial. It is well known that ROS are produced by phagocytic cells during the "oxidative" or "metabolic" burst that occurs in response to antigen opsonization. This burst is characterized by an increase in oxygen consumption by the cell and the subsequent release of ROS such as hydrogen peroxide, superoxide, and other free radicals that can damage the integrity of opsonized antigens and can thus confer protection against intracellular bacteria, viruses, or parasites. It is unclear, however, whether ROS or the oxidative burst are involved during target cell lysis via ADCC. Early experiments indicated that hypoxia and metabolic burst inhibitors (such as cation chelators) reduced the ability of monocytes to kill antibody-sensitized erythrocytes, suggesting a role for the metabolic burst in ADCC. However, these data were at odds with two observations: (1) Monocytes from patients with chronic granulomatous disease (CGD) (i.e., patients who have a genetic impairment in mounting a metabolic burst) had normal levels of ADCC; and (2) the addition of ROS scavengers (such as superoxide dismutase) did not inhibit the ADCC reaction. Similarly, experiments with granulocytes confirmed that ROS scavengers did not inhibit granulocyte-mediated ADCC and that granulocytes from CGD patients had an even greater capacity to kill compared to granulocytes from normal donors. In contrast, data from a third study indicated that both mononuclear (PBMC) and polymorphonuclear (PMN) cells from CGD patients had significantly reduced levels of ADCC against T cells. These discrepancies may be best explained by the fact that the CGD phenotype is not caused by a unique mutation; instead, it is the result of a genetically heterogeneous group of immunodeficiencies rendering cells from CGD patients differentially affected in their capacity to perform ROS-dependent ADCC. In a recent report, Horner et al. argued that the ADCC activity of PMN is neither ROS mediated nor perforin dependent but is instead related to trogocytosis, the ability of effector cells and target cells to exchange membrane lipids during the ADCC reaction in vitro. They showed that membrane lipid exchange (assessed visually by microscopy) correlated with target cell apoptosis and target cell lysis. Recently, an in vivo role of trogocytosis in ADCC has also been proposed using a mouse model. Future experiments may elucidate whether trogocytosis is a novel mechanism of ADCC or merely a consequence of the known mechanisms described earlier. In any case, the trogocytosis and ROS data summarized here illustrate how ADCC killing mechanisms for PMN and monocytes may differ substantially from the predominant perforin/granzyme pathway described for NK cells. Consistent with the notion that effector cell populations may act by different mechanisms, antibody fucosylation has been found to differentially impact ADCC mediated by NK and PMN effectors.

CONCLUSION

Antibody-dependent cell-mediated cytotoxicity (ADCC) is the killing of an antibody-coated target cell by a cytotoxic effector cell through a nonphagocytic process, characterised by the release of the content of cytotoxic granules or by the expression of cell death-inducing molecules. ADCC is triggered through interaction of target-bound antibodies (belonging to IgG or IgA or IgE classes) with certain Fc receptors (FcRs), glycoproteins present on the effector cell surface that bind the Fc region of immunoglobulins (Ig). Effector cells that mediate ADCC include natural killer (NK) cells, monocytes, macrophages, neutrophils, eosinophils and dendritic cells. ADCC is a rapid effector mechanism whose efficacy is dependent on a number of parameters (density and stability of the antigen on the surface of the target cell; antibody affinity and FcR-binding affinity). ADCC involving human IgG1, the most used IgG subclass for therapeutic antibodies, is highly dependent on the glycosylation profile of its Fc portion and on the polymorphism of Fcγ receptors.

FREQUENTLY ASKED QUESTIONS

Q1. What do you understand by Antibody dependent cell mediated cytotoxicity?

Answer: The antibody-dependent cellular cytotoxicity (ADCC), also referred to as antibody-dependent cell-mediated cytotoxicity, is a mechanism of cell-mediated immune defense whereby an effector cell of the immune system actively lyses a target cell, whose membrane-surface antigens have been bound by specific antibodies. It is one of the mechanisms through which antibodies, as part of the humoral immune response, can act to limit and contain infection.

Q2. Define and discuss opsonization?

Answer: Opsonization (also, opsonisation) is the molecular mechanism whereby molecules, microbes, or apoptotic cells are chemically modified to have a stronger attraction to the cell surface receptors on phagocytes and NK cells. With the antigen coated in opsonins, binding to immune cells is greatly enhanced. Opsonization also mediates phagocytosis via signal cascades from cell surface receptors. Opsonins aid the immune system in a number of ways. In a healthy individual, they mark dead and dying self cells for clearance by macrophages and neutrophils, activate complement proteins, and target cells for destruction through the action of natural killer (NK) cells.

Q3. Describe the role of NK cells in ADCC?

Answer: The typical ADCC involves activation of NK cells by antibodies. A NK cell expresses Fc receptors, mostly CD16. These receptors recognize, and bind to, the Fc portion of an antibody, such as IgG, which has bound to the surface of a pathogen-infected target cell. The most common Fc receptor on the surface of an NK cell is called CD16 or FcγRIII. Once the Fc receptor binds to the Fc region of IgG, the Natural Killer cell releases cytotoxic factors that cause the death of the target cell. During replication of a virus some of the viral proteins are expressed on the cell surface membrane of the infected cell. Antibodies can then bind to these viral proteins. Next, the NK cells which have Fc Receptors will bind to that antibody, inducing the NK cell to release proteins such as perforin and proteases known as granzymes, which causes the lysis of the infected cell to hinder the spread of the virus. Furthermore, NK cells are involved in killing tumor cells and other cells that may lack MHC I on their surface, indicating a non-self cell. This is because, generally, all nucleated cells (which excludes RBCs) of the body contain MHC I.

Q4. Give an account of cell mediated immunity?

Answer: Cell-mediated immunity is an immune response that does not involve antibodies. Rather, cell mediated immunity is the activation of phagocytes, antigen-specific cytotoxic T-lymphocytes, and the release of various cytokines in response to antigen. Historically, the immune system was separated into two branches: humoral immunity, for which the protective function of immunization could be found in the humor (cell-free bodily fluid or serum) and cellular immunity, for which the protective function of immunization was associated with cells. CD4 cells or helper T cells provide protection against different pathogens. Naive T cells, mature T cells that have yet to encounter an antigen, are converted into activated effector T cells after encountering antigen-presenting cells (APCs). These APCs, such as macrophages, dendritic cells, and B cells in some circumstances, load antigenic peptides onto the MHC of the cell, in turn presenting the peptide to receptors on T cells. The most important of these APCs are highly specialized dendritic cells; conceivably operating solely to ingest and present antigens.

Q5. What are the different types of Fc receptors?

Answer: there are several types of Fc receptors depending on the kind of antibody that they recognize. Among of these types, the following three types are major. One is Fc receptors recognizing Fc portion of IgG called Fc gamma receptor (FcγR). Other is Fc receptor recognizing C-terminal of IgA called Fc alpha receptor (FcαR). The last is Fc receptor recognizing C-terminal of IgE called Fc epsilon receptor (FcεR).

Q6. What is Fas-L signaling pathway?

Answer: Cross-linking of FcγRIIIA on NK cells induces transcriptional upregulation of FAS ligands (FAS-L), thereby enabling NK cells to become effectors that can kill target cells expressing FAS receptors. This may be relevant in the context of ADCC, as the killing by macrophages of antibody-coated mesanglial cells and vascular smooth muscle cells via ADCC can be inhibited by anti-FAS-L or anti-CD16 antagonistic antibodies.

Q7. What is ELISA?

Answer: ELISA (enzyme-linked Immunosorbent assay) is a plate-based assay technique designed for detecting and quantifying substances such as peptides, proteins, antibodies and hormones. Other names, such as enzyme immunoassay (EIA), are also used to describe the same technology. In an ELISA, an antigen must be immobilized on a solid surface and then complexed with an antibody that is linked to an enzyme. Detection is accomplished by assessing the conjugated enzyme activity via incubation with a substrate to produce a measureable product. The most crucial element of the detection strategy is a highly specific antibody-antigen interaction.

Q8. Give an account of Fc epsilon receptor?

Answer: Fc epsilon receptor, also known as FcεR, is an Fc receptor that binding IgE. Currently, two types of FcεR are known. The high-affinity receptor FcεRI is capable of binding monomeric IgE (it has two Ig-like domains). The classical Fc epsilon receptor I is tetrameric. FcεRI is found on epidermal Langerhans cells, eosinophils, mast cells and basophils. Based on its cellular distribution, FcεRI plays a crucial role in controlling allergic responses. The low-affinity receptor FcεRII (CD23) interacts preferentially with complex IgE. FcεRII has multiple functions as a membrane-bound or soluble receptor; it controls B cell growth and differentiation and blocks IgE-binding of eosinophils, monocytes, and basophils.

Q9. Discuss the role of antigen presenting cells in ADCC?

Answer: Macrophages are involved in killing the target cell by releasing lytic and toxic enzymes at the target site. Monocytes and macrophages express all three classes of IgG receptors, FcgRI (CD64), FcgRII (CD32) and FcgRIII (CD16), as well as the receptor for IgA (CD89). Each has been shown to be capable of triggering phagocytosis and ADCC. Although only low levels of FcgRIIIa are expressed on freshly isolated monocytes, this receptor can be markedly increased when monocytes mature into macrophages or are cultured with activating cytokines. Cytokine activation has also been shown to induce expression of FceRI. It has been noted that activated macrophages have high FcgR IIIa/IIb ratios, favoring cell activation, whereas quiescent effectors have low IIIa/IIb ratios, providing a high threshold for cell activation. Human blood dendritic cells also express IgG FcR and are potent effectors in antibody-dependent cellular cytotoxicity.

Q10. Describe briefly the function of macrophages?

Answer: Macrophages are phagocytes derived from blood monocytes. Monocytes that migrate into tissues in response to infection can differentiate into specific tissue macrophages. Macrophages are capable of ingesting and digesting exogenous antigens, such as whole microorganisms and insoluble particles, and endogenous matter, such as injured or dead host cells, cellular debris, and activated clotting factors. Macrophages are dispersed throughout the body.

Q11. How do T cells recognize antigens?

Answer: The antigen-recognition molecules of T cells are made solely as membrane-bound proteins and only function to signal T cells for activation. These T-cell receptors (TCRs) are related to immunoglobulins both in their protein structure having both V and C regions and in the genetic mechanism that produces their great variability. However, the T-cell receptor differs from the B-cell receptor in an important way: it does not recognize and bind antigen directly, but instead recognizes short peptide fragments of pathogen protein antigens, which are bound to MHC molecules on the surfaces of other cells.

Q12. What are the distinct features of granulocytes?

Answer: The granulocytic cells have irregular-shaped nuclei with two to five lobes and are often called as polymorphonuclear leukocytes. These cells include neutrophils, basophils and eosinophils and mast cells. These cells contain secretory granules in their cytoplasm which contain pharmacologically active components. Degranulation releases antimicrobial agents like defensins and cationic peptides, enzymes like acid hydrolases and lysozymes, NO etc.

Q13. How are natural killer cells involved in immune response?

Answer: Natural killer cells re the differentiated cells of common lymphoid progenitor cell. NK cells constitute 5%-10% of lymphocytes in human peripheral blood, play role in destroying cells infected with intracellular pathogens and also tumor cells in the absence of any previous immunization against the tumor. NK cells distinguish abnormalities like reduction in the display of class I MHC molecules and the unusual antigens displayed by some tumor cells and cells infected with viruses.

Q14. How do cytotoxic T lymphocytes kill target cells?

Answer: Cytotoxic T cells provide protection against intracellular pathogens such as viruses and some bacteria and parasites that multiply in the host-cell cytoplasm, where they are sheltered from attack by antibodies. When exposed to infected cells, TC cells release the cytotoxins perforin, granzymes and granulysin. Through the action of perforin, granzymes enter the cytoplasm of the target cell and their serine protease function triggers apoptosis in the target cell.

Q15. Describe briefly the role of neutrophils in immune system?

Answer: Neutrophils constitute 50%–70% of the circulating white blood cells. In response to many types of infections, the bone marrow releases more than the usual number of neutrophils and these cells generally are the first to arrive at a site of inflammation. Neutrophils are active phagocytic cells but unlike macrophages, the lytic enzymes and bactericidal substances in neutrophils are contained within primary and secondary granules. Neutrophils also employ both oxygen-dependent and oxygen-independent pathways to generate antimicrobial substances. Neutrophils are in fact much more likely than macrophages to kill ingested microorganisms.

Multiple Choice Questions

Q1. Which of the following cell has no role in immune defense?
A. B cells
B. T cells
C. RBC's
D. Monocytes

Q2. Which of the following cells secrete antibodies?
A. Neutrophils
B. B cells
C. Monocytes
D. T cells

Q3. Class II MHC molecules expressed by B cells which permit these cells to function as
A. Neutralizing the antigen
B. Development of immune response by secretion of antibodies
C. Antigen presenting cells
D. Phagocytosis

Q4. B cells recognize the antigens
A. Processed proteinaceous antigens only
B. Lipopolysaccharide
C. Unprocessed soluble proteinaceous antigens and LPS
D. Unprocessed proteinaceous antigens only

Q5. Macrophages are formed from

A. Neutrophils
B. Lymphocytes
C. Monocytes
D. Basophils

Q6. Neutrophils, eosinophils and basophils are known as:

A. Platelets
B. Granulocytes
C. Buffers
D. Astocytomas

Q7. The biological effects of antibodies can be mediated via Fc receptors (FcR) which bind to the Fc region of a particular antibody molecule. Which FcR mediates antibody dependent cellular cytotoxicity (ADCC)?

A. FcγRI (CD64)
B. FcγRII (CD32)
C. FcγRIII (CD16)
D. FcαR (CD89)

Q8. Cross-linking of FcγRIIIA on NK cells induces transcriptional upregulation of FAS ligands (FAS-L), thereby enabling NK cells to become.

A. Differentiated
B. Effector cells
C. Non effector cells
D. Secrete antibodies

Q9. ROS are produced by phagocytic cells during the?

A. Oxidative burst
B. Metabolic burst
C. In response to antigen opsonization.
D. All of the above

Q10. Which among the following affects the granule exocytosis pathway in ADCC?

A. Herbimycin A,
B. Wortmannin
C. Cholera toxin A subunit
D. All of the above

Q11. ADCC mediated by NK cells via FcgRIII is

A. Ca2+ dependent
B. Mg2+ dependent
C. Cu2+ dependent
D. Na+ dependent

Q12. ADCC mediated by monocytes or by IFN-g-activated PMN via FcgRIa is.

A. Ca2+ dependent
B. Mg2+ dependent
C. Cu2+ dependent
D. Na+ dependent

Q13. Myeloid cells are able to destroy antibody-opsonized targets by as well as by.

A. Phagocytosis
B. Extracellular lysis
C. Both A & B
D. None of the above

Q14. The basic component (s) of ADCC are recognized as

A. Being effector cells
B. Antibodies
C. Target cells coated with antigen
D. All of the above

Q15. Fc epsilon RI signaling pathways in mast cells are initiated via the interaction of antigen with

A. IgE
B. IgA
C. IgG
D. IgD

Answer Key

1(C), 2(B), 3(C), 4(C), 5(C), 6(C), 7(C), 8(B), 9(D), 10(D), 11(A), 12(B), 13(C), 14(B), 15(A)

Assignments

Q1. Give general features of antigen antibody reactions.

Q2. Give a detailed account of ADCC

Q3. Describe the different types of Fc receptors?

Q4. Discuss in detail the mechanism of ADCC?

Q5. Describe the Fas/L signaling pathway?

References

Akiyama M, Kusunoki Y, Kyoizumi S. Overview of immunological studies on A-bomb survivors. *Journal of radiation research.* 1991;32 (Supplement):301-9.

Bardos P, Altman J, Guillou PJ, Carnaud C. The cell populations mediating natural killing-(NK) and antibody-dependent cell-mediated cytotoxicity are only partially identical. *Clinical and experimental immunology*. 1981 Mar;43(3):534.

Basel J, Tripathy D, Mendelsohn J, et al. Phase II study of weekly intravenous recombinant humanized anti-p185HER2 monoclonal antibody in patients with HER2/neu-overexpressing metastatic breast cancer. *J Clin Oncol* 1996;14(3):737–44.

Bayer R, Shen A, Harris RJ, Ng D, inventors; Harris Reed J, assignee. Antibodies with enhanced adcc functions. *United States patent application US 12/852,384.* 2011 Feb 17.

Bellanti J, editor. *Immunology.* Springer Science & Business Media; 2013 Nov 11.

Berthou C, Bourge JF, Zhang Y, et al. Interferongamma-induced membrane PAF-receptor expression confers tumor cell susceptibility to NK perforin dependent lysis. *Blood* 2000;95(7):2329–36.

Bonner JA, Harari PM, Giralt J, et al. Radiotherapy plus cetuximab for squamous-cell carcinoma of the head and neck. *New Engl J Med* 2006;354(6):567–78.

Boross P, Jansen JH, Pastula A, van der Poel CE, Leusen JH. Both activating and inhibitory Fc gamma receptors mediate rituximab-induced trogocytosis of CD20 in mice. *Immunol Lett* 2012;143(1):44–52.

Boyle JJ. Human macrophages kill human mesangial cells by Fas-L-induced apoptosis when triggered by antibody via CD16. *Clin Exp Immunol* 2004;137(3):529–37.

Bruchelt G, Handgretinger R, Fierlbeck G, et al. Lysis of neuroblastoma cells by the ADCC-reaction: granulocytes of patients with chronic granulomatous disease are more effective than those of healthy donors. *Immunol Lett* 1989;22(3):217–20.

Bryceson YT, March ME, Ljunggren HG, Long EO. Activation, coactivation, and costimulation of resting human natural killer cells. *Immunological reviews.* 2006 Dec;214(1):73-91.

Capron A, Capron M, Dessaint JP. ADCC as primary mechanisms of defence against metazoan parasites. InImmunology 80. *Progress in immunology IV, Volume 2. Fourth International Congress of Immunolog*y, Paris, July 1980. Theme 12. 1980 (pp. 782-793). Academic Press, Inc..

Cartron G, Dacheux L, Salles G, et al. Therapeutic activity of humanized anti-CD20 monoclonal antibody and polymorphism in IgG Fc receptor FcgammaRIIIa gene. *Blood* 2002; 99(3):754–8.

Choudary KB. Monoclonal antibodies with ADCC and CDC enhancement for therapy. *Int J Pharma Bio Sci*. 2013;4:B588-99.

Clynes RA, Towers TL, Presta LG, Ravetch JV. Inhibitory Fc receptors modulate in vivo cytotoxicity against tumor targets. *Nat Med* 2000;6(4):443–6.

Cragg MS, French RR, Glennie MJ. Signaling antibodies in cancer therapy. *Current Opinion in Immunology*. 1999 Oct 1;11(5):541-7.

Cunningham D, Humblet Y, Siena S, et al. Cetuximab monotherapy and cetuximab plus irinotecan in irinotecan-refractory metastatic colorectal cancer. *New Engl J Med* 2004; 351(4):337–45.

Dahan R, Reiter Y. T-cell-receptor-like antibodies–generation, function and applications. *Expert Reviews in Molecular Medicine*. 2012 Feb;14.

Dall'Ozzo S, Tartas S, Paintaud G, et al. Rituximabdependent cytotoxicity by natural killer cells: influence of FCGR3A polymorphism on the concentration-effect relationship. *Cancer Res* 2004;64(13):4664–9.

de Haij S, Jansen JH, Boross P, et al. In vivo cytotoxicity of type I CD20 antibodies critically depends on Fc receptor ITAM signaling. *Cancer Res* 2010;70(8):3209–17.

Descamps B. Antibody-dependent cell-mediated cytotoxicity of ADCC. In *Annales d'Immunologie* 1980 (Vol. 131, No. 2, p. 127).

Doan T, Melvold R, Waltenbaugh C. *Concise Medical Immunology*. Lippincott Williams & Wilkins; 2005.

Eales LJ. *Immunology for Life Scientists*. J. Wiley; 2003 Aug 15.

Eischen CM, Schilling JD, Lynch DH, Krammer PH, Leibson PJ. Fc receptor-induced expression of Fas ligand on activated NK cells facilitates cell-mediated cytotoxicity and subsequent autocrine NK cell apoptosis. *J Immunol* 1996;156(8):2693–9.

Fan Z, Masui H, Altas I, Mendelsohn J. Blockade of epidermal growth factor receptor function by bivalent and monovalent fragments of 225 anti-epidermal growth factor receptor monoclonal antibodies. *Cancer Res* 1993;53(18):4322–8.

Fast LD, Hansen JA, Newman W. Evidence for T cell nature and heterogeneity within natural killer (NK) and antibody-dependent cellular cytotoxicity (ADCC) effectors: a comparison with cytolytic T lymphocytes (CTL). *The Journal of Immunology*. 1981 Aug 1;127(2): 448-52.

Ferrari G, Pollara J, Kozink D, Harms T, Drinker M, Freel S, Moody MA, Alam SM, Tomaras GD, Ochsenbauer C, Kappes JC. An HIV-1 gp120 envelope human monoclonal antibody that recognizes a C1 conformational epitope mediates potent antibody-dependent cellular cytotoxicity (ADCC) activity and defines a common ADCC epitope in human HIV-1 serum. *Journal of Virology*. 2011 Jul 15;85(14):7029-36.

Gabrielli S, Ortolani C, Del Zotto G, Luchetti F, Canonico B, Buccella F, Artico M, Papa S, Zamai L. The memories of NK cells: innate-adaptive immune intrinsic crosstalk. *Journal of Immunology Research*. 2016;2016.

Gennari R, Menard S, Fagnoni F, et al. Pilot study of the mechanism of action of preoperative trastuzumab in patients with primary operable breast tumors overexpressing *HER2. Clin Cancer Res* 2004;10(17): 5650–5.

Golub RM, Descamps-Latscha B. Role of oxygendependent mechanisms in monoclonal antibodyinduced lysis of normal T cells by phagocytes. I. Human phagocytes. *Ann Inst Pasteur Immunol* 1985;136D(1):3–18.

Gomez-Roman VR, Florese RH, Patterson LJ, et al. A simplified method for the rapid fluorometric assessment of antibody-dependent cell-mediated cytotoxicity. *J Immunol Meth* 2006;308(1–2):53–67.

Gupta S, editor. *Immunology of HIV Infection*. Springer Science & Business Media; 2013 Nov 11.

Horner H, Frank C, Dechant C, et al. Intimate cell conjugate formation and exchange of membrane lipids precede apoptosis induction in target cells during antibody-dependent, granulocyte-mediated cytotoxicity. *J Immunol* 2007;179(1):337–45.

Hubert P, Heitzmann A, Viel S, et al. Antibodydependent cell cytotoxicity synapses form in mice during tumor-specific antibody immunotherapy. *Cancer Res* 2011; 71(15): 5134–43.

Jacobs DB, Pipho C. Use of propidium iodide staining and flow cytometry to measure anti-mediated cytotoxicity: resolution of complement-sensitive and resistant target cells. *J Immunol Meth* 1983; 62(1):101–8.

Kato K, Agatsuma T, Tanabe T, Masuko T, Hashimoto Y. Release of esterase from murine 1. Antibody-Dependent Cellular Cytotoxicity (ADCC) Antibody Fc 24 lymphokine-activated killer cells in antibody-dependent cellular cytotoxic reaction. *Jpn J Cancer Res* (Gann) 1991; 82(2):206–12.

Koller CA, LoBuglio AF. Monocyte-mediated antibodydependent cell-mediated cytotoxicity: the role of the metabolic burst. *Blood* 1981; 58(2):293–9.

Kurai J, Chikumi H, Hashimoto K, et al. Antibodydependent cellular cytotoxicity mediated by cetuximab against lung cancer cell lines. *Clin Cancer Res* 2007; 13(5):1552–61.

Lanier LL. On guard—activating NK cell receptors. *Nature immunology*. 2001 Jan; 2(1): 23.

Lee SC, Srivastava RM, López-Albaitero A, Ferrone S, Ferris RL. Natural killer (NK): dendritic cell (DC) cross talk induced by therapeutic monoclonal antibody triggers tumor antigen-specific T cell immunity. *Immunologic Research*. 2011 Aug 1; 50(2-3): 248-54.

Lee-MacAry AE, Ross EL, Davies D, et al. Development of a novel flow cytometric cell-mediated cytotoxicity assay using the fluorophores PKH-26 and TO-PRO-3 iodide. *J Immunol Meth* 2001;252(1–2):83–92.

Lichtfuss GF, Meehan AC, Cheng WJ, et al. HIV inhibits early signal transduction events triggered by CD16 cross-linking on NK cells, which are important for antibody-dependent cellular cytotoxicity. *J Leukocyte Biol* 2011;89(1):149–58.

Maloney DG, Grillo-Lopez AJ, Bodkin DJ, et al. IDECC2B8: results of a phase I multiple-dose trial in patients with relapsed non-Hodgkin's lymphoma. *J Clin Oncol* 1997; 15(10):3266–74.

Marras F, Bozzano F, De Maria A. Involvement of activating NK cell receptors and their modulation in pathogen immunity. *BioMed Research International*. 2011 Aug 11;2011.

Minard-Colin V, Xiu Y, Poe JC, et al. Lymphoma depletion during CD20 immunotherapy in mice is mediated by macrophage FcgammaRI, FcgammaRIII, and FcgammaRIV. *Blood* 2008;112(4):1205–13.

Mukaida N. Immunologic tests: Antibody-dependent cell-mediated cytotoxicity (ADCC). Nihon rinsho. *Japanese Journal of Clinical Medicine*. 2005 Jul;63:34.

Musolino A, Naldi N, Bortesi B, et al. Immunoglobulin G fragment C receptor polymorphisms and clinical efficacy of trastuzumab-based therapy in patients with HER-2/neu-positive metastatic breast cancer. *J Clin Oncol* 2008; 26(11):1789–96.

Nimmerjahn F, Ravetch JV. Antibody-mediated modulation of immune responses. *Immunological Reviews*. 2010 Jul; 236(1):265-75.

Oboshi W, Watanabe T, Matsuyama Y, Kobara A, Yukimasa N, Ueno I, Aki K, Tada T, Hosoi E. The influence of NK cell-mediated ADCC: structure and expression of the CD16 molecule differ among FcγRIIIa-V158F genotypes in healthy Japanese subjects. *Human Immunology*. 2016 Feb 1; 77(2):165-71.

Ortaldo JR, Winkler-Pickett RT, Nagashima K, Yagita H, Okumura K. Direct evidence for release of poreforming protein during NK cellular lysis. *J Leukocyte Biol* 1992; 52(5):483–8.

Peipp M, Lammerts van Bueren JJ, Schneider-Merck T, et al. Antibody fucosylation differentially impacts cytotoxicity mediated by NK and PMN effector cells. *Blood* 2008; 112(6):2390–9.

Perlmann P. Antibody-dependent cellular cytotoxicity (ADCC) mediated by human killer lymphocytes (K-cells). In *Concepts in Viral Pathogenesis* 1984 (pp. 58-64). Springer, New York, NY.

Pignata C, Prasad KV, Robertson MJ, Levine H, Rudd CE, Ritz J. Fc gamma RIIIA-mediated signaling involves src-family lck in human natural killer cells. *The Journal of Immunology*. 1993 Dec 15;151(12): 6794-800.

Plater-Zyberk C, Maini RN. *Mechanisms of Cytotoxicity and the Cells Involved.*

Pollara J, Hart L, Brewer F, et al. High-throughput quantitative analysis of HIV-1 and SIV-specific ADCC mediating antibody responses. *Cytometry A J Int Soc Analyt Cytol* 2011; 79(8):603–12.

Quinnan GV, Ennis FA. Cell-mediated immunity in cytomegalovirus infections—a review. *Comparative Immunology, Microbiology and Infectious Diseases.* 1980 Jan 1; 3(3):283-90.

Ralainirina N, Poli A, Michel T, Poos L, Andrès E, Hentges F, Zimmer J. Control of NK cell functions by CD4+ CD25+ regulatory T cells. *Journal of Leukocyte Biology.* 2007 Jan; 81(1):144-53.

Riechmann L, Clark M, Waldmann H, Winter G. Reshaping human antibodies for therapy. *Nature* 1988; 332(6162):323–7.

Robertson MJ. Role of chemokines in the biology of natural killer cells. *Journal of Leukocyte Biology*. 2002 Feb; 71(2):173-83.

Rosenthal KS, Tan MJ. *Rapid Review Microbiology and Immunology E-Book: With Student Consult Online Access*. Elsevier Health Sciences; 2010 Aug 27.

Sanderson CJ. The mechanism of lymphocyte-mediated cytotoxicity. *Biological Reviews*. 1981 May; 56(2):153-96.

Schneider-Merck T, Lammerts van Bueren JJ, Berger S, et al. Human IgG2 antibodies against epidermal growth factor receptor effectively trigger antibody-dependent cellular cytotoxicity but, in contrast to IgG1, only by cells of myeloid lineage. *J Immunol* 2010; 184(1):512–20.

Shereck E, Satwani P, Morris E, Cairo MS. Human natural killer cells in health and disease. *Pediatric Blood & Cancer.* 2007 Oct 15; 49(5):615-23.

Sinkovics JG, Horvath JC. Human natural killer cells: a comprehensive review. *International Journal of Oncology*. 2005 Jul 1; 27(1):5-47.

Smyth MJ, Kelly JM, Sutton VR, et al. Unlocking the secrets of cytotoxic granule proteins. *J Leukocyte Biol* 2001; 70(1):18–29.

Spickett G, editor. *Oxford Handbook of Clinical Immunology and Allergy.* Oxford University Press; 2006 Jun 29.

Timonen T, Saksela E. A simplified isotope release assay for cell-mediated cytotoxicity against anchorage dependent target cells. *J Immunol Meth* 1977; 18(1–2):123–32.

Todd I, Spickett G, Fairclough L. *Lecture Notes: Immunology*. John Wiley & Sons; 2015 Sep 28.

Trapani JA, Smyth MJ. Functional significance of the perforin/granzyme cell death pathway. *Nat Rev Immunol* 2002; 2(10):735–47.

Vermorken JB, Trigo J, Hitt R, et al. Open-label, uncontrolled, multicenter phase II study to evaluate the efficacy and toxicity of cetuximab as a single agent in patients with recurrent and/or metastatic squamous cell carcinoma of the head and neck who failed to respond to platinum-based therapy. *J Clin Oncol* 2007; 25(16):2171–7.

Victor Raúl Gómez Romána, Joseph C. Murray and Louis M. Weinerb, Antibody-Dependent Cellular Cytotoxicity (ADCC), Antibody Fc, *Linking Adaptive and Innate Immunity.*

Vivier E. What is natural in natural killer cells? *Immunology letters*. 2006 Sep 15;107(1):1-7.

Weng WK, Levy R. Two immunoglobulin G fragment C receptor polymorphisms independently predict response to rituximab in patients with follicular lymphoma. *J Clin Oncol* 2003; 21(21):3940–7.

Wood SM, Meeths M, Chiang SC, et al. Different NK cell-activating receptors preferentially recruit Rab27a or Munc13-4 to perforin-containing granules for cytotoxicity. *Blood* 2009; 114(19):4117–27.

Yi YS, Son YJ, Ryou C, Sung GH, Kim JH, Cho JY. Functional roles of Syk in macrophage-mediated inflammatory responses. *Mediators of inflammation.* 2014; 2014.

Zhang W, Gordon M, Schultheis AM, et al. FCGR2A and FCGR3A polymorphisms associated with clinical outcome of epidermal growth factor receptor expressing metastatic colorectal cancer patients treated with single-agent cetuximab. *J Clin Oncol* 2007; 25(24):3712–8. 2014, Pages 1-27.

In: Immunoglobulins
Editor: Manzoor Ahmad Mir
ISBN: 978-1-53616-903-4

Chapter 6

MONOCLONAL ANTIBODIES IN THERAPEUTICS

Manzoor A. Mir**, *Umar Mehraj*, *Safura Nisar and Hina Qayoom
Department of Bioresources, School of Biological Sciences,
University of Kashmir, Srinagar J&K, India

ABSTRACT

The development of monoclonal antibodies using hybridoma technology was reported four decades ago and since then more than thirty therapeutic antibodies have been approved by the FDA for the treatment of autoimmune disorders, tumours and inflammatory diseases. Antibodies are highly specific and can therefore bind and affect disease-specific targets, thereby sparing normal cells, and causing fewer toxic side-effects than traditional cytotoxic chemotherapies. The specificity of therapeutic antibodies has been harnessed well in the treatment of cancer. For example, the PD-1 receptor is upregulated on activated T cells and binds two known ligands: PD-L1 and PD-L2. Through interactions with PD-L1 on the surface of tumor cells and immune cells, PD-1 signalling counters T-cell activation during the effector phase of the immune response. Humanized antibodies that target PD-L1 such as avelumab and atezolizumab demonstrate positive response in metastatic breast cancer patients. Other target molecules include CTLA-4, TNF-α and others. The monoclonal antibodies are quite specific and are very efficient, but their high cost-effectiveness, accumulating to more than one billion dollars from preclinical development to FDA approval. Because of the cost problem, therapeutic antibodies remain inaccessible to most patients in both developed and developing countries. With the advancement in science and technology and growing interest in biosimilar antibodies may provide an another alternate for treatment of diseases and furthermore the cost of therapeutic antibodies will decrease. Monoclonal antibody therapy or antibody-based drugs are revolutionizing the treatment of cancer and autoimmune diseases. The mAbs can alter the body's defence system and activate or inhibit the necessary pathways for the better

* Corresponding author: Dr Manzoor Ahmad Mir (Ph. D), Department of Bioresources, School of Biological Sciences, University of Kashmir, Srinagar J&K, India-190006, Contact:-+919622901319. E-mail: drmanzoor@kashmiruniversity.ac.in.

prognosis of the patient. In this chapter we will discuss the production of monoclonal antibodies, such as hybridoma technology, phage display technology, and the advanced next generation sequencing technology. We will also discuss different types of monoclonal antibodies approved by FDA and presently in use for the treatment of different types of cancers, and other autoimmune disorders. We will also shed some light on the mAbs presently in clinical trials or which have recently been approved by the FDA or the EMA.

Keywords: antibody, monoclonal, chimeric, humanized, phage display, hybridoma

OBJECTIVES

- Introduction to therapeutic antibodies.
- Describe the generation of monoclonal antibodies.
- Understanding the role of therapeutic antibodies in cancer.
- Discuss the use of monoclonal antibodies in the treatment of autoimmune disorders.
- Limitations of monoclonal antibodies.

INTRODUCTION

Since 1975, when Kohler and Milstein developed a procedure to efficiently produce monoclonal antibodies (mAbs) (Kohler and Milstein, 1975), it has been widely believed that these molecules would be ideal reagents for imaging and therapy, similar to the magic bullets imagined by Paul Ehrlich at the beginning of the 20th century. Antibodies are highly specific and can therefore affect and bind antigens on diseased cells specifically and efficiently remove the affected cells, thus sparing the normal healthy cells of the body. However, the early excitement was rapidly replaced by disappointment when it became clear that these molecules were facing serious problems when used as therapeutics. The first mAbs were murine molecules and were recognized as foreign when injected into patients, leading to their elimination by the patient's immune system. Moreover, in order to be effective, antibodies often need to interact with certain elements of the immune system such as receptors displayed on effector cells or the complement cascade. Because of their murine nature, these early antibodies did not interact properly with components of the human immune system and their biological efficacy was severely restricted Figure 6.1.

Over the last four decades, the biopharmaceutical market has gone up and therapeutic antibodies has become the leading product. In 2013, pharmaceutical industry sales risen up to 39 billion dollars and therapeutic antibodies represented about 50% of the sales. At present more than thirty antibodies have been approved by the FDA department of US for treatment of different types of ailments. In clinical stages, the number of therapeutic antibodies has risen to hundreds in Europe and United States. The antibodies used in therapeutics these days are no longer full length, naked mouse antibodies. With the advance in engineering technology and novel antigen discovery techniques and strategies, interest in therapeutic antibodies has greatly increased. In this chapter we will discuss the advances in the antibody therapeutics, monoclonal antibodies in the treatment of cancers and others diseases, and the growth of biobetter.

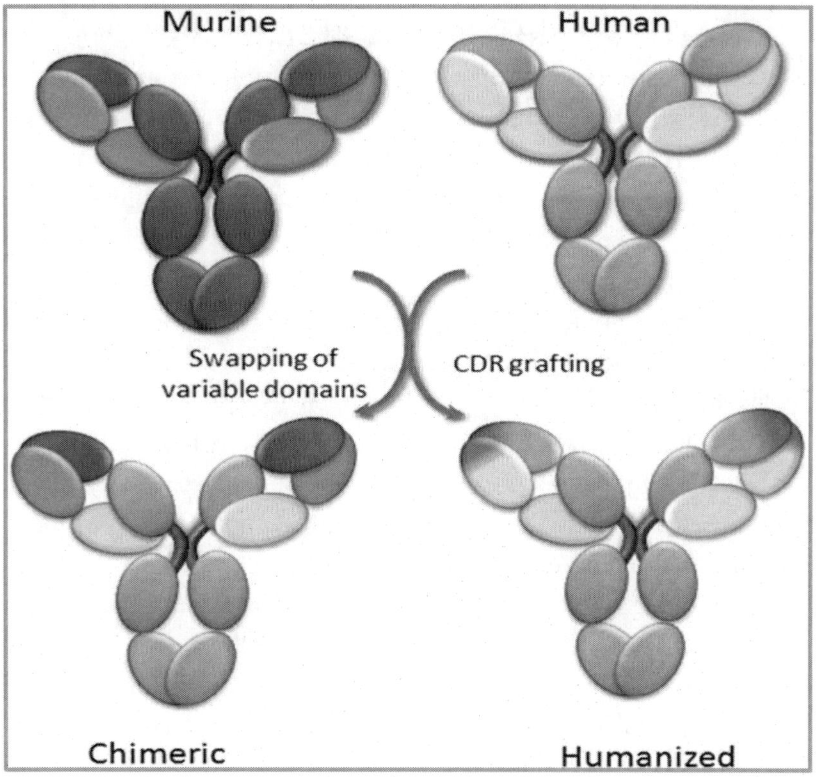

Figure 6.1. Different types of monoclonal antibodies, chimeric, mouse, humanized.

Production of Monoclonal Antibodies

In the earlier times, the generation of antibodies relied on the immunization of animals particularly mice with the desired antigen target. This process resulted in the formation of multiple antibodies produced against different epitopes of the desired target antigen, secreted by B cells with each cell secreting only one specific antibody. The antibodies produced by this strategy are polyclonal in origin. The limitation with this strategy was the replicate senescence of the B cells producing antibodies.

Hybridoma Technology

The hybridoma technology developed by Kohler and Milstein was ground-breaking as it allowed the fusion of antibody secreting cells from the spleen of immunized animals with immortal non-antibody secreting cells Figure 6.2. With this technique, the resulted fusion cells would divide continuously and produce antibodies in mass production. The first approved antibody produced with this strategy was muromonab-CD3 (OrthocloneOKT®3) in 1986 for preventing kidney transplant rejection. The technique of hybridoma had some drawbacks which include labour intensive, low yielding or genetically unstable. More importantly though, the antibody sequences originated from an immunized animal and consequently had the potential of triggering an immune response in humans. Therefore, further improvements were needed to yield antibodies more human-like and safe. These technologies have evolved from chimeric antibodies that is, grafting essential mouse amino acids needed for antigen binding onto a human antibody framework, to both *in vitro* and *in vivo* techniques for generating humanized antibodies Table 6.1. The transgenic mice viz., XenoMouse™ (Abgenix) and HuMab-Mouse® (Medarex) have endogenous murine heavy and kappa light chain genes are inactivated and replaced with the equivalent human germline sequences. Injections of antigens into the mice leads to the production of fully human antibodies. The antibodies developed by this strategy have undergone somatic hypermutation and selection to relatively high affinity. Panitumumab (Vectibix®), a fully human antibody directed against epidermal growth factor receptor (EGFR) as treatment for advanced colorectal cancer was developed using this strategy and its its approval by the regulatory authorities also led to the validation of this technology. Since then, RANK ligand-specific denosumab (Prolia®) has been approved for bone loss and TNF-specific golimumab (Simponi®) for rheumatoid arthritis.

Table 6.1. Difference between chimeric and humanized antibodies

Chimeric antibody	Humanized antibody
An antibody derived from both human and mouse DNA to which a chimeric region is also attached.	An antibody made up of a combination of human antibody and a small part of a mouse or rat antibody.
It has a large portion of a chimeric region.	It does not have a chimeric region.
Chimeric antibody helps to track down and illuminize tumor regions.	Humanized antibody is used for the immunization of humans.
INN substem is -ximab	INN substem is -zumab
For example, rituximab, infliximab and cetuximab.	For example, palivizumab, bevacizumab, and natalizumab.

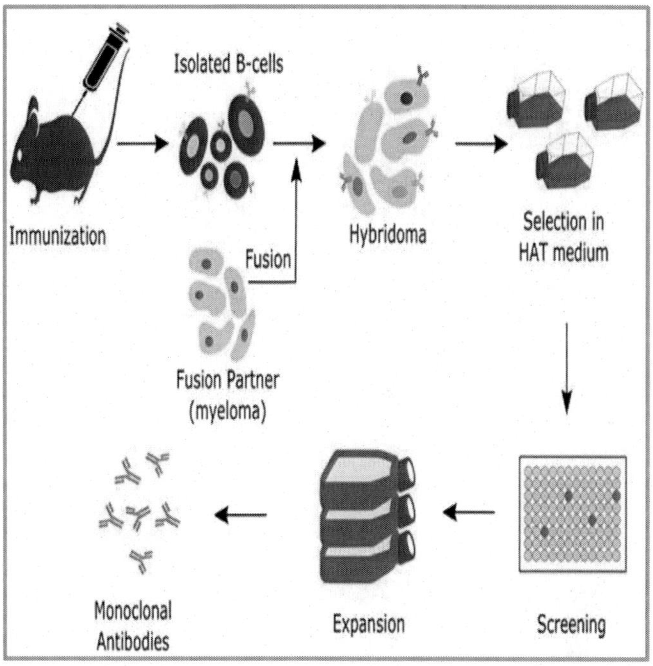

(Image source: https://www.sigmaaldrich.com/technical-documents/protocols/biology/cell-culture/hybridoma-fusion.html).

Figure 6.2. Mechanics in the development of monoclonal antibodies with hybridoma technology.

PHAGE DISPLAY TECHNOLOGY IN ANTIBODY GENERATION

Cloning and screening large libraries of sufficiently diverse human antibody genes in combination with display technology can be used for generating fully human antibodies *in vitro*. The concept of display technology provides a direct physical link between a gene (genotype) and the encoding antibody fragment (phenotype) Figure 6.3. It allows the selection of genes that encode a protein with the desired binding function. For the display of large repertoires and for the selection of high affinity antibodies to biologically relevant targets, phage display technology remains the most widely used *in vitro* method. This strategy involves the expression of proteins on the surface of filamentous phage via fusion with phage coat protein with the genetic sequence packaged within, linking phenotype to genotype selection. When combined with antibody libraries, phage display allows for rapid *in vitro* selection of antigen-specific antibodies and recovery of their corresponding coding sequence. Phage display technology is highly effective, robust and amenable to high throughput processes for screening of > 10^{10} specificities. This technology distinguishes diversity of libraries by source and design, naïve, immune, synthetic and semi-synthetic.

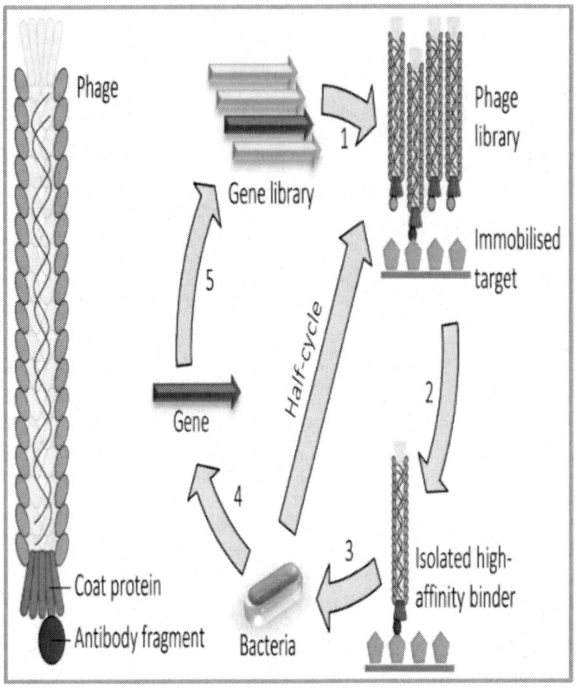

(Image source: Wikipedia).

Figure 6.3. The phage display technology, and its mechanism for the production of monoclonal antibodies.

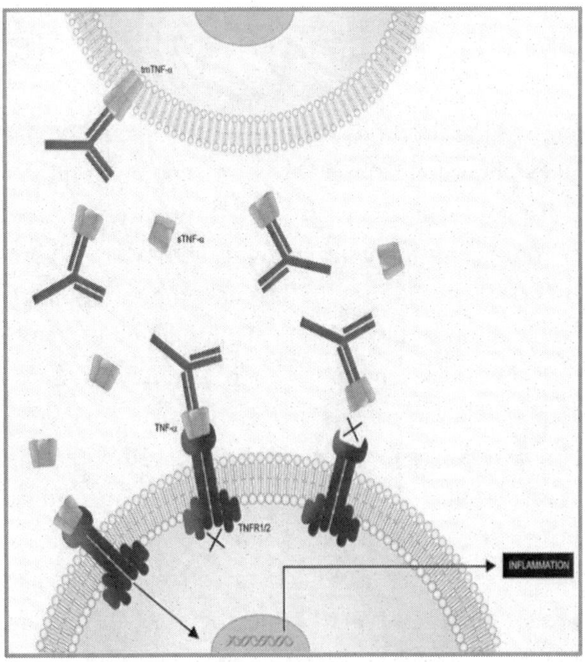

(Image source: https://www.alpco.com/classes-of-biologics-for-the-treatment -of-ibd).

Figure 6.4. Adalimumab, monoclonal antibody developed by phage display technology approved by FDA, binding the TNFα and blocks the signalling.

It was first demonstrated for single chain variable fragment (scFv) and latter it also introduced human antigen binding fragments (Fabs), domain antibodies, camelid domain antibodies, single domain shark antibodies, diabodies and even whole IgG by the screening of other formats. Adalimumab (Humira®) which binds the cytokine TNF was the first FDA approved antibody developed with this technology and was first selected as a scFv expressed on the surface of phage and was further engineered in human IgG1 format, providing major validation for phage display technologies Figure 6.4. In 2013, adalimumab generated global sales of $11 billion and remained the most lucrative antibody product of the year. The number of phage display-derived candidates currently in clinical development further demonstrates the value of phage display as an established and reliable drug discovery platform.

Bacteria, baculovirus and yeast are some other cell surface methodologies that display thousands of copies of the displayed protein on the cell surface thereby allowing for quantitative screening such as flow cytometry. The yeast display methodology is the most ideally suited method because it has the additional feature of eukaryotic protein folding pathways. Alternatively, ribosome display is used to screen large scFv and single variable domain libraries containing up to 10^{13} mutant clones. The main distinguishing feature of this method is that large libraries can be generated because the entire procedure is performed *in vitro* without the need for cell transformations. The genotype and phenotype are linked through ribosomal complexes, consisting of messenger RNA (mRNA), ribosome and encoded protein. The *in vitro* selection of antibodies from antibody libraries based on target affinity is termed "panning". The method is an iterative process whereby the population of target specific antibodies is enriched relative to the number of panning rounds.

THE NEXT-GENERATION SEQUENCING AND GENERATION OF ANTIBODIES

The analysis of natural and synthetic repertoires from which libraries have been constructed has become possible with the advent of next-generation sequencing (NGS). In one application, NGS was used to identify antigen-specific IgG antibodies from polyclonal serum of immunized rabbits and mice when used in combination with affinity chromatography coupled to LC-MS/MS Figure 6.5. The technology had demonstrated its utility in selecting antibodies with favourable properties. For example, Reddy et al. found that 21/27 (or 78%) of scFv they constructed could bind the target antigen with nanomolar affinity after using NGS data to pair together the most abundant variable heavy (VH) and variable light (VL) genes from immunized mice. Other studies have seen similar findings with natural VH:VL pairing limitations and without the limitation. Interestingly, the method of repertoire mining of VH and VL abundances using NGS of splenocytes isolated from immunized mice was compared with a phage panning approach of the same cDNA. While both methods provided completely different sets of antibodies, specificity and affinities of the antibody clones were comparable.

Applications of Antibodies in Therapeutics

Substances foreign to the body, such as disease-causing bacteria and viruses and other infectious agents, known as antigens, are recognized by the body's immune system as invaders. Our natural defences against these infectious agents are antibodies, proteins that seek out the antigens and help destroy them.

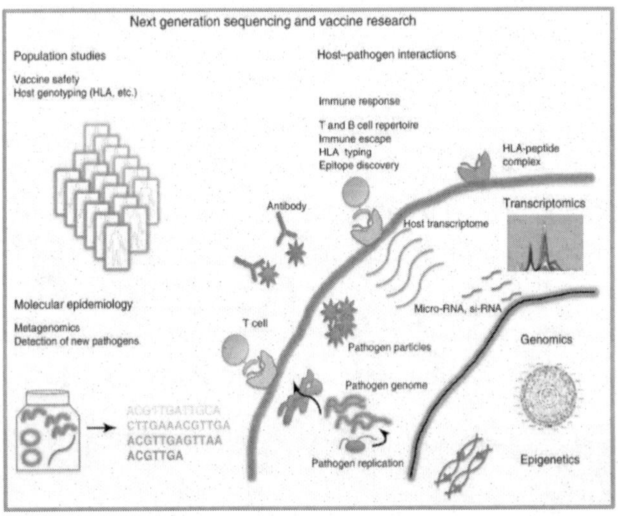

(Image source: Next generation deep sequencing and vaccine design: today and tomorrow, https://doi.org/10.1016/j.tibtech.2012.05.005).

Figure 6.5. Next generation sequencing and vaccine research for the production of monoclonal antibodies.

Antibodies have two very useful characteristics. First, they are extremely specific, that is, each antibody binds to and attacks one particular antigen. Second, some antibodies, once activated by the occurrence of a disease, continue to confer resistance against that disease; classic examples are the antibodies to the childhood diseases chickenpox and measles. It is the first trait of antibodies, their specificity, that makes monoclonal antibody technology so valuable. Not only can antibodies be used therapeutically to protect against disease but also, they can also help to diagnose a wide variety of illnesses and can detect the presence of drugs, viral and bacterial products, and other unusual or abnormal substances in the blood.

Monoclonal Antibodies in the Treatment of Cancer

Cancer immunotherapy is the strategy to employ the body's own defence system i.e., the immune system to fight against the uncontrolled population of cells. It involves the stimulation of the innate and the adaptive components of the immune system to attack the metastatic cancer cells. The stimulation can be achieved by the immunization of the patient which trains the patient's own immune system recognizes the tumor cells and fights against the it. The other way around to activate the immune system is through the administration of therapeutic antibodies as drugs, in which case the patient's immune system is recruited to destroy tumor cells by the therapeutic antibodies Figure 6.6.

Alemtuzumab

It is a DNA-derived humanized monoclonal antibody (Campath-1H) that is directed against the 21–28 kDa cell surface glycoprotein, CD52. Gemtuzumab is also used in the treatment of leukemia. It is used for the treatment of B-cell chronic lymphocytic leukemia (B-CLL) in patients who have been treated with alkylating agents and in whom fludarabine therapy has failed Figure 6.7.

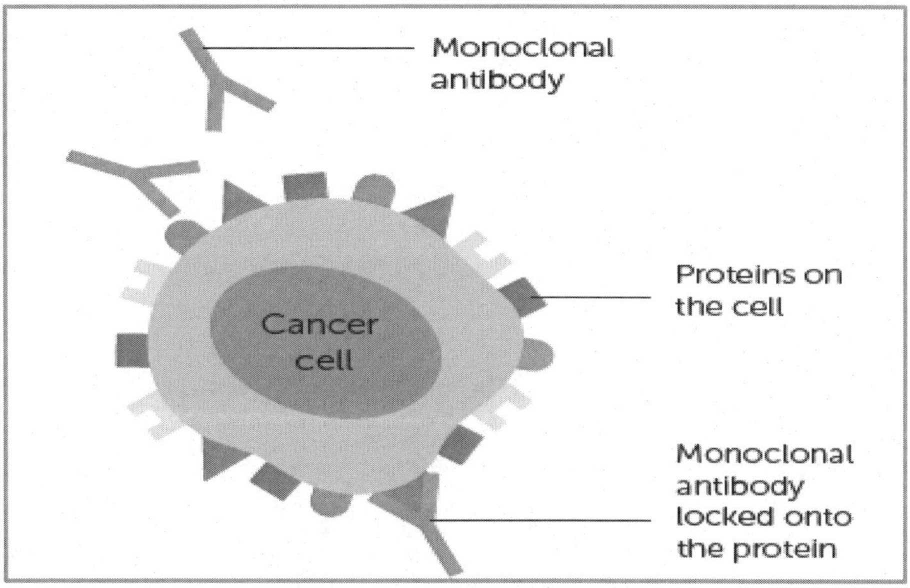

(Image source: cancer research UK https://www.cancerresearchuk.org/about-cancer/cancer-in-general/treatment/targeted-cancer-drugs/types/monoclonal-antibodies).

Figure 6.6. Monoclonal antibody therapy for the treatment of cancer.

Alemtuzumab is contraindicated in patients who have active systemic infections, underlying immunodeficiency (e.g., seropositive for HIV), or known Type I hypersensitivity or anaphylactic reactions to Campath or to any one of its components.

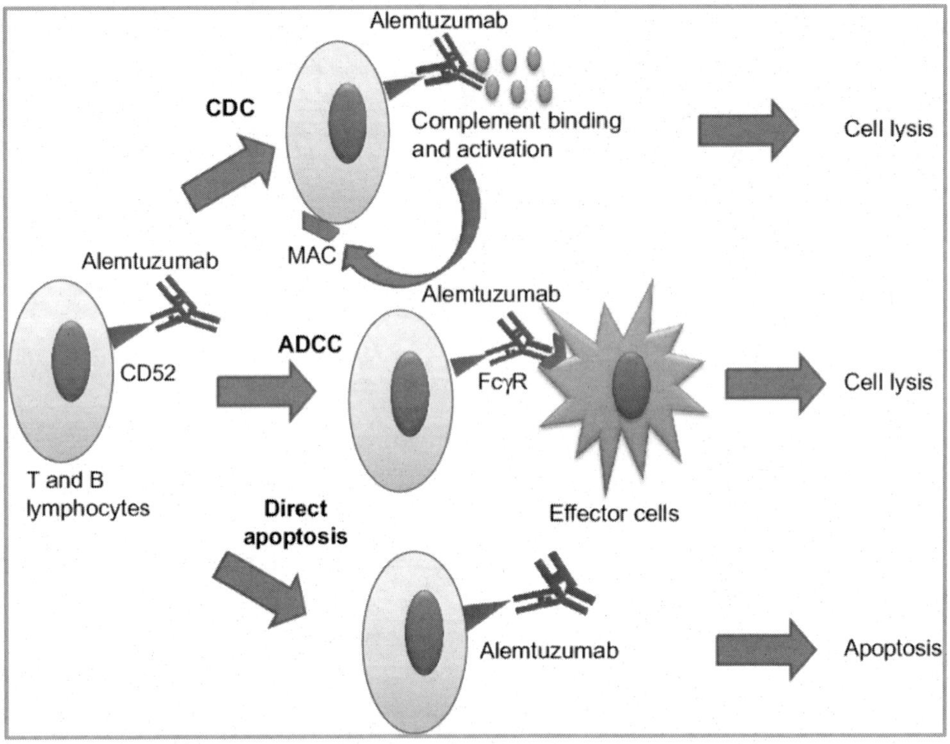

(Image source: Intractable and highly active relapsing multiple sclerosis – Role of alemtuzumab, Neuropsychiatric Disease and Treatment).

Figure 6.7. Mechanism of action of alemtuzumab.

Tremelimumab

The anti–CTLA-4 agent tremelimumab remains investigational in every tumor type. Escalating doses of tremelimumab have been tested with concurrent exemestane in 26 patients with ER^+ HER-2^{--} breast cancer. Five patients had dose-limiting toxicity (DLT), which included diarrhoea (four counts) and elevated serum transaminase levels (one count). The MTD was 6 mg/kg every 90 days. The best response was stable disease (SD) for ≥12 weeks in 42% of patients, and a significant increase in the ratio of $ICOS^+$/FoxP3^+ $CD4^+$ T cells was observed in most patients.

Ipilimumab

Ipilimumab is currently approved as a single agent for early- and late-stage melanoma, and is under investigation in multiple other tumor types both as a single agent and in combination with PD-1/PD-L1 blockade. A study in breast cancer evaluated a single dose of neoadjuvant ipilimumab alone or given with cryoablation in 12 patients with early breast cancer prior to mastectomy; six additional patients received preoperative cryoablation alone. Combination immunotherapy induced circulating Th type 1 cytokines, $ICOS^+Ki67^+CD4^+$ and $ICOS^+Ki67^+CD8^+$ T cells, and an increased $CD8^+$ T-cell/$FoxP3^+$ Treg ratio within the tumor. Clonally expanded TILs [detected by deep sequencing of T-cell receptor (TCR) DNA] correlated with the TIL score by haematoxylin and eosin. On the basis of these promising results, a follow-up study is evaluating cryoablation combined with CTLA-4 (ipilimumab 1 mg/kg) and PD-1 blockade (nivolumab 3 mg/kg; NCT02833233).

Bevacizumab

Bevacizumab is a humanized monoclonal antibody and was the first commercially available angiogenesis inhibitor. It prevents the formation of new blood vessels by targeting and inhibiting the function of a natural protein called vascular endothelial growth factor (VEGF) that stimulates new blood vessel formation thereby stops tumor growth. The drug was first developed as a genetically engineered version of a mouse antibody that contains both human and mouse components. Other drugs for treatment of colorectal cancer are cetuximab and panitumumab. Bevacizumab was approved by the Food and Drug Administration (FDA) in February 2004 for use in colorectal cancer when used with standard chemotherapy treatment. It was approved by the EMA (European medicines agency) in January 2005 for use in colorectal cancer. Israel has also approved the use of bevacizumab. Bevacizumab is usually given intravenously through the arm every 14 days. In colon cancer, it is given in combination with the chemotherapy drug 5-fluorouracil (5-FU), leucovorin, and oxaliplatin or irinotecan. Bevacizumab has also demonstrated activity in renal cell cancer and ovarian cancer when used as a single agent, and in lung cancer and breast cancer when combined with chemotherapy.

The main side effects of concern are hypertension and heightened risk of bleeding. Bowel perforation has also rarely been reported. Studies done particularly in lung cancer have shown that less than half of the patients with advanced disease qualify for treatment with this drug. The FDA updated the label on the drug on September 25, 2006, to note rare cases of brain capillary leak syndrome and nasal septum perforation.

PD-1/PD-L1 Blockade in Breast Cancer

The PD-1 receptor is upregulated on activated T cells and binds two known ligands: PD-L1 and PD-L2. Through interactions with PD-L1 on the surface of tumor cells and immune cells, PD-1 signalling counters T-cell activation during the effector phase of the immune response. Humanized antibodies that target PD-L1 such as avelumab and atezolizumab demonstrate positive response in metastatic breast cancer patients. The concerns or side effects associated with the use of these mAbs in the breast cancer patients are consistent with those expected from the drug class. It was approved by the FDA in 2017 for the treatment of Merkle-cell carcinoma, which is an aggressive type of skin cancer.

Avelumab

Avelumab is a fully human mAb developed originally for the treatment of non-small-cell lung carcinoma (NSCLC). It targets the protein programmed death ligand 1 (PD-L1). In 2017 it received the orphan drug designation by the European Medicines Agency (EMA) for the treatment of gastric cancer. The side effects of the concern are immune-mediated adverse reactions (pneumonitis, hepatitis, colitis, adrenal insufficiency, hypo- and hyperthyroidism, diabetes mellitus, and nephritis) and life-threatening infusion reactions.

Atezolizumab

It is a fully humanized, engineered mAb of IgG1 isotype developed to target the protein programmed death ligand 1 (PD-L1). In May 2016 FDA granted accelerated approval to atezolizumab for locally advanced or metastatic urothelial carcinoma treatment after failure of cisplatin-based chemotherapy. Atezolizumab is also used to treat extensive stage small cell lung cancer. Atezolizumab blocks the interaction of PD-L1 with programmed cell death protein 1 (PD-1) and CD80 receptors (B7-1Rs). PD-L1 can be highly expressed on certain tumors, which is thought to lead to reduced activation of immune cells (cytotoxic T-cells in particular) that might otherwise recognize and attack the cancer. Inhibition of PD-L1 by atezolizumab can remove this inhibitor effect and thereby engender an anti-tumor response.

Monoclonal Antibodies in the Treatment of Autoimmune Diseases

Monoclonal antibodies used for autoimmune diseases include infliximab and adalimumab, which are effective in rheumatoid arthritis, Crohn's disease, ulcerative colitis and ankylosing spondylitis by their ability to bind to and inhibit TNF-α. Basiliximab and daclizumab inhibit IL-2 on activated T cells and thereby help prevent acute rejection of kidney transplants [45]. Omalizumab inhibits human immunoglobulin E (IgE) and is useful in treating moderate-to-severe allergic asthma.

Basiliximab

It is a chimeric mouse-human mAb targeting the alpha chain of the IL-2 receptor of T cells. Basiliximab is used in the prevention of graft rejection or organ transplantation, particularly in kidney transplants. It acts as an immunosuppressant and acts to prevent immediate transplant rejection in people with kidney transplants, in combinations with other agents.

Infliximab

It is a purified recombinant DNA-derived chimeric IgG monoclonal antibody that consists of mouse heavy and light chain variable regions combined with human hevy and light chain constant regions. It targets the TNF-α both in the soluble form in the blood and transmembrane form on T cells. TNF-α is a chemical messenger (cytokine) and a key part of the autoimmune reaction. Infliximab is used for the treatment of a number of autoimmune disorders such as Crohn's disease, ulcerative colitis, rheumatoid arthritis, ankylosing spondylitis, psoriasis, psoriatic arthritis, and Behçet's disease. In rheumatoid arthritis, infliximab aids by preventing TNF-α from binding to its receptor in the cell. It is given by slow injection into a vein, typically at six- to eight-week intervals. Common side effects include infections, acute infusion reactions, and abdominal pain.

Bavituximab

It is a human mouse chimeric mAb against phosphatidylserine, which is a component of cell membranes that is exposed when a cell is transformed into solid tumor cancer cell or dies, and when cells are infected with hepatitis C. The process of cell death is highly controlled and so there usually no immune response to phosphatidylserine but when bavituximab binds to it, the conjugate appears to stimulate an immune response in humans.

MONOCLONAL ANTIBODIES IN THE TREATMENT OF ALZHEIMER'S DISEASE

Alzheimer's disease (AD) is a multi-faceted, age-dependent, progressive neurodegenerative disorder, and is a major cause of dementia. According to the Amyloid hypothesis, the accumulation of extracellular amyloid beta peptides (Aβ) into plaques via oligomerization leads to hallmark symptomatic conditions of AD through synaptic dysfunction and neurodegeneration. Immunotherapy via exogenous monoclonal antibody (MAB) administration has been known to treat various central nervous disorders, such as AD, by inhibiting Aβ-oligomerization thereby preventing neurotoxicity. However, MABs are large for passive protein channels and are therefore inefficient due to the blood-brain barrier preventing MAB passage into the brain. However, the Peripheral Sink hypothesis proposes a mechanism where MABs may not need to cross the blood-brain barrier. Therefore, many research studies are being conducted from failed attempts to treat AD in the past.

However, anti-Aβ vaccines can promote antibody-mediated clearance of Aβ plaques in transgenic mice models with amyloid precursor proteins (APP), and can reduce cognitive impairments. Vaccines can stimulate the immune system to produce its own antibodies, in this case by introducing Aβ into transgenic animal models, known as active immunization. They can also introduce antibodies into animal models, known as passive immunization. In mice expressing APP, both active and passive immunization of anti-Aβ antibodies has been shown to be effective in clearing plaques, and can improve cognitive function. Therefore, several clinical trials using passive and active immunization approaches by development of certain drugs approved by the FDA are currently underway, and are expected to yield results in a couple of years. The implementation of these drugs is during the onset of AD. Other research and drug development for early intervention and AD prevention is ongoing. Various drugs that are under research to treat AD include Bapineuzumab, Solanezumab, and Gautenerumab.

ANTIBODIES IN APPROVAL

As of now more than fifty monoclonal antibodies have been approved by the EU EMA and FDA department for use in the treatment of different diseases (Table 6.2). While full-length and IgG1 antibodies still dominate, a humanized Fab also gained approval. There is expected to be an influx of antibodies in the coming years for the treatment of inflammatory skin conditions including psoriasis, encompassing various cytokine targets or their respective receptors; IL-17a, IL-17R, IL-1α, IL-4R and IL-23.

BIOBETTER ANTIBODIES

Despite its success, antibody-based therapy still presents a long list of important shortcomings that need to be overcome to fully exploit their full therapeutic potential. Typical drawbacks in antibody-based therapy include limited efficacy due to poor tissue and tumor penetration, low *in vivo* efficacy, cumbersome administration, antibody aggregation, solubility as well as high production costs. The use of antibody engineering to improve the properties of therapeutic antibodies has advanced greatly in the last decades giving rise to a varied set of novel formats that offer enhanced attributes for therapeutic and research purposes. These novel molecules are often referred to as biobetters or next-generation antibodies and include platforms such as: engineered antibodies for enhanced effector functions, antibody drug-conjugates (ADC), multi-specific antibodies and single-domain antibody fragments (sdAb or nanobodies®).

A biobetter can have modifications to its chemical structure such as humanization, fusion/conjugation or be glyco-engineered to be less immunogenic or more efficacious. For instance, anti-CD20 ocrelizumab which is currently under review for FDA approval for MS is a humanized version of rituximab (Rituxan®). A biobetter may be modified to have an improved formulation for improved treatment regimen so treatment is less evasive or have a simplified manufacturing process. This was demonstrated with novel subcutaneous (SC) formulations of trastuzumab (Herceptin®) and rituximab (Mabthera®) developed by co-formulating the antibody with hyaluronidase, an enzyme that increases the absorption and distribution of the injected product. A biobetter may be engineered to have higher target affinity, bind at a different epitope or stronger effector function to enhance efficacy and potentially reduce any off-target side effects. Any such improvement means that the therapeutic has been modified and incomparable to the original therapeutic; it is therefore classified as a new biological therapeutic and requires more laborious testing to obtain regulatory approval before market entry.

Table 6.2. List of therapeutic monoclonal antibodies developed in recent years and their clinical status

International Non-proprietary Name (INN)	Target	Type	Year of first EU EMA approval	Year of first FDA approval	Cell line	Therapeutic indication(s)
Adalimumab	TNFα	Human IgG1	2017	2016	CHO	Arthritis; juvenile rheumatoid arthritis; psoriatic arthritis; rheumatoid colitis; ulcerative Crohn's disease; psoriasis; spondylitis; ankylosing
Bezlotoxumab	C. difficile toxin B	Human monoclonal antitoxin antibody	2017	2016	CHO	Enterocolitis; pseudomembranous
Avelumab	PD-L1	Human IgG1/κ	2017	2017	CHO	Metastatic Merkel cell carcinoma
Dupilumab	IL-4Rα	Human IgG4	Not approved	2017	CHO	Asthma; dermatitis
Durvalumab	PD-L1	Human IgG1/κ	Not approved	2017	CHO	Metastatic urothelial carcinoma
Ocrelizumab	CD20	Humanized IgG1κ	Not approved	2017	CHO	Multiple sclerosis
Brodalumab	IL-17RA	Human IgG2/κ	Not approved	2017	CHO	Psoriasis
Reslizumab	IL-5	Human IgG4/κ	2016	2016	NSO	Asthma
Olaratumab	PDGFR-α	Human IgG1	2016	2016	CHO	Sarcoma
Daratumumab	CD38	Human IgG1/κ	2016	2015	CHO	Multiple myeloma
Elotuzumab	SLAMF7	Human IgG1	2016	2015	NS0	Multiple myeloma
Necitumumab	EGFR	Human IgG1	2016	2015	NS0	Carcinoma, non-small-cell lung
Infliximab	TNFα	Chimeric human-murine IgG1	2013	2016	Sp2/0-Ag14	Spondylitis; ankylosing; arthritis; rheumatoid colitis; ulcerative arthritis; psoriatic Crohn's Disease; psoriasis
Obiltoxaximab	PA component of B. anthracis toxin	Chimeric (mouse/human) IgG1/κ	Not approved	2016	NS0	Anthrax infection
Atezolizumab	PD-L1	Human IgG1	Not approved	2016	CHO	Metastatic non-small cell lung cancer
Secukinumab	interleukin-17A	Human IgG1/κ	2015	2015	CHO	Arthritis; psoriatic psoriasis; spondylitis; ankylosing
Mepolizumab	IL-5	Human IgG1/κ	2015	2015	CHO	Asthma
Nivolumab	PD-1	Human IgG4	2015	2015	CHO	Carcinoma; non-small-cell lung carcinoma; renal cell Hodgkin disease melanoma
Alirocumab	PCSK9	Human IgG1	2015	2015	CHO	Dyslipidemias
Idarucizumab	dabigatran etexilate	Human FaB	2015	2015	CHO	Hemorrhage
Evolocumab	LDL-C / PCSK9	Human IgG2	2015	2015	CHO	Dyslipidemias; hypercholesterolemia
Dinutuximab	GD2	Human IgG1/κ	2015 (1)	2015	Sp2/0	Neuroblastoma

International Non-proprietary Name (INN)	Target	Type	Year of first EU EMA approval	Year of first FDA approval	Cell line	Therapeutic indication(s)
Bevacizumab	CD19	BiTEs	2015	2014	CHO	Precursor cell lymphoblastic leukemia-lymphoma
Pembrolizumab	PD-1	Human IgG4	2015	2014	CHO	Melanoma
Ramucirumab	VEGF	Human IgG1	2014	2014	NS0	Stomach neoplasms
Vedolizumab	Integrin-α4β7	HumanizedIgG1	2014	2014	CHO	Colitis; ulcerative Crohn's disease
Siltuximab	cCLB8	Chimeric IgG1κ	2014	2014	CHO	Giant lymph node hyperplasia
Alemtuzumab	CD52	Humanized IgG1	2013	2014	CHO	Multiple sclerosis
Trastuzumab emtansine	HER2	Humanized IgG1 as ADC	2013	2013	CHO	Breast cancer
Pertuzumab	HER2	Humanized IgG1	2013	2012	CHO	Breast cancer
Infliximab	TNF-alpha	Chimeric IgG1 Ab	2013	Not approved	CSC-Ps0006	Spondylitis; ankylosing arthritis; rheumatoid colitis; ulcerative Crohn's disease; arthritis; psoriatic psoriasis
Obinutuzumab	CD20	Humanized IgG1	Not approved	2013	CHO	CLL
Brentuximab	CD30 (conjugate of Mab and MMAE)	Chimeric IgG1 as ADC (antibody drug conjugate)	2012	2011	CHO	Hodgkin lymphoma (HL), systemic anaplastic large cell lymphoma (ALCL)
Raxibacumab	Bacillus anthracis protective antigen	Human IgG1	Not approved	2012	NS0	Prevention and treatment of inhalation anthrax
Belimumab	BLyS	Human IgG1	2011	2011	NS0	Systemic lupus erythematosus (SLE)
Ipilimumab	CTLA-4	Human IgG1	2011	2011	CHO	Melanoma
Denosumab	RANKL	Human IgG2	2011	2011	CHO	Prevention of SREs in patients with bone metastases from solid tumours
Denosumab	RANKL	Human IgG2	2010	2010	CHO	Osteoporosis
Ofatumumab	CD20	Human IgG1	2010	2009	NS0	Chronic lymphocytic leukemia
Besilesomab	NCA-95	Murine IgG1	2010	Not approved	Hybridoma	*In vivo* diagnosis of inflammation / infection sites via scintigraphic imaging —> non-therapeutic
Tocilizumab	IL-6 receptor	Humanized IgG1	2009	2010	CHO	Rheumatoid arthritis
Canakinumab	IL-1ß	Human IgG1	2009	2009	Sp2/0	Cryopyrin-associated periodic syndromes including familial cold autoinflammatory syndrome and Muckle-Wells syndrome; tumor necrosis factor receptor associated periodic syndrome (TRAPS); hyperimmunoglobulin D Syndrome (HIDS)/mevalonate kinase deficiency (MKD) and familial mediterranean fever (FMF)

Table 6.2. (Continued)

International Non-proprietary Name (INN)	Target	Type	Year of first EU EMA approval	Year of first FDA approval	Cell line	Therapeutic indication(s)
Golimumab	TNF*a*	Human IgG1	2009	2009	Sp2/0	Rheumatoid arthritis; psoriatic arthritis; ankylosing spondylitis
Ustekinumab	IL-12 / IL-23	Human IgG1	2009	2009	Sp2/0	Plaque psoriasis
Certolizumab pegol	TNF*a*	Humanized IgG Fab fragment	2009	2008	E. coli	Chron's disease; rheumatoid arthritis
Catumaxomab	EpCAM and CD3	Trifunctional MAb IgG2a / IgG2b	2009	Not approved	Rat-Mouse hybrid hybridoma	Malignant ascites in patients with EpCAM-positive carcinomas
Eculizumab	Complement C5	Humanized IgG2/4	2007	2007	NS0	Paroxysmal nocturnal hemoglobinuria
Ranibizumab	VEGF-A	Humanized IgG1 Fab fragment	2007	2006	E. coli	Neovascular (wet) age-related macular degeneration; macular edema following retinalvein occlusion
Panitumumab	EGFR	Human IgG2	2007	2006	CHO	Metastatic colorectal carcinoma
Natalizumab	VLA-4	Humanized IgG4	2006	2004	Murine myeloma	Multiple sclerosis (relapsing); Crohn's disease
Catumaxomab	EpCAM	Humanized MAb	2005	2005	CHO	Head and neck cancer
Bevacizumab	VEGF	Humanized IgG1	2005	2004	CHO	Metastatic colorectal cancer; non-small cell lung cancer; metastatic breast cancer; hlioblastoma multiforme; metastatic renal cell carcinoma
Omalizumab	IgE	Humanized IgG1	2005	2003	CHO	Asthma
Cetuximab	EGFR	Chimeric IgG1	2004	2004	Sp2/0	Head and neck cancer; colorectal cancer
Efalizumab	CD11a	Humanized IgG1	2004 (2)	2003 (2)	CHO	Psoriasis
Ibritumomab tiuxetan	CD20	Murine IgG1	2004	2002	CHO	Non-Hodgkin's lymphoma
Fanolesomab	CD15	Murine MAb	Not approved	2004	Hybridoma	Imaging of equivocal appendicitis –> non-therapeutic
Adalimumab	TNFα	Human IgG1	2003	2002	CHO	Rheumatoid arthritis; juvenile idiopathic arthritis; psoriatic arthritis; ankylosing spondylitis; Crohn's disease, plaque psoriasis
Tositumomab and iodine 131 tositumomab	CD20	Murine IgG2a	Not approved	2003	Hybridoma	Non-Hodgkin's lymphoma
Alemtuzumab	CD52	Humanized IgG1	2001	2001	CHO	B-cell chronic lymphocytic leukemia
Trastuzumab	HER-2	Humanized IgG1	2000	1998	CHO	Breast cancer; metastatic gastric or gastroesophageal junction adenocarcinoma

International Non-proprietary Name (INN)	Target	Type	Year of first EU EMA approval	Year of first FDA approval	Cell line	Therapeutic indication(s)
Gemtuzumab ozogamicin	CD33	Humanized IgG4 / toxin conjugate	Not approved	2000 (3)	NS0	Acute myeloic leucemia (AML)
Infliximab	TNFα	Chimeric IgG1	1999	1998	Sp2/0	Crohn's disease; ulcerative colitis; rheumatoid arthritis; ankylosing spondylitis; psoriatic arthritis; plaque psoriasis
Palivizumab	F-protein of RS virus	Humanized IgG1	1999	1998	NS0	Respiratory syncytial virus (RSV)
Necitumumab	CD25 (a chain of IL2 receptor)	Humanized IgG1	1999 (4)	1997 (4)	Sp2/0	Reversal of transplantation rejection
Basiliximab	CD25 (a chain of IL2 receptor)	Chimeric IgG1	1998	1998	NS0	Reversal of transplantation rejection
Rituximab	CD20	Chimeric IgG1	1998	1997	CHO	Non-Hodgkin's lymphoma; chronic lymphocytic leukemia; rheumatoid arthritis
Votumumab	Cytokeratintumor-associated antigen		1998 (5)	Not approved	Human lympho-blastoid cell line transformed with EBV	Detection of carcinoma of the colon or rectum –> non-therapeutic
Sulesomab	NCA90	Murine Fab fragment	1997	Not approved	NS0	Diagnostic imaging forosteomyelitis –> non-therapeutic
Arcitumomab	Human CEA (carcinoembryonic antigen)	Murine Fab fragment	1996 (6)	1996	Hybridoma	Detection of colorectal cancer –> non-therapeutic
Imiciromab	Human cardiac myosin	Murine Fab fragment	Not approved	1996 (7)	Murine ascites	Myocardial infarction imaging agent –> non-therapeutic
Capromab	Tumor surface antigen PSMA	Murine MAb	Not approved	1996	Hybridoma	Detection of prostate adenocarcinoma –> non-therapeutic
Nofetumomab	Carcinoma-associated antigen	Murine Fab fragment	Not approved	1996	Hybridoma	Diagnostic imaging of small-cell lung cancer –> non-therapeutic
Abciximab	GPIIb/IIIa	Chimeric IgG1 Fab	1995 (8)	1994	Sp2/0	High risk angioplasty (prevention of blood clots)
Satumomab	TAG-72	Murine MAb	Not approved	1992	Hybridoma	Detection of colorectal and ovarian cancers –> non-therapeutic
Muromonab-CD3	CD3	Murine IgG2a	1986 (8)	1986	Hybridoma	Transplantation rejection

FREQUENTLY ASKED QUESTIONS

Q1. What are antibodies?

Answer: Antibody, also called immunoglobulin, a protective protein produced by the immune system in response to the presence of a foreign substance, called an antigen. Antibodies recognize and latch onto antigens in order to remove them from the body. A wide range of substances are regarded by the body as antigens, including disease-causing organisms and toxic materials such as insect venom. Antibodies have a Y shaped structure made up of two heavy and two light chains. Antibodies have set of complimentary determining regions (CDRs) in the variable region which form paratope and binds specifically with the epitope of the antigen during antibody antigen interaction. Antibody can be polyclonal or monoclonal. There are five classes of antibodies namely IgA, IgD, IgE, IgG and IgM depending upon the presence of heavy chain in the antibody molecule.

Q2. What is meant by therapeutic antibodies?

Answer: Development of therapeutic antibodies boomed in the 1980's, and the first therapeutic antibody, a mouse antibody, was launched in 1986 as an immunosuppressive agent used during organ transplantation. Antibodies can swiftly provide therapeutics to target the disease-related molecules that have been discovered in genomic research because 1) the high level of specificity and affinity to the target molecule or antigen achieves a high level of efficacy and fewer adverse events, 2) their ability to target diverse molecules and the modes of action of the antibodies allow them to be applied to a wide range of therapeutic targets, and 3) modification and refinement by genetic engineering technology and the establishment of recombinant manufacturing technology has made industrial manufacturing possible.

Q3. What do you understand by antibody dependent cell mediated cytotoxicity?

Answer: The antibody-dependent cell-mediated cytotoxicity also known as antibody-dependent cellular cytotoxicity (ADCC), is a mechanism of cell-mediated immune defense whereby an effector cell of the immune system actively lyses a target cell, whose membrane-surface antigens have been bound by specific antibodies. It can also be determined a lytic process in which antibody coated target cells are killed by the cells of the immune system such as NK cells and macrophages. The significance of this ADCC is that virus-infected or other diseased cells are targeted for destruction by components of the cell-mediated immune system, such as natural killer cells. ADCC is independent of the immune complement system that also lyses targets but does not require any other cell. ADCC requires an effector cell which classically is known to be natural killer (NK) cells that typically interact with IgG antibodies. However, macrophages, neutrophils and eosinophils can also mediate ADCC. ADCC is part of the adaptive immune response due to its dependence on a prior antibody response.

Q4. What are monoclonal antibodies?

Answer: Monoclonal antibodies (mAb or moAb) are antibodies that are made by identical immune cells that are all clones of a unique parent cell. Monoclonal antibodies can have monovalent affinity, in that they bind to the same epitope. In contrast, polyclonal antibodies bind to multiple epitopes and are usually made by several different plasma cell lineages. Bispecific monoclonal antibodies can also be engineered, by increasing the therapeutic targets of one single monoclonal antibody to two epitopes. Monoclonal antibodies are also known as magic bullets and are produced by hybridoma technology which involves identifying antigen-specific plasma/plasmablast cells that produce antibodies specific to an antigen of interest and fusing these cells with myeloma cells using polyethylene glycol as fusinogen and selecting the hybrid cells in HAT media followed by selection, purification, screening and identification. These monoclonal antibodies have diagnostic, analytical and therapeutic uses.

Q5. How are Monoclonal Antibodies (mAbs) produced?

Answer: mAbs are produced through hybridoma technology. The production of monoclonal antibodies involves several steps. In the first instance a laboratory animal is injected with a desired target to stimulate their immune system. Following this, antibody producing cells, B lymphocytes, are harvested from the animal's spleen and fused with an immortal myeloma cell line to create hybrid cells, or hybridomas. The hybrid cells are then screened to find those that secrete antibodies with the desired specificity for a particular target. Once identified the hybrid cell is cloned to establish a hybridoma colony. This colony is then maintained in a culture medium to provide a continual supply of monoclonal antibodies. Each of the antibodies secreted by the hybrid cell is identical (monoclonal) and has the capacity to bind to a specific receptor found on the surface of a cell.

Q6. How are Monoclonal Antibodies (mAbs) useful?

Answer: Monoclonal antibodies (mAb) are important reagents used in biomedical research, in diagnosis of diseases, and in treatment of such diseases as infections and cancer. The use of monoclonal antibodies (mAb) in biomedical research has been and will continue to be important for the identification of proteins, carbohydrates, and nucleic acids. Their use has led to the elucidation of many molecules that control cell replication and differentiation, advancing our knowledge of the relationship between molecular structure and function. These advances is basic biologic sciences have improved our understanding of the host response to infectious-disease agents and toxins produced by these agents, to transplanted organs and tissues, to spontaneously transformed cells (tumours), and to endogenous antigens (involved in autoimmunity). In addition, the exquisite specificity of mAb allows them to be used in humans and animals for disease diagnosis and treatment. Under the appropriate conditions, mAb-producing hybridomas survive indefinitely, so continued production of mAb is associated with the use of fewer animals, especially when production involves the use of *in vitro* methods.

Q7. Describe cancer immunotherapy?

Answer: Cancer immunotherapy (sometimes called immuno-oncology) is the artificial stimulation of the immune system to treat cancer, improving on the immune system's natural ability to fight cancer. It is an application of the fundamental research of cancer immunology and a growing subspeciality of oncology. It exploits the fact that cancer cells often have tumor antigens, molecules on their surface that can be detected by the antibody proteins of the immune system, binding to them. The tumor antigens are often proteins or other macromolecules (e.g., carbohydrates). Normal antibodies bind to external pathogens, but the modified immunotherapy antibodies bind to the tumor antigens marking and identifying the cancer cells for the immune system to inhibit or kill. In 2018, James P. Allison and Tasuku Honjo received the Nobel Prize in Physiology or Medicine for their discovery of cancer therapy by inhibition of negative immune regulation.

Q8. What do you understand by phage display technology?

Answer: Phage display is a laboratory technique for the study of protein–protein, protein–peptide, and protein–DNA interactions that uses bacteriophages (viruses that infect bacteria) to connect proteins with the genetic information that encodes them. In this technique, a gene encoding a protein of interest is inserted into a phage coat protein gene, causing the phage to "display" the protein on its outside while containing the gene for the protein on its inside, resulting in a connection between genotype and phenotype. These displaying phages can then be screened against other proteins, peptides or DNA sequences, in order to detect interaction between the displayed protein and those other molecules. In this way, large libraries of proteins can be screened and amplified in a process called *in vitro* selection, which is analogous to natural selection.

Q9. Describe hybridoma technology?

Answer: Hybridoma technology is a method for producing large numbers of identical antibodies (also called monoclonal antibodies). This process starts by injecting a mouse (or other mammal) with an antigen that provokes an immune response. A type of white blood cell, the B cell, produces antibodies that bind to the injected antigen. These antibody producing B-cells are then harvested from the mouse and, in turn, fused with immortal B cell cancer cells, a myeloma, to produce a hybrid cell line called a hybridoma, which has both the antibody-producing ability of the B-cell and the exaggerated longevity and reproductivity of the myeloma. The hybridomas can be grown in culture, each culture starting with one viable hybridoma cell, producing cultures each of which consists of genetically identical hybridomas which produce one antibody per culture (monoclonal) rather than mixtures of different antibodies (polyclonal). The myeloma cell line that is used in this process is selected for its ability to grow in tissue culture and for an absence of antibody synthesis. In contrast to polyclonal antibodies, which are mixtures of many different antibody molecules, the monoclonal antibodies produced by each hybridoma line are all chemically identical.

Q10. Discuss Alemtuzumab?

Answer: Alemtuzumab is a medication used to treat chronic lymphocytic leukemia (CLL) and multiple sclerosis. In CLL has been used as both a first line and second line treatment. In MS it is generally only recommended if other treatments have not worked. It is given by injection into a vein. It is a monoclonal antibody that binds to CD52, a protein present on the surface of mature lymphocytes, but not on the stem cells from which these lymphocytes are derived. After treatment with alemtuzumab, these CD52-bearing lymphocytes are targeted for destruction. Alemtuzumab was approved for medical use in the United States in 2001. Alemtuzumab is used for the treatment of B-cell chronic lymphocytic leukemia (B-CLL) in people who have been treated with alkylating agents and who have failed fludarabine therapy. It is an unconjugated antibody, thought to work via the activation of antibody-dependent cell-mediated cytotoxicity (ADCC).

Q11. Describe the properties and function of infliximab?

Answer: Infliximab, sold under the brand name Remicade and is a medication used to treat a number of autoimmune diseases. This includes Crohn's disease, ulcerative colitis, rheumatoid arthritis, ankylosing spondylitis, psoriasis, psoriatic arthritis, and Behçet's disease. It is given by slow injection into a vein, typically at six- to eight-week intervals. Common side effects include infections, acute infusion reactions, and abdominal pain. Infliximab is a chimeric monoclonal antibody biologic. It works by binding to TNF-α. TNF-α is a chemical messenger (cytokine) and a key part of the autoimmune reaction. In rheumatoid arthritis, infliximab seems to work by preventing TNF-α from binding to its receptor in the cell. Infliximab was approved for medical use in the United States in 1998. It was originally developed in mice as a mouse antibody. Because humans have immune reactions to mouse proteins, the mouse common domains were replaced with similar human antibody domains. They are monoclonal antibodies and have identical structures and affinities to the target. Because they are a combination of mouse and human antibody amino acid sequences, they are called as "chimeric monoclonal antibody".

Q12. Describe the properties and function of avelumab?

Answer: Avelumab (trade name Bavencio) is a fully human monoclonal antibody developed by Merck KGaA and Pfizer[citation needed] as a pharmaceutical drug for use in immunotherapy, originally for the treatment of non-small-cell lung carcinoma (NSCLC). Avelumab targets the protein programmed death-ligand 1 (PD-L1). Formation of a PD-1/PD-L1 receptor/ligand complex leads to inhibition of CD8+ T cells, and therefore inhibition of an immune reaction. Immunotherapy aims at ceasing this immune blockage by blocking those receptor ligand pairs. In the case of avelumab, the formation of PD-1/PDL1 ligand pairs is blocked and CD8+ T cell immune response should be increased. PD-1 itself has also been a target for immunotherapy. Therefore, avelumab belongs to the group of immune checkpoint blockade cancer therapies. It has received orphan drug designation by the European Medicines Agency (EMA) for the treatment of gastric cancer in January 2017. The US Food and Drug Administration (FDA) approved it in March 2017 for Merkel-cell carcinoma, an aggressive type of skin cancer. The EMA approved it in September 2017 for the same indication.

Q13. Describe the properties and function of ipilimumab?

Answer: Ipilimumab is a type of cancer treatment called immunotherapy. It is used to treat melanoma that can't be removed with surgery or has spread to another part of the body (advanced melanoma) and advanced kidney cancer (renal cell carcinoma) with another drug called nivolumab. Ipilimumab works to activate the immune system by targeting CTLA-4, a protein receptor that downregulates the immune system. Cytotoxic T lymphocytes (CTLs) can recognize and destroy cancer cells. However, an inhibitory mechanism interrupts this destruction. Ipilimumab turns off this inhibitory mechanism and allows CTLs to function.

Q14. Describe the properties and function of bevacizumab?

Answer: Bevacizumab, sold under the trade name Avastin, is a medication used to treat a number of types of cancers and a specific eye disease. For cancer it is given by slow injection into a vein and used for colon cancer, lung cancer, glioblastoma, and renal-cell carcinoma. For age-related macular degeneration it is given by injection into the eye.

Common side effects when used for cancer include nose bleeds, headache, high blood pressure, and rash. Other severe side effects include gastrointestinal perforation, bleeding, allergic reactions, blood clots, and an increased risk of infection. When used for eye disease side effects can include vision loss and retinal detachment. Bevacizumab is in the angiogenesis inhibitor and monoclonal antibody families of medication. It works by slowing the growth of new blood vessels by inhibiting vascular endothelial growth factor A (VEGF-A). Bevacizumab was approved for medical use in the United States in 2004.

Q15. Describe the properties and function of basiliximab?

Answer: Basiliximab (trade name Simulect) is a chimeric mouse-human monoclonal antibody to the α chain (CD25) of the IL-2 receptor of T cells. It is used to prevent rejection in organ transplantation, especially in kidney transplants. Basiliximab is an immunosuppressant agent used to prevent immediate transplant rejection in people who are receiving kidney transplants, in combination with other agents. It has been reported that some cases of lichen planus have been successfully treated with basiliximab as an alternative therapy to cyclosporin. No short-term side effects have been reported.

Q16. Discuss biobetter antibodies?

Answer: The use of antibody engineering to improve the properties of therapeutic antibodies has advanced greatly in the last decades giving rise to a varied set of novel formats that offer enhanced attributes for therapeutic and research purposes. These novel molecules are often referred to as biobetters or next-generation antibodies and include platforms such as: engineered antibodies for enhanced effector functions, antibody drug-conjugates (ADC), multi-specific antibodies and single-domain antibody fragments.

A biobetter can have modifications to its chemical structure such as humanization, fusion/conjugation or be glyco-engineered to be less immunogenic or more efficacious. For instance, anti-CD20 ocrelizumab which is currently under review for FDA approval for MS is a humanized version of rituximab (Rituxan®). A biobetter may be modified to have an improved formulation for improved treatment regimen so treatment is less evasive or have a simplified manufacturing process.

Multiple Choice Questions

Q1. Antibodies are produced by which cells in the human body?

A. Lymphocytes
B. Phagocytes
C. Monocytes
D. Thrombocytes

Q2. Antigen binding sites of an immunoglobulin are located in

A. Light chain alone
B. Heavy chain alone
C. Fc region of the antibody
D. Fab regions of the antibody

Q3. The class of immunoglobulins which can cross the placenta is

A. Immunoglobulin IgM
B. Immunoglobulin IgG
C. Immunoglobulin IgA
D. Immunoglobulin IgD

Q4. Hybridoma technology is a new biotechnological approach for commercial production of:

A. Monoclonal antibodies
B. Interferons
C. Alcohol
D. Antibiotics

Q5. Which of the following is incorrect?

A. mAbs can be used to diagnose or treat diseases
B. mAbs can be used in tumour scanning
C. mAbs can be used in cancer diagnosis
D. mAbs cannot be used in such non-infectious diseases as those of endocrine system

Q6. What is the first step in producing monoclonal antibodies?

A. Injecting a mouse with an antibody
B. Injecting a mouse with an antigen
C. Removing spleen cells from a mouse
D. Inducing tumour formation

Q7. All are mAbs except

A. Rituximab
B. Transtuzumab
C. Infliximab
D. tamoxifen

Q8. Which of the following is incorrect?

A. mAbs can be used to diagnose or treat diseases
B. mAbs can be used in tumour scanning
C. mAbs can be used in cancer diagnosis
D. mAbs cannot be used in such non-infectious diseases as those of endocrine system

Q9. The EBV-hybridoma technique

A. Immortalizes the donor B cells
B. Facilitates the proliferation of antigen specific B cells
C. Gives much higher hybridization frequencies
D. All of the above

Q10. Generally, antibodies produced against a pathogen is

A. Monoclonal

B. Homogenous

C. Polyclonal

D. All of same specificity

Q11. Which among the following statements about CTLA-4 is incorrect?

A. CTLA-4 is a member of the immunoglobulin (Ig) gene superfamily present on T cells.

B. CTLA-4 has only one N-linked glycosylation site

C. CTLA-4 can be found on the surface of both CD4+ T cell and CD8+ T cell within 24h after activation

D. It is highly expressed in naïve T cell

Q12. The cross linkage of antigens by antibodies is known as

A. Agglutination

B. Complement fixation

C. Across reaction

D. All of these

Q13. T cells are the source of

A. Interleukin

B. Interferon

C. Lymphotoxin

D. All of these

Q14. In monoclonal antibody technology, tumour cells that can replicate endlessly are fused with mammalian cells that produce an antibody. The result of this cell fusion is a

A. Hybridoma

B. Myeloma

C. Natural killer cell

D. Lymphoblast

Q15. Which type of cell actually secrets antibodies?

A. Plasma cells

B. T cells

C. Macrophages

D. Dendritic cells

ANSWER KEY

1(A), 2(D), 3(B), 4(A), 5(D), 6(B), 7(D), 8(D), 9(D), 10(C), 11(D), 12(A), 13(D), 14(A), 15(A)

ASSIGNMENTS

Q1. Write a brief note on the discovery on monoclonal antibodies.

Q2. Why are mAbs referred to as magic bullets?

Q3. Explain in detail about the use of mAbs in Cancer therapy?

Q4. Write about the various applications of mAbs in the field of medicine.

Q5. Describe in detail Biobetter antibodies?

REFERENCES

Barbas, CF; III. Kang, AS; Lerner, RA; Benkovic, SJ. Assembly of combinatorial antibody libraries on phage surfaces: the gene III site, *Proc. Natl. Acad. Sci.*, 88, (1991), 7978–7982.

Barbas, CFI; Bain, JD; Hoekstra, DM; Lerner, RA. Semisynthetic combinatorial antibody libraries: a chemical solution to the diversity problem, *Proc. Natl. Acad. Sci.*, 89, (1992), 4457–4461.

Berghout, A. Clinical programs in the development of similar biotherapeutic products: rationale and general principles, *Biologicals*, 39, (2011), 293–296.

Blackstone, EA; Joseph, PF. The economics of biosimilars, *Am. Health Drug Benefits*, 6, (2013), 469–478.

Brown, J; Kudrin, A. Switching to biosimilar infliximab (CT-P13): evidence of clinical safety, effectiveness and impact on public health, *Biologicals*, 44, (2016), 257–266.

Chames, P; Van Regenmortel, M; Weiss, E; Baty, D. Therapeutic antibodies: successes, limitations and hopes for the future, *Br. J. Pharmacol.*, 157, (2009), 220–233.

Cheung, WC; Beausoleil, SA; Zhang, X; Sato, S; Schieferl, SM; Wieler, JS; Beaudet, JG; Ramenani, RK; Popova, L; Comb, MJ; Rush, J; Polakiewicz, RD. A proteomics approach for the identification and cloning of monoclonal antibodies from serum, *Nat. Biotechnol.*, 30, (2012), 447–452.

Committee for Medicinal Products for Human Use, *Guideline on Immunogenicity Assessment of Monoclonal Antibodies Intended for In Vivo Clinical Use*, European Medicines Agency, London (UK), 2012.

Committee for Medicinal Products for Human Use, *Guideline on Similar Biological Medicinal Products Containing Monoclonal Antibodies—Non-clinical and Clinical Issues*, European Medicines Agency, London (UK), 2012.

Committee for Medicinal Products for Human Use, *Guideline on Similar Biological Medicinal Products Containing Biotechnology-derived Proteins as Active Substance: Quality Issues (Revision 1)*, European Medicines Agency, London (UK), 2014.

Committee for Medicinal Products for Human Use, *Guideline on Similar Biological Medicinal Products*, European Medicines Agency, London (UK), 2014.

Committee for Medicinal Products for Human Use, *Guideline on Similar Biological Medicinal Products Containing Biotechnology-derived Proteins as Active Substance: Non-clinical and Clinical Issues*, European Medicines Agency, London (UK), 2014.

Daugherty, PS; Chen, G; Olsen, MJ; Iverson, BL; Georgiou, G. Antibody affinitymaturation using bacterial surface display, *Protein Eng.*, 11, (1998), 825–832.

Davies, J; Riechmann, L. An antibody variable heavy domain with a lox-Cre site integrated into its coding region: bacterial recombination within a single polypeptide chain, *FEBS Lett.*, 377, (1995), 92–96.

de Labry, AO; Giménez, E; Lindner, L; García, L; Espín, J; Rovira, J. Biosimilars in the European market, *Generics Biosimilars Initiative J. (GaBI J.)*, 2, (2013), 30–35.

Drake, PM; Rabuka, D. An emerging playbook for antibody-drug conjugates: lessons from the laboratory and clinic suggest a strategy for improving efficacy and safety, *Curr. Opin. Chem. Biol.*, 28, (2015), 174–180.

Ecker, DM; Jones, SD; Levine, HL. The therapeutic monoclonal antibody market, *MAbs*, 7, (2015), 9–14.

European Medicines Agency, Committee for Medicinal Products for Human Use (CHMP), *Guideline on Developement, Production, Characterisation and Specification For Monoclonal Antibodies and Related Products*, European Medicines Agency, London (UK), 2008.

Food and Drug Administration, *Guidance for Industry, Clinical Pharmacology Data to Support a Demonstration of Biosimilarity to a Reference Product (Draft Guidance)*, 2014.

Food and Drug Administration, *Quality Considerations in Demonstrating Biosimilarity of a Therapeutic Protein Product to a Reference Product, Guidance for Industry*, Silver Spring (MD), 2015.

Food and Drug Administration, *Scientific Considerations in Demonstrating Biosimilarity to a Reference Product, Guidance for Industry*, Silver Spring (MD), 2015.

Garrone, P; Djossou, O; Fossiez, F; Reyes, J; Ait-Yahia, S; Maat, C; Ho, S; Hauser, T; Dayer, JM; Greffe, J; Miossec, P; Lebecque, S; Rousset, F; Banchereau, J. Generation and characterization of a human monoclonal autoantibody that acts as a high affinity interleukin-1 alpha specific inhibitor, *Mol. Immunol.*, 33, (1996), 649–658.

Georgiou, G; Ippolito, GC; Beausang, J; Busse, CE; Wardemann, H; Quake, SR. The promise and challenge of high-throughput sequencing of the antibody repertoire, *Nat. Biotechnol.*, 32, (2014), 158–168.

Hanes, J; Schaffitzel, C; Knappik, A; Pluckthun, A. Picomolar affinity antibodies from a fully synthetic naive library selected and evolved by ribosome display, *Nat. Biotechnol.*, 18, (2000), 1287–1292.

Harmsen, MM; De Haard, HJ. Properties, production, and applications of camelid single-domain antibody fragments, *Appl. Microbiol. Biotechnol.*, 77, (2007), 13–22.

Hong, DS; Hui, D; Bruera, E; Janku, F; Naing, A; Falchook, GS; Piha-Paul, S; Wheler, JJ; Fu, S; Tsimberidou, AM; Stecher, M; Mohanty, P; Simard, J; Kurzrock, R. MABp1, a first-in-class true human antibody targeting interleukin-1alpha in refractory cancers: an open-label, phase 1 dose-escalation and expansion study, *Lancet Oncol.*, 15, (2014), 656–666.

Hoogenboom, HR; Griffiths, AD; Johnson, KS; Chiswell, DJ; Hudson, P; Winter, G. Multi-subunit proteins on the surface of filamentous phage: methodologies for displaying antibody (Fab) heavy and light chains, *Nucleic Acids Res.*, 19, (1991), 4133–4137.

Hoogenboom, HR; Winter, G. By-passing immunisation. Human antibodies from synthetic repertoires of germline VH gene segments rearranged *in vitro*, *J. Mol. Biol.*, *227*, (1992), 381–388.

Huse, WD; Sastry, L; Iverson, SA; Kang, AS; Alting-Mees, M; Burton, DR; Benkovic, SJ; Lerner, RA. Generation of a large combinatorial library of the immunoglobulin repertoire in phage lambda, *Science*, 246, (1989), 1275–1281.

Huse, WD; Sastry, L; Iverson, SA; Kang, AS; Alting-Mees, M; Burton, DR; Benkovic, SJ; Lerner, RA. Generation of a large combinatorial library of the immunoglobulin repertoire in phage lambda, *Science*, 246, (1989), 1275–1281.

Jakobovits, A; Amado, R; Yang, X; Roskos, L; Schwab, G. From XenoMouse technology to panitumumab, the first fully human antibody product fromtransgenicmice, *Nat. Biotechnol.*, 25, (2007), 1134–1143.

Jespers, LS; Roberts, A; Mahler, SM; Winter, G; Hoogenboom, HR. Guiding the selection of human antibodies from phage display repertoires to a single epitope of an antigen, *Biotechnology* (N Y), 12, (1994), 899–903.

Jones, PT; Dear, PH; Foote, J; Neuberger, MS; Winter, G. Replacing the complementarity-determining regions in a human antibody with those from a mouse, *Nature*, 321, (1986), 522–525.

Kim, YS; Choi, BW; Yang, SW; Shin, SM; Nam, SW; Roh, YS; Lee, JY; Lee, KJ; Kim, YJ; Kwon, JY; Kim, DI. Biosimilars: challenges and path forward, Biotechnol. *Bioprocess Eng.*, 19, (2014), 755–765.

Kohler, G; Milstein, C. Continuous cultures of fused cells secreting antibody of predefined specificity, *Nature*, 256, (1975), 495–497.

Kohler, G; Milstein, C. Derivation of specific antibody-producing tissue culture and tumor lines by cell fusion, *Eur. J. Immunol.*, 6, (1976), 511–519.

Kontermann, RE; Brinkmann, U. Bispecific antibodies, *Drug Discov. Today*, 20, (2015), 838–847.

Kudrin, A; Knezevic, I; Joung, J; Kang, HN. Case studies on clinical evaluation of biosimilar monoclonal antibody: scientific considerations for regulatory approval, *Biologicals*, 43, (2015), 1–10.

Lehmann-Horn, K; Kronsbein, HC; Weber, MS. Targeting B cells in the treatment of multiple sclerosis: recent advances and remaining challenges, Ther. *Adv. Neurol. Disord.*, 6, (2013), 161–173.

Li, W; Caberoy, NB. New perspective for phage display as an efficient and versatile technology of functional proteomics, *Appl. Microbiol. Biotechnol.*, 85, (2010), 909–919.

Lonberg, N. Human antibodies fromtransgenic animals, *Nat. Biotechnol.*, 23, (2005), 1117–1125.

Marks, JD; Hoogenboom, HR; Bonnert, TP; McCafferty, J; Griffiths, AD; Winter, G. By-passing immunization. Human antibodies from V-gene libraries displayed on phage, *J. Mol. Biol.*, 222, (1991), 581–597.

Mathonet, P; Ullman, CG. The application of next generation sequencing to the understanding of antibody repertoires, *Front. Immunol.*, 4, (2013).

Mazor, Y; Van Blarcom, T; Mabry, R; Iverson, BL; Georgiou, G. Isolation of engineered, full-length antibodies from libraries expressed in Escherichia coli, *Nat. Biotechnol.*, 25, (2007), 563–565.

McCafferty, J; Griffiths, AD; Winter, G; Chiswell, DJ. Phage antibodies: filamentous phage displaying antibody variable domains, *Nature*, 348, (1990), 552–554.

Mottershead, DG; Alfthan, K; Ojala, K; Takkinen, K; Oker-Blom, C. Baculoviral display of functional scFv and synthetic IgG-binding domains, *Biochem. Biophys. Res. Commun.*, 275, (2000), 84–90.

Nakano, K; Kojima, T; Kasutani, K; Senoh, C; Natori, O; Ishii, S; Tsunoda, H; Hattori, K. Effective screening method of agonistic diabodies based on autocrine growth, *J. Immunol. Methods*, 347, (2009), 31–35.

Niwa, R; Satoh, M. The current status and prospects of antibody engineering for therapeutic use: focus on glycoengineering technology, *J. Pharm. Sci.*, 104, (2015), 930–941.

Nixon, AE; Sexton, DJ; Ladner, RC. Drugs derived from phage display: from candidate identification to clinical practice, *MAbs*, 6, (2014), 73–85.

Nuttall, SD; Krishnan, UV; Doughty, L; Pearson, K; Ryan, MT; Hoogenraad, NJ; Hattarki, M; Carmichael, JA; Irving, RA; Hudson, PJ. Isolation and characterization of an IgNAR variable domain specific for the human mitochondrial translocase receptor Tom70, *Eur. J. Biochem.*, 270, (2003), 3543–3554.

Ravn, U; Didelot, G; Venet, S; Ng, KT; Gueneau, F; Rousseau, F; Calloud, S; Kosco-Vilbois, M; Fischer, N. Deep sequencing of phage display libraries to support antibody discovery, *Methods*, 60, (2013), 99–110.

Ravn, U; Gueneau, F; Baerlocher, L; Osteras, M; Desmurs, M; Malinge, P; Magistrelli, G; Farinelli, L; Kosco-Vilbois, MH; Fischer, N. By-passing *in vitro* screening–next generation sequencing technologies applied to antibody display and in silico candidate selection, *Nucleic Acids Res.*, 38, (2010), e193.

Reddy, ST; Ge, X; Miklos, AE; Hughes, RA; Kang, SH; Hoi, KH; Chrysostomou, C; Hunicke-Smith, SP; Iverson, BL; Tucker, PW; Ellington, AD; Georgiou, G. Monoclonal antibodies isolated without screening by analyzing the variable-gene repertoire of plasma cells, *Nat. Biotechnol.*, 28, (2010), 965–969.

Riechmann, L; Clark, M; Waldmann, H; Winter, G. Reshaping human antibodies for therapy, *Nature*, 332, (1988), 323–327.

Saggy, I; Wine, Y; Shefet-Carasso, L; Nahary, L; Georgiou, G; Benhar, I. Antibody isolation from immunized animals: comparison of phage display and antibody discovery via V gene repertoire mining, *Protein Eng. Des. Sel.*, 25, (2012), 539–549.

Scott, CT. Mice with a human touch, *Nat. Biotechnol.*, 25, (2007), 1075–1077.

Smits, LJ; Derikx, LA; de Jong, DJ; Boshuizen, RS; van Esch, AA; Drenth, JP; Hoentjen, F. Clinical outcomes following a switch from remicade to the biosimilar CT P13 in inflammatory bowel disease patients: a prospective observational cohort study, *J. Crohns Colitis*, 10, (2016), 1287–1293.

Thimmaraju, PK; Rakshambikai, R; Farista, R; Juluru K., Legislations on biosimilar interchangeability in the US and EU – developments far from visibility, *Generics Biosimilars Initiative*, (2015).

Vazquez, AM; Hernandez, AM; Macias, A; Montero, E; Gomez, DE; Alonso, DF; Gabri, MR; Gomez, RE. Racotumomab: an anti-idiotype vaccine related to Nglycolyl-containing gangliosides - preclinical and clinical data, *Front. Oncol.*, 2, (2012), 150.

WHO, Guidelines on evaluation of similar biotherapeutic products (SBPs), *World Health Organization (WHO) Expert Committee on Biological Standardization*, 2009.

In: Immunoglobulins
Editor: Manzoor Ahmad Mir

ISBN: 978-1-53616-903-4

Chapter 7

ALEXA FLUORE TAGGING OF ANTIBODIES AND FLOW CYTOMETRY

Manzoor A. Mir**, Umar Mehraj, Safura Nisar and Hina Qayoom
Department of Bioresources, School of Biological Sciences,
University of Kashmir, Srinagar, J&K India

ABSTRACT

Flow-cytometry is a popular cell biology technique that utilizes laser-based technology to count, sort, and profile cells in a heterogeneous fluid mixture using fluorescent dyes and alexa-floures. Using a flow cytometer machine, cells or other particles tagged with flour-chromes suspended in a liquid stream are passed through a laser light beam in single file fashion, and interaction with the light is measured by an electronic detection apparatus as light scatter and fluorescence intensity. The forward scatter tells the size of the cells in a FACS Dot-plot whereas, the side scatter tells the colour intensity of the tagged flourochome. If a fluorescent label, or fluorochrome, is specifically and stoichiometrically bound to a cellular component, the fluorescence intensity will ideally represent the amount of that particular cell component. The Alexa Fluor family of fluorescent dyes is a series of dyes invented by Molecular Probes, now a part of Thermo Fisher Scientific, and sold under the Invitrogen brand name. Alexa Fluor dyes are frequently used as cell and tissue labels in fluorescence microscopy and cell biology. Alexa Fluor dyes can be conjugated directly to primary antibodies or to secondary antibodies to amplify signal and sensitivity or other biomolecules. Flow-cytometry is a powerful tool because it allows simultaneous multiparametric analysis of the physical and chemical characteristics of up to thousands of particles per second. This makes it a rapid and quantitative method for analysis and purification of cells in suspension. Using flow-cytometery, we can determine the phenotype and function of the cells and even sort live cells. The use of flow-cytometry in

* Corresponding author: Dr Manzoor Ahmad Mir (Ph. D), Department of Bioresources, School of Biological Sciences, University of Kashmir, Srinagar J&K, India-190006, Contact:-+919622901319. E-mail: drmanzoor@kashmiruniversity.ac.in.

the clinical laboratory has grown substantially in the past decade. This is an attribute to the development of smaller, user-friendly, less-expensive instruments and a continuous increase in a number of clinical applications.

Keywords: flow-cytometry, fluorochrome, fluorescein isothyocynate, phycoerythrin, flouroscent dye, alexa-fluor, lymphocyte

OBJECTIVES

- To understand the concept of Alexa fluor tagging and flow cytometry
- To learn briefly about the history of this technique
- To understand the basic principles of this technique
- To describe applications of flow-cytometry in clinical diagnostics
- A brief look at future prospectus of this technique

INTRODUCTION

The *Alexa Fluor* family of *fluorescent* dyes is a series of dyes invented by *Molecular Probes,* now a part of *Thermo Fisher Scientific,* and sold under the Invitrogen brand name. Alexa Fluor dyes are frequently used as cell and tissue labels in *fluorescence microscopy* and *cell biology.* Alexa Fluor dyes can be conjugated directly to *primary antibodies* or to *secondary antibodies* to amplify signal and sensitivity or other biomolecules. The *excitation* and *emission* spectra of the Alexa Fluor series cover the *visible spectrum* and extend into the *infrared.* The individual members of the family are numbered roughly according to their excitation maxima (in *nm*).

History

Richard and Rosaria Haugland the founders of Molecular Probes, are well known in biology and chemistry for their research into fluorescent dyes to be used in biological applications. At the time when Molecular Probes were founded, these products were essentially unavailable commercially. Now a number of well-known fluorescent dyes come out of the labs as molecular Probes, dyes such as: Texas Red, Cascade Blue, Oregon Green, Marina Blue and the Alexa Fluor family. Probably the most famous, Alexa Fluor family of dyes, were intended to improve upon the properties of previously developed biological fluorescent dye families and solve some of the issues that they possessed. Structurally, the Alexa Fluor dyes are generated through the sulfonation and additional modification of certain well known dye families. In particular, the dye families of coumarin, rhodamine, xanthene (of which the industry-standard fluorescein is a member), and cyanine dyes were used. Sulfonation makes the Alexa Fluor dyes negatively charged and more hydrophilic than their precursors, while additional modification were used to improve the performance in other areas. For example, Alexa Fluor 488, a sulfonated and chemically modified form of fluorescein, was designed to solve the well-known issues of rapid photo-bleaching and pH dependent fluorescent intensity common in the industry-standard FITC dye. The Alexa Fluor dyes were named after Alex Haugland, son of Richard and Rosaria Haugland. Molecular Probes were acquired in 2003, by Invitrogen who further worked to expand the Alexa Fluor family adding new dyes to fill gaps in the spectrum. In 2008, Invitrogen became Life Technologies when they merged with Applied Biosystems, bringing the Alexa Fluor family of dyes under the new Life Technologies title. In 2014, Life Technologies was acquired by Thermo Fisher Scientific who revitalized the Invitrogen name and brought the Alexa Fluor family of dyes back under the Invitrogen brand.

Antibody labelling is an important molecular biological technique. Antibodies are widely used in immunoassays to detect and quantify antigens. The antibody that recognizes the antigen is referred to as the 'primary' antibody and confers specificity to the assay. The labeling of proteins allows following specific molecules using fluorescent microscopy. Fluorescence results from a process that occurs when certain molecules (generally polyaromatic hydrocarbons or heterocycles) called fluorophores, fluorochromes, or fluorescent dyes absorb light. The absorption of light by a population of these molecules raises their energy level to a brief excited state. As they decay from this excited state, they emit fluorescent light.

Flow-Cytometry

The flow cytometer was developed in the 1970's and rapidly became an essential instrument for the biologic sciences Flow-cytometry is a technique of quantitative single cell analysis (See Figure 7.1 and 7.2). This technology is used to analyze the physical and chemical characteristics of particles in a fluid as it passes through at least one laser. Spurred by the HIV pandemic and a plethora of discoveries in hematology, specialized flow cytometers for use in the clinical laboratory were developed by several manufacturers.

The major clinical application of flow-cytometry is diagnosis of hematologic malignancy, but a wide variety of other applications exist, such as reticulocyte enumeration and cell function analysis. The technique of analyzing individual cells in a fluidic channel was first described by Wallace Coulter in the 1950s, and applied to automated blood cell counting. Subsequent developments in the field of computer science, laser technology, monoclonal antibody production, cytochemistry, and fluorochrome chemistry led to the development of the flow cytometer almost two decades later. Because the first commercial flow cytometers were large, complex, expensive, and difficult to operate and maintain, they were primarily used in the research laboratory. However, the enormous value of the flow cytometer in the medical and biologic sciences was quickly appreciated, and its cost and complexity gradually decreased as its analytic capability increased. The present "state-of-the art" flow cytometers are capable of analyzing up to 13 parameters (forward scatter, side scatter, 11 colors of immunofluorescence) per cell at rates up to 100,000 cells per second. Automation and robotics is increasingly being applied to flow-cytometry to reduce analytic cost and improve efficiency.

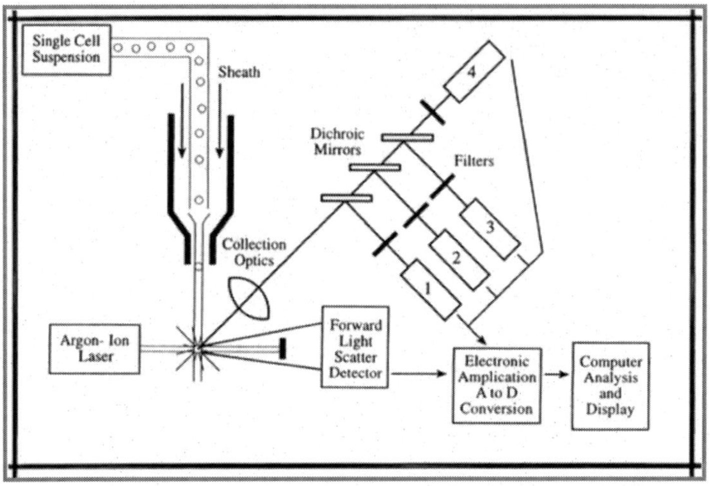

Figure 7.1. Schematic representation of Flow cytometer showing internal architecture.

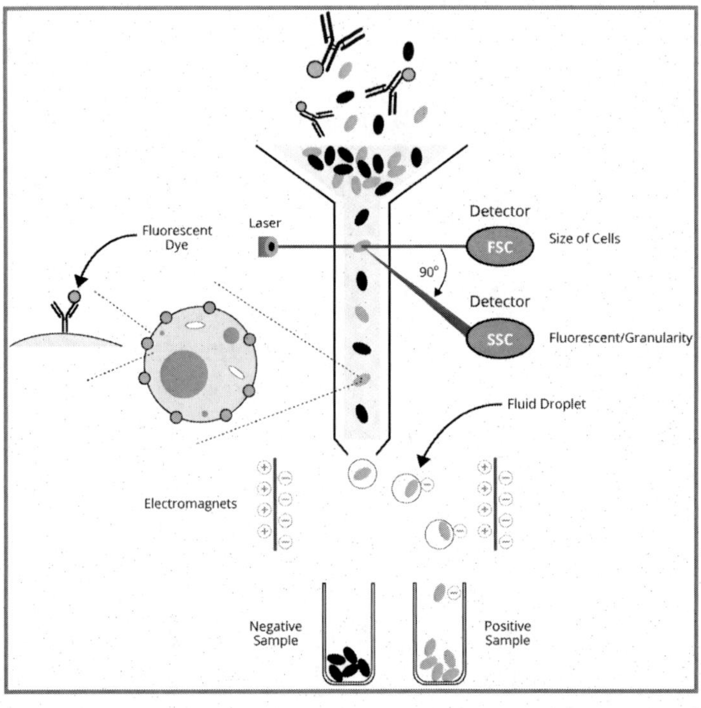

Figure 7.2. Alexa-fluore tagging of antibodies and flow cytometry.

Basic Principles of Flow-Cytometry and Flow Cytometer

For flow cytometric analysis, single cell or particle suspensions are necessary. Various immunofluorescent dyes or antibodies can be attached to the antigen or protein of interest. The suspension of cells or particles is aspirated into a flow of cells which are surrounded by a narrow fluid stream as they pass one at a time through a focused laser beam. The light is either scattered or absorbed when it strikes a cell (See Figure 7.3). Absorbed light of the appropriate wavelength may be re-emitted as fluorescence if the cell contains a naturally fluorescent substance or one or more fluorochrome-labeled antibodies are attached to surface or internal cell structures. Light scatter is dependent on the internal structure of the cell and its size and shape. Fluorescent substances absorb light of an appropriate wavelength and reemit light of a different wavelength. Fluorescein isothiocyanate (FITC), Texas red, and phycoerythrin (PE) are the most common fluorescent dyes used in the biomedical sciences (See Table 7.1).

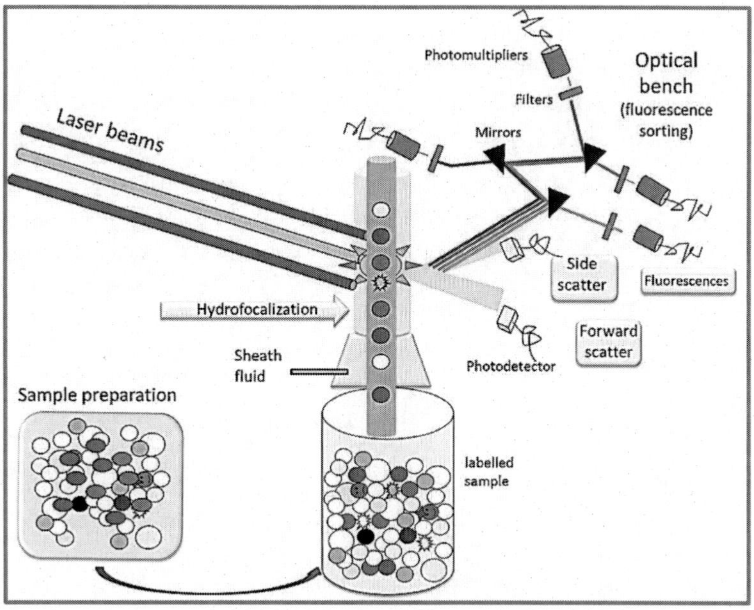

Image Source: Depince-Berger et al. 2016 DOI: 10.1016/j.morpho.2016.05.001.

Figure 7.3. Showing basic principle of a flowcytometre.

Light and/or fluorescence scatter signals are detected by a series of photodiodes and amplified. Optical filters are essential to block unwanted light and permit light of the desired wavelength to reach the photodetector. The resulting electrical pulses are digitized, and the data is stored, analyzed, and displayed through a computer system (See Figure 7.4). The end results are quantitative information about every cell analyzed. Since large numbers of cells are analyzed in a short period of time (>1,000/sec), statistically valid information about cell populations is quickly obtained.

Table 7.1. Some Flourochromes used to label proteins

Flourochrome	Abbreviation	Absorption Min nm	Emmision Max nm
violet laser diode			
Cascade blue		380 401	419
Cascade yellow		399	549
Pacific Blue		405	456
argon-ion laser			
fluorescein	FITC	495	520
Alexa 488		488	520
R-phycoerythrin	PE	495 564	576
PE-Texas Red	ECD	495 564	620
PE-cyanine 5	PC5	495 564	670
Peridinin-chlorophyll	PerCP	490	677
PE-cyanine-5.5	PC5.5	495 564	696
PE-cyanine 7	PC7	495 564	767
He-Ne laser			
allophycocyanin	APC	650	660
APC-cyanine 5.5	APC-Cy5.5	650	710
APC-cyanine 7	APC-Cy7	650	767

The flow cytometer instrument consists of three core systems: fluidics, optics, and electronics. The fluidics system includes a flow cell, where the sample fluid is injected. The flow cell requires sheath fluid to carry and align the cells or particles so that they pass through a narrow channel and into the laser intercept (light beam) in a single file. This hydrodynamic focusing allows the analysis of one cell at a time by laser interrogation (See Figure 7.5).

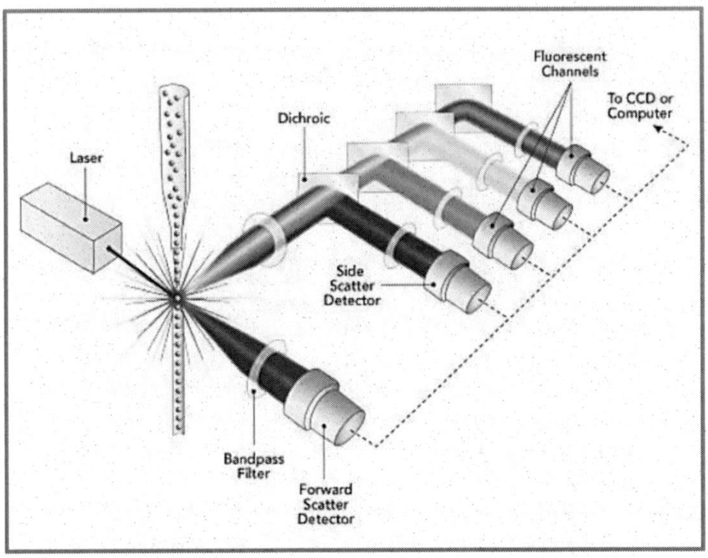

Figure 7.4. Showing side scatter, forward scatter and fluorescent channels of a flow cytometer.

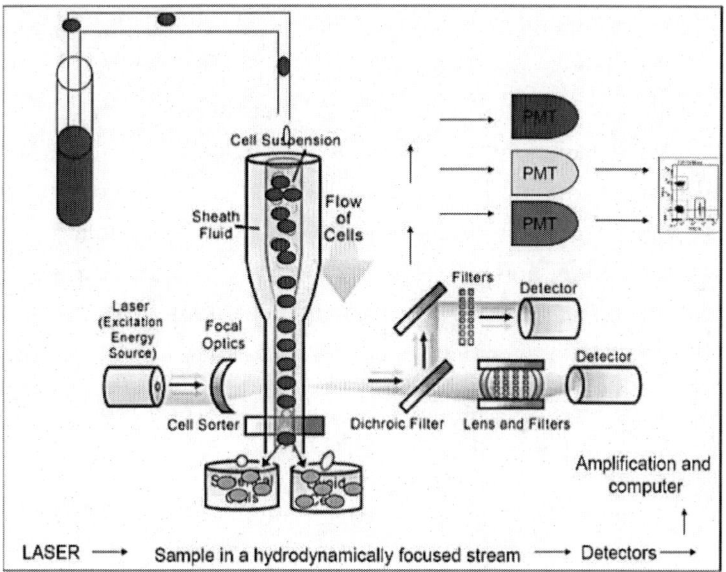

Figure 7.5. Laser optics, hydrodynamics (fluidics) and electronics of flow cytometer.

The optics system consists of various filters, light detectors, and the light source, which is usually a laser line producing a single wavelength of light at a specific frequency. This is where the particles are passed through at least one laser beam. Lasers are available at different wavelengths ranging from ultraviolet to far red and have a variable range of power levels as well (photon output/time). Interrogation by the laser beam excites any compatible fluorescent probes that are conjugated to antibodies, causing the probes to emit light (or fluoresce) at specified wavelengths. A detector in front of the light beam measures forward scatter light signals (FSC) and detectors to the side measure side scatter light signals (SSC). Fluorescence detectors measure the fluorescence signal intensity emitted from positively stained cells and particles. Within the flow cytometer, all of these different light signals are split into defined wavelengths and channelled by a set of filters and mirrors so that each sensor will detect fluorescence only at a specified wavelength. These sensors are called photomultiplying tubes (PMTs). Various filters are used in the flow cytometer to direct photons of the correct wavelength to each PMT. Short pass (SP) filters allow transmission of photons below a specific wavelength while long pass (LP) filters allow transmission above a specific wavelength. Band pass (BP) filters allow transmission of photons that have wavelengths within a narrow range. Each PMT will also detect any other fluorophores emitting at a similar wavelength to the fluorophore it is detecting. These light signals are converted by the electronics system to data that can be visualized and interpreted by software.

The key parameters which are measured by a flow cytometer include forward light scatter (FSC), side light scatter (SSC), and fluorescence emission signals. Forward scattered light (FSC) is light that is refracted by a cell in the forward direction, and continues in the same direction that the light was already traveling (typically up to 20°

offset from the laser beams axis). This signal is collected by a PMT called the forward scatter channel (FSC), and is commonly used to determine particle size. Usually, bigger particles will produce more forward scattered light than smaller ones, and larger cells will have a stronger forward scatter signal. Side scattered light (SSC) is light that is refracted by cells and travels in a different direction than its original path (measured at a 90° angle to the excitation line). It usually provides information about the granularity and complexity of the cells. Cells with a low granularity and complexity will produce less side scattered light, while highly granular cells with a high degree of internal complexity (such as neutrophils) will result in a higher side scatter signal. Though it is affected by shape and size of the cells, it's more sensitive to membranes, cytoplasm, nucleus etc. Thus, by using forward and side scattered light detection, cell populations can often be distinguished based on characteristic differences in cell size and granularity. Both FSC and SSC measurements are influenced by multiple factors and also depends on the quality of sample preparation.

Applications of Flow Cytometry

Diagnosis of Haematological Malignancies

The identification and quantitation of cellular antigens with fluorochrome-labeled monoclonal antibodies ("immunophenotyping") is one of the most important applications of the flow cytometer. Immunophenotypic analysis is critical to the initial diagnosis and classification of the acute leukemias, chronic lymphoproliferative diseases, and malignant lymphomas since treatment strategy often depends upon antigenic parameters. In addition, immunophenotypic analysis provides prognostic information not available by other techniques and provides a sensitive means to monitor the progress of patient after chemotherapy or bone marrow transplantation. It often permits the detection of minimal residual diseases. Flow cytometric analysis of apoptosis, multidrug resistance, leukemia-specific chimeric proteins, cytokine receptors and other parameters may provide additional diagnostic or prognostic information in the near future. Leukemias represent abnormal proliferations of hematopoietic cells that are arrested at a discrete stage of differentiation. Leukemias are classified into acute and chronic forms based on a constellation of clinical and laboratory findings.

Analysis of DNA Ploidy, the Cell Cycle, and Cell Death

The significant medical discovery of understanding the human cell cycle in combination with flow cytometric technology allowed the development of DNA analysis of neoplasia by flow cytometry. A defined number of cells are stained with a known saturating amount of DNA-specific fluorescent dye under controlled conditions of temperature, pH, and ionic strength. The cells are then analyzed using a flow cytometer where, upon excitation by a light beam of the appropriate wavelength, the amount and intensity of fluorescent emission of the dye bound to DNA of each cell, is measured based on a statistically significant number of cells (i.e., ≥10,000) in a period of a few minutes. The relative total DNA content in an unknown cell population is determined when compared to cells analyzed with known, constant DNA content. In addition, small populations of cells can be detected in a heterogenous mixture, and cell populations with small variations in DNA content (~ 4% with a CV of 2%) can be detected. The advent of univariate flow cytometric DNA analysis in the late 1970's was soon followed by reports of the independent prognostic significance of tumor cell DNA content and/or proliferative activity in a number of human malignancies. Numerous additional publications soon appeared, many of which did not substantiate the independent prognostic significance of DNA analysis. In addition, interlaboratory variation in specimen preparation, analysis, data interpretation, and quality control lead to questions about the validity of some results. In 1996 a committee of the American Society for Clinical Oncology concluded that the existing data did not warrant the routine application of measurements of DNA ploidy or proliferation analysis. Since 1996, the utilization of DNA analysis has significantly decreased, and it is most often performed in patients with node-negative breast carcinoma and other tumors where the clinical correlation is strongest. However, recent technological innovations may lead to a revival of interest in clinical DNA analysis. In this regard, Bagwell modified DNA analysis to optimize the accuracy of DNA ploidy and S phase in node-negative breast cancer, eliminate spurious technical inconsistencies, and apply standardized modeling rules to data analysis and interpretation. Furthermore, the authors developed a prognostic model that combines DNA ploidy and the S phase into a Relative Risk Index (RRI). Multiparametric DNA analysis using cytokeratin expression to exclude normal background cells has also been shown to improve the predictive value of the measurements.

Cell Function Analysis

The analysis of cell function can provide relevant clinical information that cannot be acquired from static cellular parameters, such as the expression of surface antigens. Since these considerations are particularly relevant in transplantation medicine and diseases of the immune system, many investigations have focused on functional analysis of the lymphocyte. Virtually every event that occurs during the process of lymphocyte activation can be measured by flow cytometry, but determinations of tyrosine phosphorylation, calcium flux, oxidative metabolism, neoantigen expression, and cellular proliferation have the greatest clinical potential at this time. Tyrosine phosphorylation can be measured within the cell by multiparametric flow-cytometry and labelled ant-iphosphotyrosine monoclonal antibodies, or within activated cell lysates with colored microbeads labelled with monoclonal antibodies specific for different tyrosine kinase substrates (multiplex bead technology). Intracellular calcium flux is measured with ratiometric Ca^{2+} indicators whose spectral characteristics change with Ca^{2+} binding. Calcium flux has been used to study platelet activation in response to different agonists and lymphocyte activation in viral infection and other diseases. Flow cytometric measurement of the oxidative burst in neutrophils has been used as a screening test for chronic granulomatous disease. The most common technique for this purpose uses a nonfluorescing dye (dihydrorhodamine-123) that is selectively concentrated in the mitochondria and is oxidized to a brightly fluorescent compound (rhodamine-123) during the normal oxidative burst. Lymphocyte neoantigens are surface or intracellular proteins, including cytokines that are up-regulated during lymphocyte activation. The detection of these substances may become one of the most important flow cytometric assays. CD11b/CD18 and CD154 are present examples of diagnostically significant neoantigens.

Future Expectations

The flow cytometer is a versatile tool with enormous potential for the study of cells and particles. Because of its unique analytic capabilities, the flow cytometer has become an integral part of the medical research laboratory during the past two decades. The ingress of the flow cytometer into the clinical laboratory has been slower and more controversial. Immunophenotypic analysis and lymphocyte subset analysis is widely performed in the clinical laboratory, but most of the other applications are limited to larger and/or specialized laboratories due to economic considerations. However, no other laboratory instrument provides multiparametric analysis at the single cell level, and the flow cytometer or application-variants of the flow cytometer will become more valuable as medical diagnosis and therapy changes. New fluorochromes, including UV-excited, complex of dyes ("tandem dyes"), and nanocrystals are under development, as well as a new generation of modular flow cytometers using small, solid state lasers, robotics, and advanced, innovative bioinformatics software. An example of the new, specialized flow cytometers of the future is the High Throughput Pharmacological System (HTPS). This system, designed for automated high throughput analysis of novel bioresponse-modifying drugs, permits analysis of 9-10 cell samples/minute from 96-well microplates. Another interesting offshoot of the flow cytometer is the laser scanning cytometer (LSC), a microscope slide-based technology capable of acquiring multiparametric data from selected cells from a heterogeneous population which is proving particularly useful for the analysis of fine needle aspirate and body fluid specimens.

Frequently Asked Questions

Q1. What Is Flow Cytometry?

Answer: Flow-cytometryis a popular cell biology technique that utilizes laser-based technology to count, sort, and profile cells in a heterogeneous fluid mixture. Using a flow cytometer machine, cells or other particles suspended in a liquid stream are passed through a laser light beam in single file fashion, and interaction with the light is measured by an electronic detection apparatus as light scatter and fluorescence intensity. If a fluorescent label, or fluorochrome, is specifically and stoichiometrically bound to a cellular component, the fluorescence intensity will ideally represent the amount of that particular cell component.

Q2. What Is the Principle Behind Flow Cytometry?

Answer: The basic principle of flow-cytometryis the passage of cells in single file in front of a laser so they can be detected, counted and sorted. Cell components are fluorescently labelled and then excited by the laser to emit light at varying wavelengths. The fluorescence can then be measured to determine the amount and type of cells present in a sample.

Q3. What Do You Understand by Alexa Fluor Antibody Tagging?

Answer: The Alexa Fluor family of fluorescent dyes is a series of dyes invented by Molecular Probes, now a part of Thermo Fisher Scientific, and sold under the Invitrogen brand name. Alexa Fluor dyes are frequently used as cell and tissue labels in fluorescence microscopy and cell biology.

Q4. What Is a Flow Cytometer?

Answer: A flow cytometer is an instrument that provides quantifiable data from a sample. Other instruments using flow-cytometryinclude cell sorters which physically separate and thereby purify cells of interest based on their optical properties. A flow cytometer is similar to a microscope, except that, instead of producing an image of the cell, flow-cytometry offers high-throughput, automated quantification of specified optical parameters on a cell-by-cell basis. To analyze solid tissues, a single-cell suspension must first be prepared.

Q5. What Are the Components of a Typical Flow Cytometer?

Answer: A flow cytometer has five main components: a flow cell, a measuring system, a detector, an amplification system, and a computer for analysis of the signals. The flow cell has a liquid stream (sheath fluid), which carries and aligns the cells so that they pass single file through the light beam for sensing. The measuring system commonly use measurement of impedance (or conductivity) and optical systems - lamps (mercury, xenon); high-power water-cooled lasers (argon, krypton, dye laser); low-power air-cooled lasers (argon (488 nm), red-HeNe (633 nm), green-HeNe, HeCd (UV)); diode lasers (blue, green, red, violet) resulting in light signals. The detector and analog-to-digital conversion (ADC) system converts analog measurements of forward-scattered light (FSC) and side-scattered light (SSC) as well as dye-specific fluorescence signals into digital signals that can be processed by a computer. The amplification system can be linear or logarithmic.

Q6. What Does the Term Acquisition Mean?

Answer: The process of collecting data from samples using the flow cytometer is termed 'acquisition'. Acquisition is mediated by a computer physically connected to the flow cytometer, and the software which handles the digital interface with the cytometer. The software is capable of adjusting parameters (e.g., voltage, compensation) for the sample being tested, and also assists in displaying initial sample information while acquiring sample data to ensure that parameters are set correctly. Early flow cytometers were, in general, experimental devices, but technological advances have enabled widespread applications for use in a variety of both clinical and research purposes. Due to these developments, a considerable market for instrumentation, analysis software, as well as the reagents used in acquisition such as fluorescently labeled antibodies has developed.

Q7. What Is Cell Sorting by Flow Cytometry?

Answer: Cell sorting is a method to purify cell populations based on the presence or absence of specific physical characteristics. In flow cytometers with sorting capabilities, the instrument detects cells using parameters including cell size, morphology, and protein expression, and then droplet technology to sort cells and recover the subsets for post-experimental use.

Q8. How Do Flow-Cytometry Cell Sorters Work?

Answer: Flow-cytometry cell sorters have a collection system unlike flow-cytometry analyzers. The collection process starts when a sample is injected into a stream of sheath fluid that passes through the flow cell and laser intercepts. The stream then carries the cell through a vibrating nozzle. The disturbance in the stream causes it to break into a droplet containing ideally one cell. An electrical charging ring is placed just at the point where the stream breaks into droplets. A charge is placed on the ring based immediately prior to fluorescence intensity being measured, and the opposite charge is trapped on the droplet as it breaks from the stream. The charged droplets then fall through an electrostatic deflection system that diverts droplets into containers based upon their charge. In some systems, the charge is applied directly to the stream, and the droplet breaking off retains charge of the same sign as the stream. The stream is then returned to neutral after the droplet breaks off.

If collected under sterile conditions, these cells can be further cultured, manipulated, and studied.

Q9. What Is Fluorescent Labelling of Antibodies?

Answer: Antibody labelling is an important molecular biological technique. Antibodies are widely used in immunoassays to detect and quantify antigens. The antibody that recognizes the antigen is referred to as the 'primary' antibody and confers specificity to the assay. The labelling of proteins allows to follow specific molecules using fluorescent microscopy. Fluorescence results from a process that occurs when certain molecules (generally polyaromatic hydrocarbons or heterocycles) called fluorophores, fluorochromes, or fluorescent dyes absorb light. The absorption of light by a population of these molecules raises their energy level to a brief excited state. As they decay from this excited state, they emit fluorescent light.

Q10. What Are Some of the Uses of Flow Cytometry?

Answer: The most common application performed on the cytometer is immunophenotyping. This technique identifies and quantifies populations of cells in a heterogeneous sample - usually blood, bone marrow or lymph. Apart from this flow-cytometry is used in cell sorting, Cell Cycle Analysis, Cell Proliferation Assays, clinical diagnostics etc.

Q11. What Is Fluorescence?

Answer: Fluorescence is the emission of light by a substance that has absorbed light or other electromagnetic radiation. It is a form of luminescence. In most cases, the emitted light has a longer wavelength, and therefore lower energy, than the absorbed radiation.

Q12. How Is Fluorescence Excited in a Flow Cytometer?

Answer: Fluorescent molecules are the tools of the trade in flow cytometry. When a fluorophore absorbs light, its electrons become excited and move from a resting state (S0, Figure 8A) to a maximal energy level called the excited electronic singlet state (S2). The amount of energy required for this transition will differ for each fluorophore. The duration of the excited state depends on the fluorophore and typically lasts for 1-10 nanoseconds. The fluorophore then undergoes a conformational change, the electrons fall to a lower, more stable energy level called the electronic singlet state (S1), and some of the absorbed energy is released as heat. The electrons subsequently fall back to their resting state (S0) releasing the remaining energy (EEmission) as fluorescence. The difference between wavelengths of the emission and excitation maxima is called the Stokes shift. This cycle can repeat several thousand times for a single fluorophore, which allows recycling of fluorophores and thus amplification of the signal. Emitted light typically contains less energy than was originally put into the fluorophore to excite it. Therefore, the emission wavelength of any fluorophore is longer (lower energy) than its excitation wavelength and thus a different color.

Q13. What Are a Few Advantages of Using Flow Cytometry?

Answer: A few advantages of flow-cytometryare:

1. High speed analyses (100.000 events/seg) depending on the flow rate;
2. Measures single cells and a large number of cells;
3. Simultaneous analysis multiple parameters;
4. Identifies small populations;
5. Quantification of flurescence intensities;
6. Sorting of predefined cells populations (up to 70.000/s);
7. Portable equipments.

Q14. What Are Fluorescent Proteins?

Answer: Fluorescent proteins can be categorized into two groups:

The first group is made of natural proteins that encode their fluorescence in the protein structure. These fluorescent proteins include green fluorescent protein (GFP), YFP, RFP, and various derivatives that have been mutated and validated. Many fluorescent proteins are commercially available in expression vectors to allow researchers to clone them into a vector expressing their favorite targets and follow the expression of the target by using the fluorescent surrogate.

A second class of fluorescent proteins are those derived from the phycobiliproteins found in algae and plants. These proteins use phycobilin cofactors to harvest light and include phycoerythrin (PE), allophycocyanin (APC), andperidinin chlorophyll (PerCP). Phycobiliproteins, especially PE, are among the brightest fluorescent molecules currently available to researchers (see Figure 2 for the structure of phycoerythrin and its cofactor). It should be noted that, while very good for flow cytometry, these molecules are not well-suited for fluorescence microscopy because they rapidly photobleach.

Q15. What Are the 5 Classes of Fluorochromes?

Answer: The five Classes of fluorochromes are as:

1. Fluorescent Proteins.
2. Synthetic Small Molecules.
3. Quantum Dots.
4. Polymer Dyes.
5. Tandem Dyes.

MULTIPLE CHOICE QUESTIONS

Q1. Flow-cytometry uses which among the following:

A. Heavy isotope
B. Radioactive elements
C. Immunological techniques
D. Energy content

Q2. In flow-cytometry which of the combination is impossible?

A. Both the labeled antigens in same particle
B. None of the labeled antigens on a particle
C. One of the labeled antigen
D. All can be true

Q3. In human normal B and T cells develop the CDC4+ and CDC8+ first and then they differentiate to CDC4+ CDC- and CDC4- CDC+. If there is a defect in development which quadrant in the flow cytometer graph will show more density?

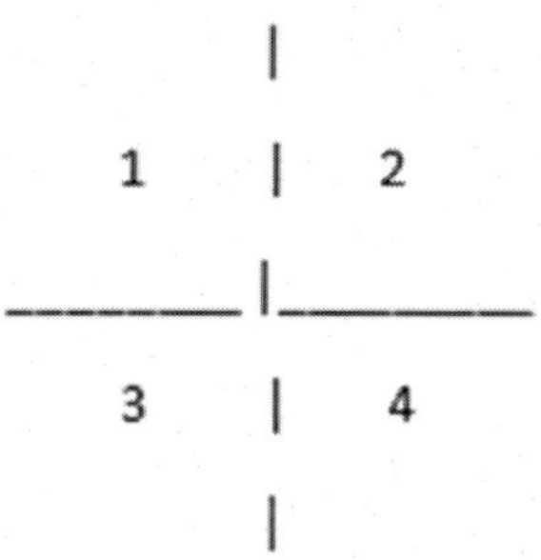

A. 1
B. 2
C. 3
D. 4

Q4. How are the cells sorted?

A. By dilution plating until there are only single cell in each well
B. By the differential weight
C. By electrostatic force
D. By magnetic force

Q5. In a flow cytometer you will see the cells lacking both the labels in which quadrant?

A. 1st quadrant
B. 2nd quadrant
C. 3rd quadrant
D. 4th quadrant

Q6. Which fluorescent dye can be used for red fluorescence?

A. Rhodamine
B. Fluorescein
C. Carmine
D. DAPI

Q7. Define the main purpose of flow cytometry.

A. Amplifies short sequences of targeted DNA or RNA
B. Captures and evaluates colors in a solid or liquid to ensure color accuracy
C. Analyzes light to determine cell or protein properties
D. Replicates cells suspended in fluid

Q8. Which "tag"/conduit does flow-cytometryuse to identify its targets?

A. Glucose
B. Plasmids
C. Protein scaffolds
D. Antibodies

Q9. Which 2 sources of energy does flow-cytometryuse, as mentioned in the article?

A. Lasers and vibrational energy
B. Infrared beams and wind
C. Thermal energy and heat
D. Lasers and impedance

Q10. Flow-cytometryallows simultaneous multiparametric analysis of the physical and/or ________ characteristics of up to thousands of particles per second.

A. Oxygen
B. Carbon
C. Chemistry
D. Chemical substance

Q11. The data generated by flow-cytometers can be plotted in a single ________, to produce a histogram, or in two-dimensional dot plots or even in three dimensions.

A. Polytope
B. Five-dimensional space
C. Dimension
D. Fourth dimension

Q12. Flow-cytometryhas broad application in ________ (especially in transplantation, hematology, tumor immunology and chemotherapy, genetics and sperm sorting for sex preselection).

A. Physician
B. Specialty
C. Surgery
D. Medicine

Q13. A flow cytometer is similar to a ________, except that, instead of producing an image of the cell, flow-cytometryoffers "high-throughput" (for a large number of cells) automated quantification of set parameters.

A. Microscope
B. Cuvette
C. Agar plate
D. Eudiometer

Q14. Intracellular antigens (various ________, secondary mediators, etc.

A. Interleukin
B. Chemokine
C. Cytokine
D. CD154

Q15. In flow cytometry, a beam of light (usually _______ light) of a single wavelength is directed onto a hydrodynamically-focused stream of fluid.

A. X-ray
B. Laser applications
C. Helium
D. Laser

ANSWER KEY

1(C), 2(D), 3(B), 4(C), 5(C), 6(A), 7(C), 8(D), 9(D), 10(D), 11(C), 12(D), 13(A), 14(C), 15(B)

References

Avril M, Gamain B, Lépolard C, Viaud N, Scherf A, Gysin J. Characterization of anti-var2CSA-PfEMP1 cytoadhesion inhibitory mouse monoclonal antibodies. *Microbes and infection*. 2006 Nov 1; 8(14-15):2863-71.

Basford C, Forraz N, McGuckin C. Optimized multiparametric immunophenotyping of umbilical cord blood cells by flow cytometry. *Nature protocols.* 2010 Jul; 5(7):1337.

Bennett JL, Lam C, Kalluri SR, Saikali P, Bautista K, Dupree C, Glogowska M, Case D, Antel JP, Owens GP, Gilden D. Intrathecal pathogenic anti–aquaporin-4 antibodies in early neuromyelitis optica. Annals of Neurology: *Official Journal of the American Neurological Association and the Child Neurology Society*. 2009 Nov; 66(5):617-29.

Berlier JE, Rothe A, Buller G, Bradford J, Gray DR, Filanoski BJ, Telford WG, Yue S, Liu J, Cheung CY, Chang W. Quantitative comparison of long-wavelength Alexa Fluor dyes to Cy dyes: fluorescence of the dyes and their bioconjugates. *Journal of Histochemistry & Cytochemistry*. 2003 Dec; 51(12):1699-712.

Campos CB, Paim BA, Cosso RG, Castilho RF, Rottenberg H, Vercesi AE. Method for monitoring of mitochondrial cytochrome c release during cell death: Immunodetection of cytochrome c by flow-cytometryafter selective permeabilization of the plasma membrane. Cytometry Part A: *The Journal of the International Society for Analytical Cytology.* 2006 Jun; 69(6):515-23.

Cochran JR, Kim YS, Olsen MJ, Bhandari R, Wittrup KD. Domain-level antibody epitope mapping through yeast surface display of epidermal growth factor receptor fragments. *Journal of Immunological Methods*. 2004 Apr 1; 287(1-2):147-58.

Diermeier-Daucher S, Clarke ST, Hill D, Vollmann-Zwerenz A, Bradford JA, Brockhoff G. Cell type specific applicability of 5-ethynyl-2′-deoxyuridine (EdU) for dynamic proliferation assessment in flow cytometry. Cytometry Part A: *The Journal of the International Society for Advancement of Cytometry*. 2009 Jun; 75(6):535-46.

Dimitriadis G, Maratou E, Boutati E, Psarra K, Papasteriades C, Raptis SA. Evaluation of glucose transport and its regulation by insulin in human monocytes using flow cytometry. Cytometry Part A: *The Journal of the International Society for Analytical Cytology*. 2005 Mar; 64(1):27-33.

Even-Desrumeaux K, Baty D, Chames P. Strong and oriented immobilization of single domain antibodies from crude bacterial lysates for high-throughput compatible cost-effective antibody array generation. *Molecular BioSystems*. 2010; 6(11):2241-8.

Feldhaus MJ, Siegel RW, Opresko LK, Coleman JR, Feldhaus JM, Yeung YA, Cochran JR, Heinzelman P, Colby D, Swers J, Graff C. Flow-cytometric isolation of human antibodies from a nonimmune Saccharomyces cerevisiae surface display library. *Nature Biotechnology*. 2003 Feb; 21(2):163.

Feldhaus MJ, Siegel RW. Yeast display of antibody fragments: a discovery and characterization platform. *Journal of Immunological Methods*. 2004 Jul 1; 290(1-2):69-80.

Ferrante EA, Pickard JE, Rychak J, Klibanov A, Ley K. Dual targeting improves microbubble contrast agent adhesion to VCAM-1 and P-selectin under flow. *Journal of Controlled Release*. 2009 Dec 3; 140(2):100-7.

Ferullo DJ, Cooper DL, Moore HR, Lovett ST. Cell cycle synchronization of Escherichia coli using the stringent response, with fluorescence labeling assays for DNA content and replication. *Methods*. 2009 May 1; 48(1):8-13.

Giepmans BN, Adams SR, Ellisman MH, Tsien RY. The fluorescent toolbox for assessing protein location and function. *Science*. 2006 Apr 14; 312(5771):217-24.

Grate JW, Bruckner-Lea CJ, Jarrell AE, Chandler DP. Automated sample preparation method for suspension arrays using renewable surface separations with multiplexed flow-cytometryfluorescence detection. *Analytica Chimica Acta*. 2003 Feb 12; 478(1):85-98.

Gunn J, Wallen H, Veiseh O, Sun C, Fang C, Cao J, Yee C, Zhang M. A multimodal targeting nanoparticle for selectively labeling T cells. *Small*. 2008 Jun; 4(6):712-5.

Holmes KL, Lantz LM. Protein labeling with fluorescent probes. *Methods in Cell Biology*. 2001 Jan 1; 63:185-204.

Hsu ER, Anslyn EV, Dharmawardhane S, Alizadeh-Naderi R, Aaron JS, Sokolov KV, El-Naggar AK, Gillenwater A, Richards-Kortum RR. A far-red fluorescent contrast agent to image epidermal growth factor receptor expression. *Photochemistry and Photobiology*. 2004 Mar; 79(3):272-9.

Jaye DL, Geigerman CM, Fuller RE, Akyildiz A, Parkos CA. Direct fluorochrome labeling of phage display library clones for studying binding specificities: applications in flow-cytometryand fluorescence microscopy. *Journal of Immunological Methods*. 2004 Dec 1; 295(1-2):119-27.

Krutzik PO, Nolan GP. Fluorescent cell barcoding in flow-cytometryallows high-throughput drug screening and signaling profiling. *Nature Methods*. 2006 May; 3(5):361.

Krutzik PO, Clutter MR, Trejo A, Nolan GP. Fluorescent cell barcoding for multiplex flow cytometry. *Current Protocols in Cytometry*. 2011 Jan; 55(1):6-31.

Krutzik PO, Nolan GP. Intracellular phospho-protein staining techniques for flow cytometry: Monitoring single cell signaling events. Cytometry Part A: *The Journal of the International Society for Analytical Cytology*. 2003 Oct; 55(2):61-70.

Letestu R, Rawstron A, Ghia P, Villamor N, Leuven NB, Boettcher S, Buhl AM, Duerig J, Ibbotson R, Kroeber A, Langerak A. Evaluation of ZAP-70 expression by flow-cytometryin chronic lymphocytic leukemia: A multicentric international harmonization process. Cytometry Part B: Clinical Cytometry: *The Journal of the International Society for Analytical Cytology*. 2006 Jul 15; 70(4):309-14.

Lee G, Papapetrou EP, Kim H, Chambers SM, Tomishima MJ, Fasano CA, Ganat YM, Menon J, Shimizu F, Viale A, Tabar V. Modelling pathogenesis and treatment of familial dysautonomia using patient-specific iPSCs. *Nature.* 2009 Sep; 461(7262):402.

Li F, Yang M, Wang L, Williamson I, Tian F, Qin M, Shah PK, Sharifi BG. Autofluorescence contributes to false-positive intracellular Foxp3 staining in macrophages: a lesson learned from flow cytometry. *Journal of Immunological Methods*. 2012 Dec 14; 386(1-2):101-7.

Mie M, Takahashi F, Funabashi H, Yanagida Y, Aizawa M, Kobatake E. Intracellular delivery of antibodies using TAT fusion protein A. *Biochemical and Biophysical Research Communications*. 2003 Oct 24; 310(3):730-4.

Miller JN. Fluorescence energy transfer methods in bioanalysis. *Analyst.* 2005; 130(3):265-70.

Orozco AF, Lewis DE. Flow cytometric analysis of circulating microparticles in plasma. *Cytometry Part A*. 2010 Jun; 77(6):502-14.

Oflazoglu E, Stone IJ, Brown L, Gordon KA, Van Rooijen N, Jonas M, Law CL, Grewal IS, Gerber HP. Macrophages and Fc-receptor interactions contribute to the antitumour activities of the anti-CD40 antibody SGN-40. *British Journal of Cancer*. 2009 Jan; 100(1):113.

Perez OD, Nolan GP. Simultaneous measurement of multiple active kinase states using polychromatic flow cytometry. *Nature Biotechnology*. 2002 Feb; 20(2):155.

Piepers S, De Vliegher S, Demeyere K, Lambrecht BN, de Kruif A, Meyer E, Opsomer G. Flow cytometric identification of bovine milk neutrophils and simultaneous quantification of their viability. *Journal of Dairy Science*. 2009 Feb 1; 92(2):626-31.

Phan TG, Green JA, Gray EE, Xu Y, Cyster JG. Immune complex relay by subcapsular sinus macrophages and noncognate B cells drives antibody affinity maturation. *Nature Immunology*. 2009 Jul; 10(7):786.

Quah BJ, Parish CR. New and improved methods for measuring lymphocyte proliferation *in vitro* and *in vivo* using CFSE-like fluorescent dyes. *Journal of Immunological Methods*. 2012 May 31; 379(1-2):1-4.

Ramdzan YM, Polling S, Chia CP, Ng IH, Ormsby AR, Croft NP, Purcell AW, Bogoyevitch MA, Ng DC, Gleeson PA, Hatters DM. Tracking protein aggregation and mislocalization in cells with flow cytometry. *Nature Methods*. 2012 May; 9(5):467.

Shapiro HM. *Practical Flow Cytometry*. John Wiley & Sons; 2005 Feb 25.

Santos AN, Simm A, Holthoff V, Boehm G. A Method for the Detection of Amyloid-β 1-40, Amyloid-β 1-42 and Amyloid-β Oligomers in Blood Using Magnetic Beads in Combination with Flow-cytometryand its Application in the Diagnostics of Alzheimer's Disease. *Journal of Alzheimer's Disease*. 2008 Jan 1; 14(2):127-31.

Sawa M, Hsu TL, Itoh T, Sugiyama M, Hanson SR, Vogt PK, Wong CH. Glycoproteomic probes for fluorescent imaging of fucosylated glycans *in vivo*. *Proceedings of the National Academy of Sciences*. 2006 Aug 15; 103(33):12371-6.

Sirk SJ, Olafsen T, Barat B, Bauer KB, Wu AM. Site-specific, thiol-mediated conjugation of fluorescent probes to cysteine-modified diabodies targeting CD20 or HER2. *Bioconjugate Chemistry*. 2008 Nov 18; 19(12):2527-34.

Ståhl S, Frejd FY, Orlova A. Affibody molecules for epidermal growth factor receptor targeting *in vivo*: aspects of dimerization and labeling chemistry. *Journal of Nuclear Medicine*. 2009 Feb 1; 50(2):274-83.

Tacken PJ, de Vries IJ, Gijzen K, Joosten B, Wu D, Rother RP, Faas SJ, Punt CJ, Torensma R, Adema GJ, Figdor CG. Effective induction of naive and recall T-cell responses by targeting antigen to human dendritic cells via a humanized anti–DC-SIGN antibody. *Blood.* 2005 Aug 15; 106(4):1278-85.

Vira S, Mekhedov E, Humphrey G, Blank PS. Fluorescent-labeled antibodies: Balancing functionality and degree of labeling. *Analytical Biochemistry*. 2010 Jul 15; 402(2):146-50.

Vos AC, Wildenberg ME, Duijvestein M, Verhaar AP, van den Brink GR, Hommes DW. Anti–tumor necrosis factor-α antibodies induce regulatory macrophages in an Fc region-dependent manner. *Gastroenterology*. 2011 Jan 1; 140(1):221-30.

Waggoner A. Fluorescent labels for proteomics and genomics. *Current Opinion in Chemical Biology*. 2006 Feb 1; 10(1):62-6.

Wang L, Abbasi F, Ornatsky O, Cole KD, Misakian M, Gaigalas AK, He HJ, Marti GE, Tanner S, Stebbings R. Human $CD4^+$ lymphocytes for antigen quantification: characterization using conventional flow-cytometryand mass cytometry. *Cytometry Part A*. 2012 Jul; 81(7):567-75.

Warnes G. Measurement of autophagy by flow cytometry. *Current Protocols in Cytometry*. 2014 Apr; 68(1):9-45.

Wittrup A, Zhang SH, Gerdy B, van Kuppevelt TH, Bengtson P, Johansson M, Welch J, Mörgelin M, Belting M. ScFv Antibody-induced Translocation of Cell-surface Heparan Sulfate Proteoglycan to Endocytic Vesicles Evidence for Heparan Sulfate Epitope Specificity and Role of Both Syndecan and Glypican. *Journal of Biological Chemistry*. 2009 Nov 20; 284(47):32959-67.

Yang X, Li X, Prow TW, Reece LM, Bassett SE, Luxon BA, Herzog NK, Aronson J, Shope RE, Leary JF, Gorenstein DG. Immunofluorescence assay and flow-cytometry selection of bead-bound aptamers. *Nucleic Acids Research*. 2003 May 15; 31(10): e54-.

Ye F, Wu C, Jin Y, Wang M, Chan YH, Yu J, Sun W, Hayden S, Chiu DT. A compact and highly fluorescent orange-emitting polymer dot for specific subcellular imaging. *Chemical Communications*. 2012; 48(12):1778-80.

Zeng Y, Ramya TN, Dirksen A, Dawson PE, Paulson JC. High-efficiency labeling of sialylated glycoproteins on living cells. *Nature Methods*. 2009 Mar; 6(3):207.

In: Immunoglobulins
Editor: Manzoor Ahmad Mir

ISBN: 978-1-53616-903-4

Chapter 8

HYBRIDOMA TECHNOLOGY

Manzoor A. Mir*, Umar Mehraj, Safura Nisar and Hina Qayoom
Department of Bioresources, School of Biological Sciences,
University of Kashmir, Srinagar, J&K, India

ABSTRACT

Hybridomas are cells that have been engineered to produce a desired antibody in large amounts. The development of monoclonal antibodies using hybridoma technology was reported four decades ago and since then more than thirty therapeutic antibodies have been approved by the FDA for the treatment of autoimmune disorders, tumours and inflammatory diseases. Antibodies are highly specific and can therefore bind and affect disease-specific targets, thereby sparing normal cells, and causing fewer toxic side-effects than traditional cytotoxic chemotherapies. To produce monoclonal antibodies, B-cells are removed from the spleen of an animal that has been challenged with the relevant antigen. These B-cells are then fused with myeloma tumor cells that can grow indefinitely in culture (myeloma is a B-cell cancer). This fusion is performed by making the cell membranes more permeable. The fused hybrid cells (called hybridomas), being cancer cells, will multiply rapidly and indefinitely and will produce large amounts of the desired antibodies. They have to be selected and subsequently cloned by limiting dilution. Supplemental media containing Interleukin-6 (such as briclone) are essential for this step. The production of monoclonal anti-bodies was first invented by Cesar Milstein, Georges J. F. Köhler and Niels Kaj Jerne in 1975. Selection occurs via culturing the newly fused primary hybridoma cells in selective-media, specifically media containing 1x concentration HAT for roughly 10–14 days. After using HAT it is often desirable to use HT containing media. Cloning occurs after identification of positive primary hybridoma cells. Clone by limited dilution.

* Corresponding author: Dr Manzoor Ahmad Mir (Ph. D), Department of Bioresources, School of Biological Sciences, University of Kashmir, Srinagar J&K, India-190006, Contact:-+919622901319. E-mail: drmanzoor@kashmiruniversity.ac.in.

Keywords: monoclonal antibody, polyclonal antibody, hybridoma cells, immunization, HGPRT

Objectives

- To understand polyclonal and monoclonal antibodies (MAbs).
- To know the history and early research about discovery of MAbs
- To describe and discuss about Hybridoma.
- To discuss regarding Hybridoma technology of MAb production.
- To describe and understand various applications of MAbs.

Introduction

Antibodies are a part of the defense system to protect the body against the invading foreign substances namely antigens. Also called as immunoglobulins, these are protein molecules produced by a specialized group of cells called B-lymphocytes (plasma cells) in mammals.

Each antigen has specific antigen determinants (epitopes) located on it. The antibodies have complementary determining regions (CDRs) which are mainly responsible for the antibody specificity. In response to an antigen (with several different epitopes), B-lymphocytes gear up and produce many different antibodies. These types of antibodies which can react with the same antigen are designated as polyclonal antibodies. The polyclonal antibody production is variable and is dependent on factors such as epitopes, response to immunity etc (Figure 8.1). Due to lack of specificity and heterogenic nature, there are several limitations on the utility of polyclonal antibodies for therapeutic and diagnostic purposes.

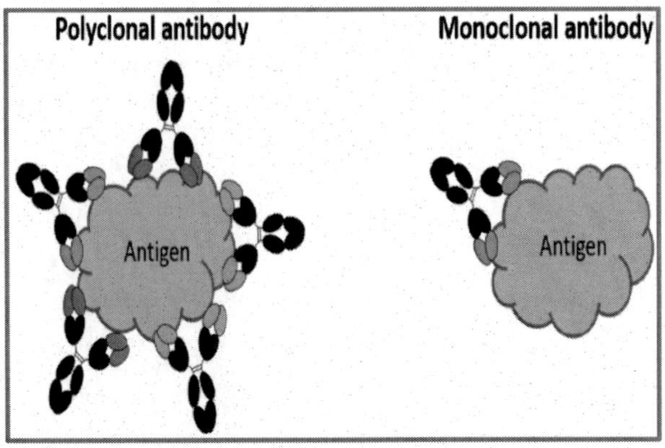

Figure 8.1. Representing the variability of polyclonal antibody in comparison to monoclonal antibody.

HISTORY

From the time the first monoclonal antibody was generated in 1975 and the first monoclonal antibody fully licenced in 1986, the field of monoclonal antibody development represents a novel way in which to target specific mutations and defects in protein structure and expression in a wide range of diseases and conditions. The term hybridoma was coined by Leonard Herzenberg in the laboratory of César Milstein's in 1975 when they used this technology to generate monoclonal antibodies in mice.

After completing his doctoral in 1957, Milstein joined the laboratory of Malcolm Dixon in Cambridge for a second doctoral study, working with Nobel Prize laureate Frederick Sanger on enzymology. Sanger's work on protein and nucleic acid sequencing became one of the foundations of Milstein's future scientific exploits. After finishing his second thesis, he became head of the Department of Molecular Biology at the Malbrán Institute for two years, but political turmoil and persecution again prompted Milstein to emigrate to Europe. It was only after his return to Cambridge that Milstein began working on antibodies. He spent the first months at the Laboratory of Molecular Biology (LMB) investigating the Bence-Jones protein, a protein complex formed by two antibody light chains, found in the urine of some multiple myeloma patients. At the time, protein sequencing was still tedious work, but Milstein and his students succeeded in elucidating the position and structural function of the disulphide bonds: First in the Bence-Jones protein and later in the subunits of normal antibodies. In the early 1970s, Milstein changed his focus from the genetic origin of antibody diversity to analysing the mRNA of antibody secreting tumours. In light of the rather meager results it became apparent that to this end, Milstein's group had to move away from solid tumours and use tissue culture cells. Milstein and his colleague, the pathologist George Brownlee, chose the mouse myeloma cell line, MOPC21, both for its easy growth and high immunoglobulin production.

POLYCLONAL ANTIBODIES

In general, an antigen molecule has antigenic determinants of more than one specificity, i.e; different determinants of an antigen will interact with different antibodies. If an antigen has determinants specific for two antibodies. Each distinct antigenic determinant of the antigen will bind to a distinct mature B-cell, whose specific Ig matches the specificity represented by the concerned determinant. As a result, such a single antigen will activate two or more different types of B-cells; each B-cell type will have sIgs specific to a distinct epitope of the antigen and will divide and differentiate to give rise to a clone of plasma cells producing antibodies of the same specificity. Thus, a single antigen would induce more than one distinct clone of plasma cells, which will produce antibodies of different specifies. Therefore, the serum of an animal immunised by a single antigen will contain antibodies of different specificities that will react to the same antigen; these are called *Polyclonal antibodies* since they are produced by several different plasma cell clones.

PRODUCING POLYCLONAL ANTIBODIES

Antibodies used for research and diagnostic purposes are often obtained by injecting a lab animal such as a rabbit or a goat with a specific *antigen.* Within a few weeks, the animal's immune system will produce high levels of antibodies specific for the antigen. These antibodies can be harvested in an *antiserum,* which is whole serum collected from an animal following exposure to an antigen. Because most antigens are complex structures with multiple epitopes, they result in the production of multiple antibodies in the lab animal. This so-called *polyclonal antibody* response is also typical of the response to infection by the human immune system. Antiserum drawn from an animal will thus contain antibodies from multiple clones of B cells, with each B cell responding to a specific epitope on the antigen (Figure 8.2).

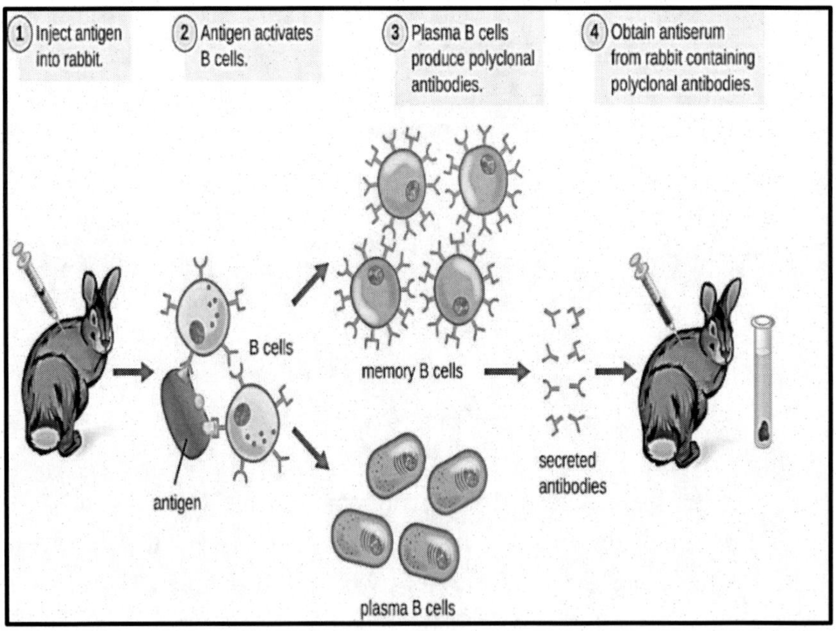

Source: NYS.

Figure 8.2. This diagram illustrates the process for harvesting polyclonal antibodies produced in response to an antigen.

Lab animals are usually injected at least twice with antigen when being used to produce antiserum. The second injection will activate memory cells that make class *IgG* antibodies against the antigen. The memory cells also undergo *affinity maturation*, resulting in a pool of antibodies with higher average affinity. Affinity maturation occurs because of mutations in the immunoglobulin gene variable regions, resulting in B cells with slightly altered antigen-binding sites. On re-exposure to the antigen, those B cells capable of producing antibody with higher affinity antigen-binding sites will be stimulated to proliferate and produce more antibody than their lower-affinity peers. An *adjuvant,* which is a chemical that provokes a generalized activation of the immune system that stimulates greater antibody production, is often mixed with the antigen prior to injection.

Antiserum obtained from animals will not only contain antibodies against the antigen artificially introduced in the laboratory, but it will also contain antibodies to any other antigens to which the animal has been exposed during its lifetime. For this reason, antisera must first be "purified" to remove other antibodies before using the antibodies for research or diagnostic assays.

MONOCLONAL ANTIBODIES

Originally discovered by Emil von Bering and Paul Erhlich, antibodies are proteins produced by the immune system that target a specific foreign object (antigen). They are called monoclonal Antibodies (mAbs) when they are produced by clones derived from a single parent cell. Monoclonal antibodies are monovalent antibodies which bind to the same epitope and are produced from a single B-lymphocyte clone. The production of monoclonal antibodies was invented by César Milstein and Georges J. F. Köhler in 1975. They shared the Nobel Prize of 1984 for Medicine and Physiology with Niels Kaj Jerne, who made other contributions to immunology.

MAbs are widely used in laboratories for research and diagnostics purposes. They can be produced in large quantities as immunotherapeutics against a range of inflammatory diseases, infection and cancer. MAbs are used in immunotherapy in order to boost the body's natural defenses.

There are four types of monoclonal antibodies (Figure 8.3):

- Murine (-omab): entirely derived from a murine source. They can lead to an allergic reaction in humans.
- Chimeric (-ximab): the variable regions are of murine origins whereas the constant regions are human. These can also cause an allergy.
- Humanized (-zumab): mostly derived from a human source except for the part of the antibody which binds to its target.
- Human (-umab) : entirely derived from a human source

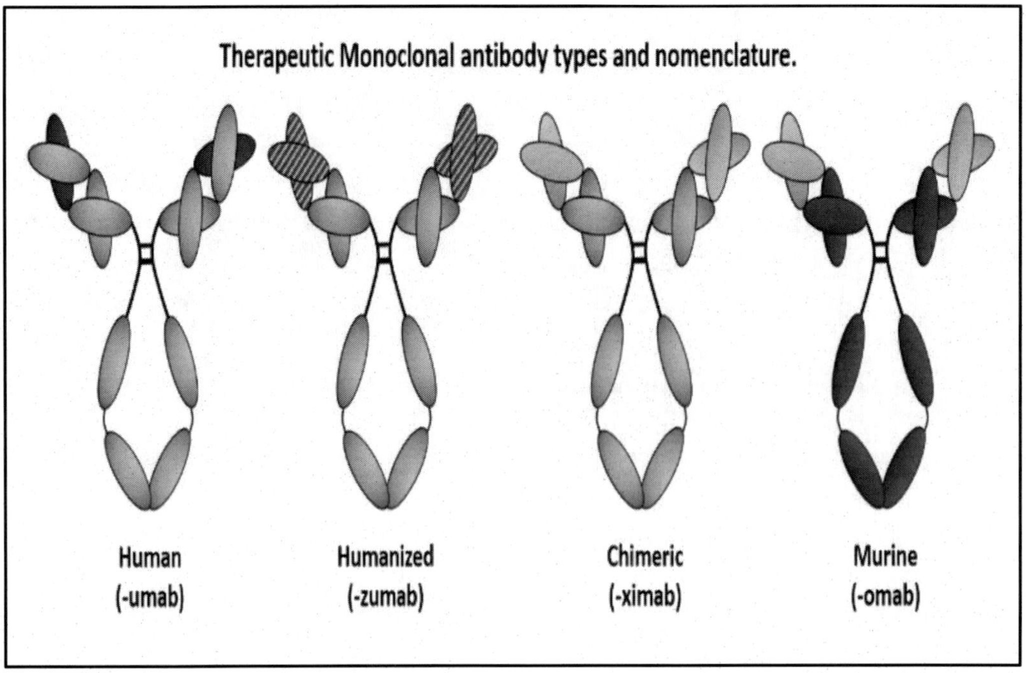

Figure 8.3. Showing monoclonal antibodies and their types.

PRODUCTION OF MONOCLONAL ANTIBODIES

The traditional monoclonal antibody (mAb) production process usually starts with generation of mAb-producing cells (i.e., hybridomas) by fusing myeloma cells with desired antibody-producing splenocytes (e.g., B cells). These B cells are typically sourced from animals, usually mice. After cell fusion, large numbers of clones are screened and selected on the basis of antigen specificity and immunoglobulin class. Once candidate hybridoma cell lines are identified, each "hit" is confirmed, validated, and characterized using a variety of downstream functional assays. Upon completion, the clones are scaled up where additional downstream bioprocesses occur.

The production of monoclonal antibodies involves several steps (Figure 8.4). In the first instance a laboratory animal is injected with a desired target to stimulate their immune system. Following this, antibody producing cells, B lymphocytes, are harvested from the animal's spleen and fused with an immortal myeloma cell line to create hybrid cells, or hybridomas. The hybrid cells are then screened to find those that secrete antibodies with the desired specificity for a particular target. Once identified the hybrid cell is cloned to establish a hybridoma colony. This colony is then maintained in a culture medium to provide a continual supply of monoclonal antibodies. Each of the antibodies secreted by the hybrid cell is identical (monoclonal) and has the capacity to bind to a specific receptor found on the surface of a cell. The entire process of Mab production is known as, *hybridoma technology.*

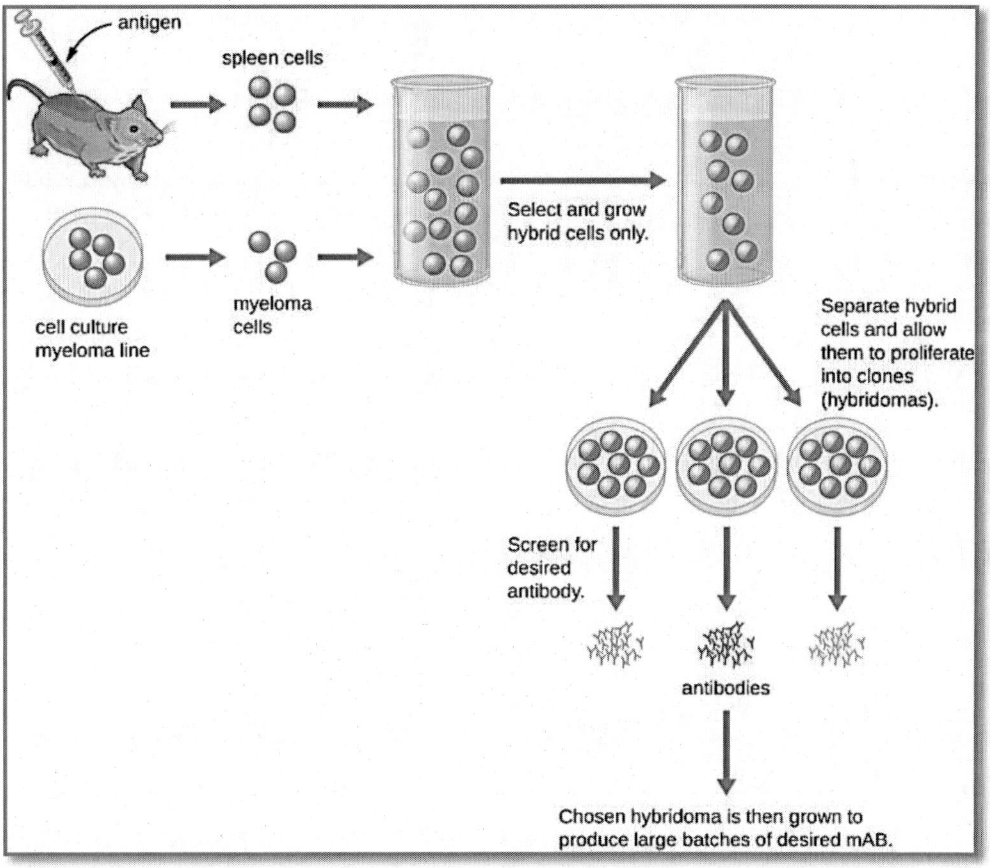

Source: NYS.

Figure 8.4. Monoclonal antibodies (mAbs) are produced by introducing an antigen to a mouse and then fusing polyclonal B cells from the mouse's spleen to myeloma cells. The resulting hybridomas cells are cultured and continue to produce antibodies to the antigen. Hybridomas producing the desired mAb are then grown in large numbers on a selective medium that is periodically harvested to obtain the desired mAbs.

APPLICATIONS AND DISADVANTAGES OF MONOCLONAL ANTIBODIES

One of the major advantages of monoclonal antibodies is that they can be produced very cheaply and on a large-scale. They are now routinely used as probes in laboratories to unravel the pathways of numerous diseases, whether it be cancer, Alzheimer's disease or autoimmune disorders.

Monoclonal antibodies are also vital to the typing of tissue for organ transplants and blood for transplants. Every day they are also used to analyse bodily fluids for diagnosis and to monitor hospital infections. They are also instrumental in the detection of heart attacks.

Beyond the clinic, monoclonal antibodies are important components in home-testing kits for determining ovulation, pregnancy, menopause and many other conditions.

In addition to their use in diagnostics, monoclonal antibodies have proven useful as drugs. Since their emergence, monoclonal antibodies have helped improve the treatment of over 50 major diseases. They are used, for example, to prevent the rejection of organ transplants and for the treatment of cancer, inflammatory and autoimmune disorders, cardiovascular disease, infectious diseases, allergies, osteoporosis and ophthalmic problems. They have also been shown to help with migraines and reduce high cholesterol.

Governments are also reliant on monoclonal antibodies for protecting public health. The antibodies are used, for example, in tests to detect and monitor infectious agents like HIV, the cause of AIDS, or the influenza virus causing pandemic flu and for identifying diseases released by biological weapons such as smallpox and anthrax. In addition, monoclonal antibodies help in checks for salmonella and bacteria that cause food poisoning. They are also instrumental in detecting viruses in animal livestock or plants and monitoring environmental pollution.

From Myeloma to Hybridoma Technology

Over the course of the next two to three years, Milstein and his students, most notably David Secher and Dick Cotton, established protocols for myeloma cell culturing and cloning. Of particular importance was Cotton's discovery that by fusing different myeloma cells, he was able to generate hybrid cells that expressed both sets of antibody subunits.

In 1974, Georges Köhler joined Milstein in Cambridge. Köhler began his postdoctoral research by looking for V-region mutants of their myeloma. The results were rather disappointing and even in the rare event of finding a V-mutant, there was no practical way of determining its specificity. Cotton's previous experiments, as well as a 1974 publication by J. Schwaber and E. Cohen on fusing human lymphocytes with myeloma cells, now came into play. Milstein and Köhler wondered whether they could create immortal hybrid cells by fusing normal B cells from immunized mice with their myeloma cells (Figure 8.5). For incidental reasons, they had all the requirements fulfilled and it worked in the first attempt. By cloning individual hybrid cells, they established the first hybridoma cell lines. Their 1975 paper on these experiments has become a citation classic, and in the course of a few years, hundreds of different monoclonal antibodies became available to researchers worldwide.

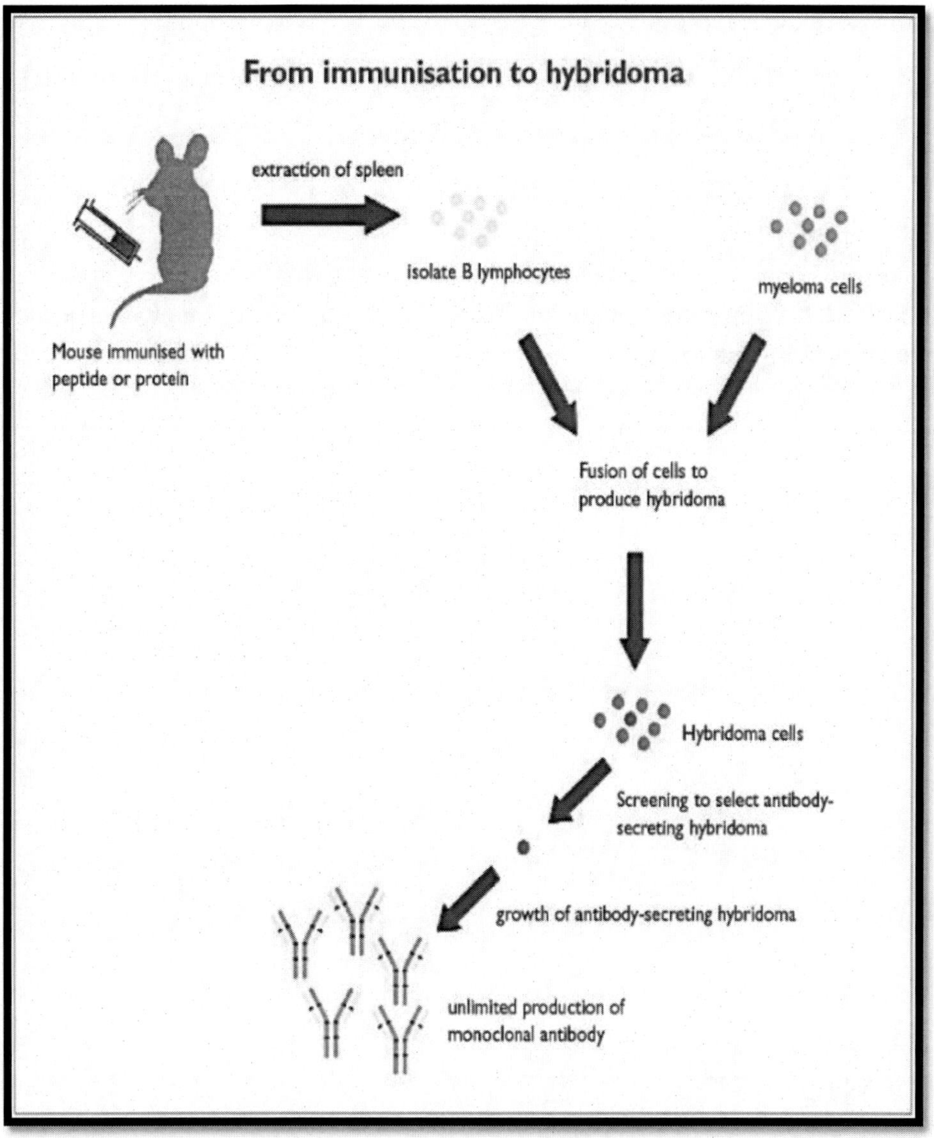

Source: Google scholar.

Figure 8.5. Antibody producing cells isolated from the spleen of a mouse immunised with the antigen of interest are fused with immortal myeloma cells. The fusion product is an immortal hybridoma which can produce unlimited quantities of monoclonal antibody.

PRINCIPLE FOR CREATION OF HYBRIDOMA CELLS

The myeloma cells used in hybridoma technology for the generation of hybrid cells must not be capable of synthesizing their own antibodies. The selection of hybridoma cells is based on inhibiting the nucleotide (consequently the DNA) synthesizing machinery. The mammalian cells can synthesize nucleotides by two pathways—de novo synthesis and salvage pathway which will be discussed next.

PATHWAYS FOR THE SYNTHESIS OF NUCLEOTIDES

The *de novo synthesis* of nucleotides requires tetrahydrofolate which is formed from dihydrofolate. The formation of tetrahydrofolate (and therefore nucleotides) can be blocked by the inhibitor aminopterin.

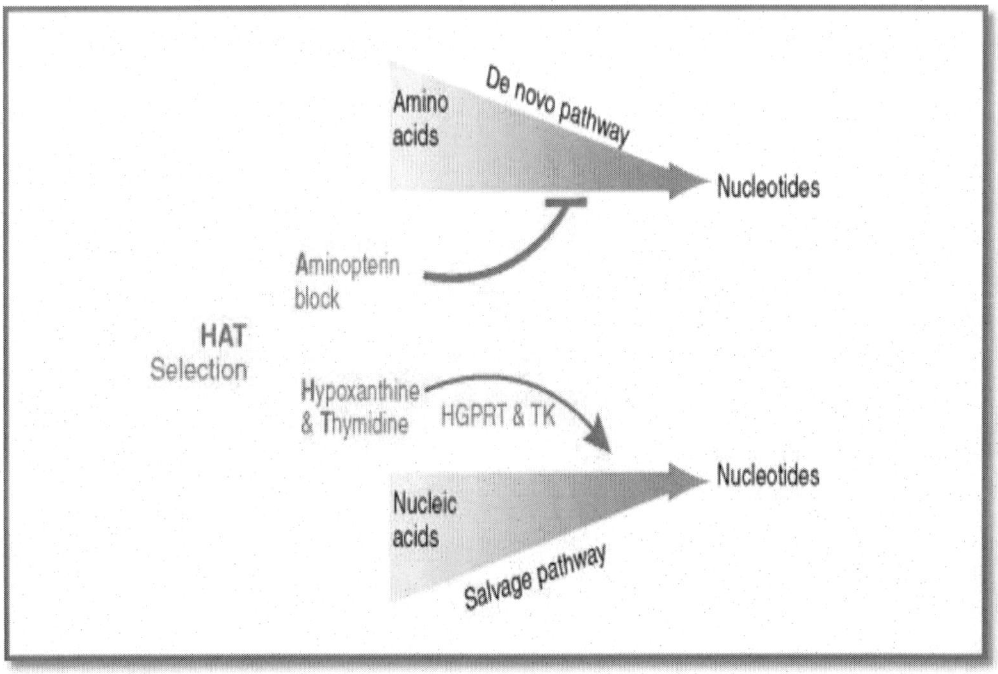

Figure 8.6. Pathways for the synthesis of nucleotides (HGPRT-Hypoxanthine guanine phosphoribosyl transferase; TK- Thymidine kinase) (Science direct, hybrid cell line).

The *salvage pathway* involves the direct conversion of purines and pyrimidine's into the corresponding nucleotides. Hypoxanthine guanine phosphoribosyl transferase (HGPRT) is a key enzyme in the salvage pathway of purines. It converts hypoxanthine and guanine respectively to inosine monophosphate and guanosine monophosphate. Thymidine kinase (TK), involved in the salvage pathway of pyrimidine's converts thymidine to thymidine monophosphate (TMP). Any mutation in either one of the enzymes (HGPRT or TK) blocks the salvage pathway. When cells deficient (mutated cells) in HGPRT are grown in a medium containing hypoxanthine aminopterin and Thymidine (HAT medium), they cannot survive due to inhibition of de novo synthesis of purine nucleotides (Salvage pathway is not operative due to lack of HGPRT). Thus, cells lacking HGPRT, grown in HAT medium die (Figure 8.6).

The hybridoma cells possess the ability of myeloma cells to grow in vitro with a functional HGPRT gene obtained from lymphocytes (with which myeloma cells are fused). Thus, only the hybridoma cells can proliferate in HAT medium, and this procedure is successfully used for their selection.

METHODOLOGY

A *hybridoma* is a hybrid cell obtained by fusing a B-lymphocyte with usually a tumour cell of the antibody forming system or B-lymphocytes (these are called *myelomas*). The hybrid cells thus produced possess the ability to produce antibodies due to the B-lymphocyte genome and the capacity for indefinite growth in vitro due to the tumour (myeloma) cell involved in the fusion. Therefore, specific hybridomas are either cultured in vitro or passage through mouse peritoneal cavity to obtain monoclonal antibodies, this is called *hybridoma technology*.

This technology is based on the fusion between myeloma cells (malignant plasma cells) and spleen cells from a suitably immunized animal. Spleen cells die in a short period under ordinary tissue culture conditions while myeloma cells are adopted to grow permanently in culture. Mutants of myeloma cells lack the enzyme hypoxanthine guanine phosphoribosyl transferase (azaquinine resistant) or thymidine kinase (bromodeoxy-uridine resistant). These mutants cannot grow in a medium containing aminopterin, supplemented with hypoxanthine and thymidine (HAT medium). Hybrids between the mutant myeloma cells and spleen cells can be selected and cultured in HAT medium.

From the growing hybrids, individual clones can be chosen that secrete the desired antibodies (monoclonal origin). The selected clones like ordinary myeloma cells can be maintained indefinitely. In short, the hybridoma technology for the production of monoclonal antibodies involves the following steps (Figure 8.10):

- Immunization
- Cell fusion
- Selection of hybridomas
- Screening the products
- Cloning and propagation.
- Characterisation and storage

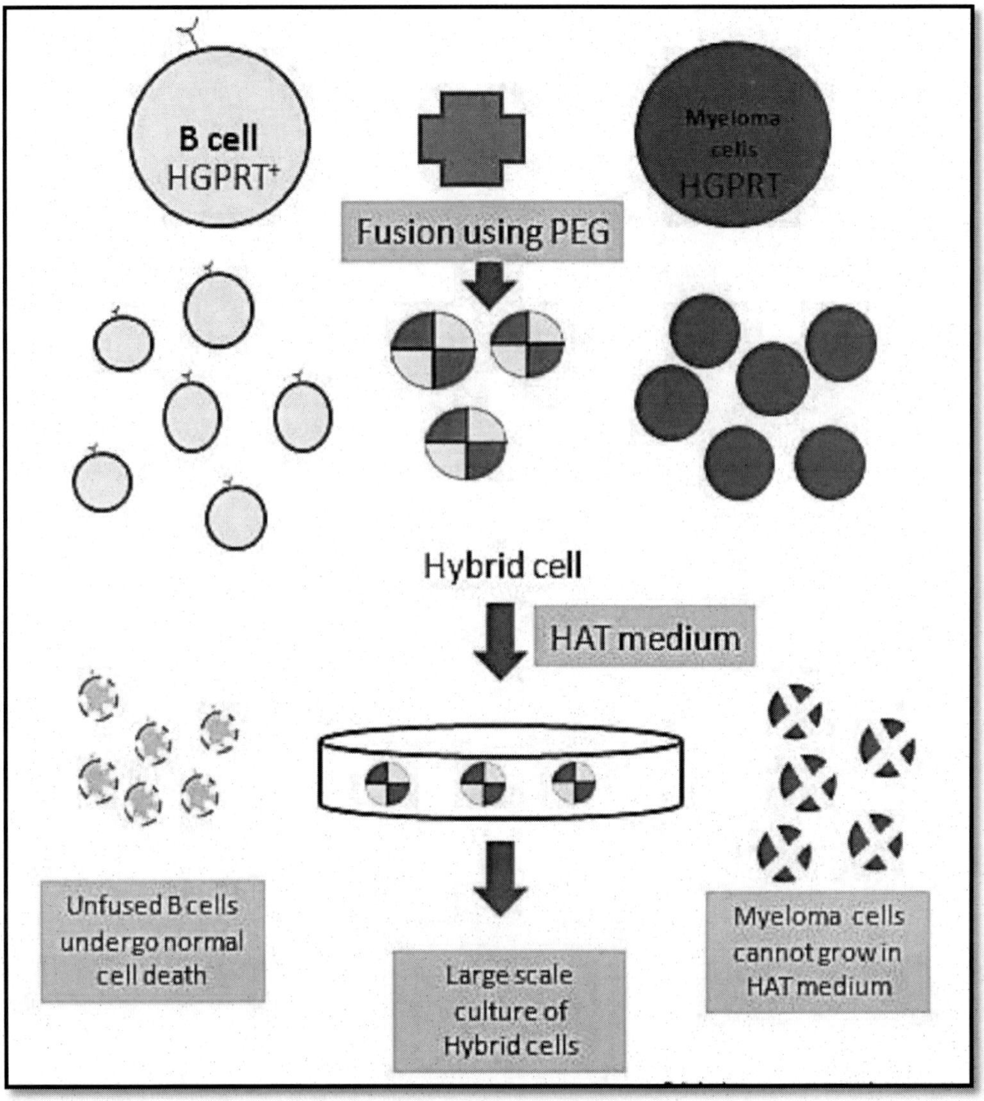

Figure 8.7. A schematic representation of the production of hybridoma clones.

Immunization

The very first step in hybridoma technology is to immunize an animal (usually a mouse), with appropriate antigen. The antigen, along with an adjuvant like Freund's complete or incomplete adjuvant is injected subcutaneously (adjuvants are non-specific potentiators of specific immune responses). The injections at multiple sites are repeated several times. This enables increased stimulation of B-lymphocytes which are responding to the antigen.

Three days prior to killing of the animal, a final dose of antigen is intravenously administered. The immune-stimulated cells for synthesis of antibodies have grown maximally by now. The concentration of the desired antibodies is assayed in the serum of the animal at frequent intervals during the course of immunization. When the serum concentration of the antibodies is optimal, the animal is sacrificed. The spleen is aseptically removed and disrupted by mechanical or enzymatic methods to release the cells. The lymphocytes of the spleen are separated from the rest of the cells by density gradient centrifugation.

Cell Fusion

The thoroughly washed lymphocytes are mixed with HGPRT defective myeloma cells. The mixture of cells is exposed to polyethylene glycol (PEG) for a short period (a few minutes), since it is toxic. PEG is removed by washing and the cells are kept in a fresh medium. These cells are composed of a mixture of hybridomas (fused cells), free myeloma cells and free lymphocytes (Figure 8.8).

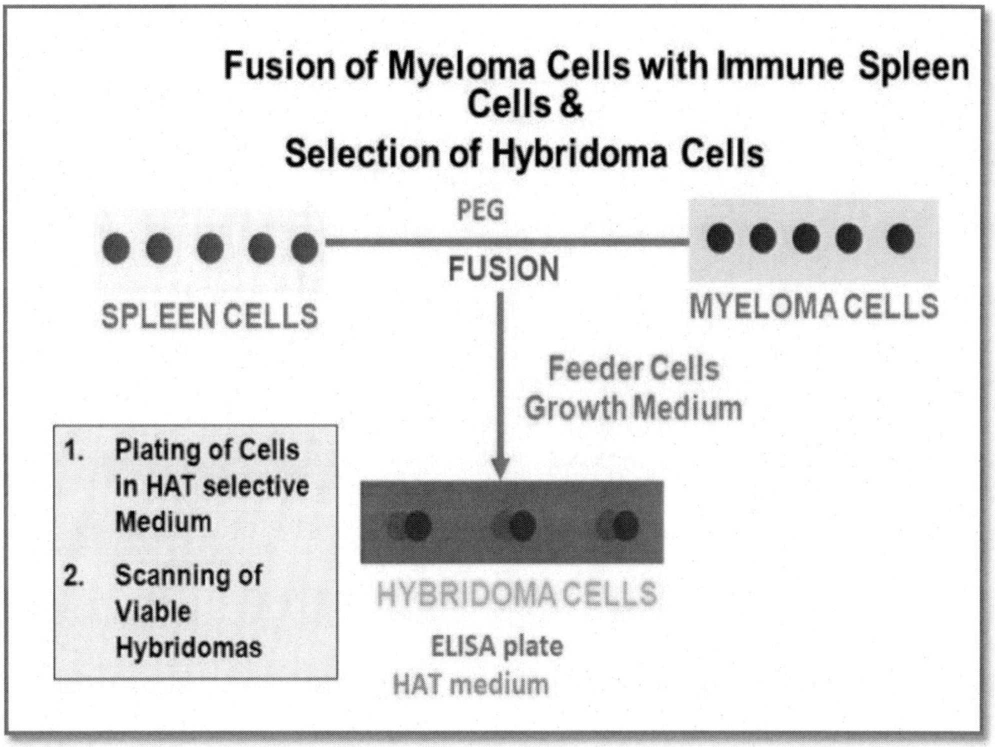

Figure 8.8. Representing the fusion of myeloma cells and spleen cells in HAT medium.

Selection of Hybridomas

When the cells are cultured in HAT medium (the principle described above), only the hybridoma cells grow, while the rest will slowly disappear. This happens in 7-10 days of culture. Selection of a single antibody producing hybrid cells is very important. This is possible if the hybridomas are isolated and grown individually. The suspension of hybridoma cells is so diluted that the individual aliquots contain on an average one cell each. These cells, when grown in a regular culture medium, produce the desired antibody (Figure 8.9).

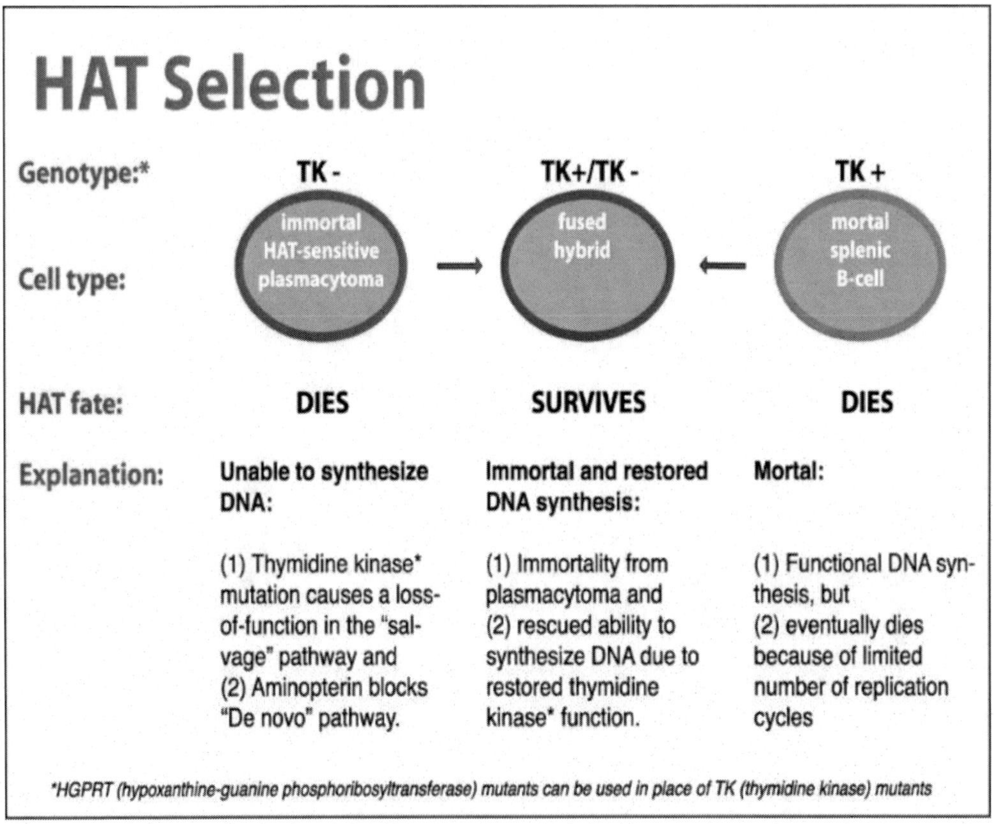

Figure 8.9. Showing the selection of hybridomas from HAT medium.

Screening the Products

The hybridomas must be screened for the secretion of the antibody of desired specificity. The culture medium from each hybridoma culture is periodically tested for the desired antibody specificity. The two techniques namely ELISA and RIA are commonly used for this purpose. In both the assays, the antibody binds to the specific antigen (usually coated to plastic plates) and the unbound antibody and other components of the medium can be washed off. Thus, the hybridoma cells producing the desired antibody can be identified by screening. The antibody secreted by the hybrid cells is referred to as monoclonal antibody.

Cloning and Propagation

The single hybrid cells producing the desired antibody are isolated and cloned. Two techniques are commonly employed for cloning hybrid cells-limiting dilution method and soft agar method.

Limiting Dilution Method

In this procedure, the suspension of hybridoma cells is serially diluted and the aliquots of each dilution are put into micro culture wells. The dilutions are so made that each aliquot in a well contains only a single hybrid cell. This ensures that the antibody produced is monoclonal.

Soft Agar Method

In this technique, the hybridoma cells are cultured in soft agar. It is possible to simultaneously grow many cells in semisolid medium to form colonies. These colonies will be monoclonal in nature. In actual practice, both the above techniques are combined and used for maximal production of MAbs.

Characterization and Storage

The monoclonal antibody has to be subjected to biochemical and biophysical characterization for the desired specificity. It is also important to elucidate the MAb for the immunoglobulin class or sub-class, the epitope for which it is specific and the number of binding sites it possesses.

The stability of the cell lines and the MAbs are important. The cells (and MAbs) must be characterized for their ability to withstand freezing, and thawing. The desired cell lines are frozen in liquid nitrogen at several stages of cloning and culture.

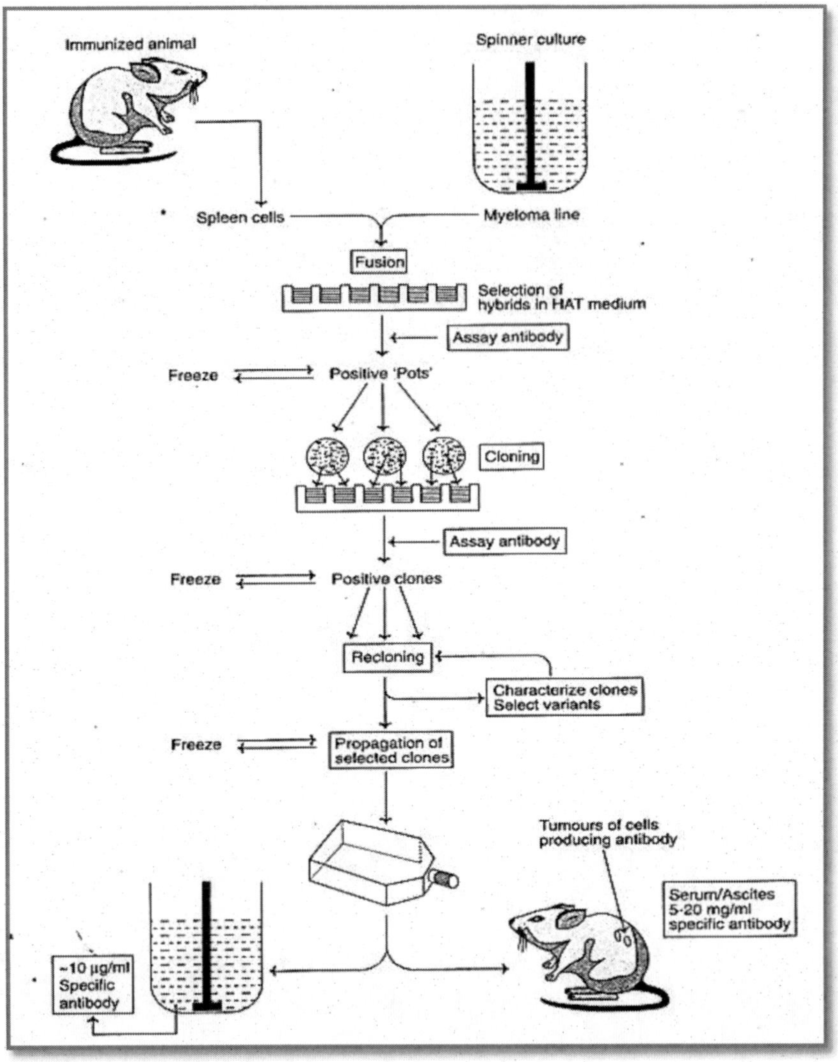

Figure 8.10. Basic protocol for the derivation of monoclonal antibodies from hybrid myelomas (Adapted from: monoclonal antibodies: production, advantages and limitations by Nandkishore Jha).

LARGE SCALE PRODUCTION OF MABS

The production of MAbs in vitro in the culture bottles is rather low (5-10 (µg/ml). The yield can be increased by growing the hybrid cells as ascites in the peritoneal cavity of mice. The ascitic fluid contains about 5-20 mg of mAb/ml. This is far superior than the in vitro cultivation techniques but collection of mAb from ascitic fluid is associated with the heavy risk of contamination by pathogenic organisms of the animal. In addition, several animals have to be sacrificed to produce mAb. Hence, many workers prefer in vitro techniques rather than the use of animals.

Encapsulated Hybridoma Cells for Commercial Production of mAbs

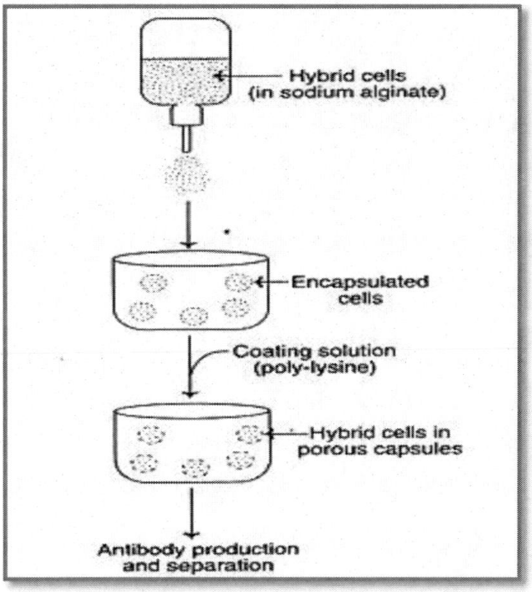

Figure 8.11. Production of monoclonal antibodies by microencapsulation
(Adapted from: monoclonal antibodies: production, advantages and limitations by Nandkishore Jha).

The yield of MAb production can be substantially increased by increasing the hybridoma cell density in suspension culture. This can be done by encapsulating the hybridomas in alginate gels and using a coating solution containing poly-lysine. These gels allow the nutrients to enter in and antibodies to come out. By this approach, a much higher concentration of mAb production (10-100 μg/ml) can be achieved. Damon Biotech Company and Cell-Tech use encapsulated hybridoma cells for large-scale production of mAbs. They employ 100-liter fermenters to yield about 100g of mAbs in about 2 weeks period (Figure 8.11).

Drawbacks of Early Monoclonal Antibodies and Possible Developmental Alternatives

The first licenced monoclonal antibody was Orthoclone OKT3 (muromonab-CD3) which was approved in 1986 for use in preventing kidney transplant rejection. It is a monoclonal mouse IgG2a antibody whose cognate antigen is CD3. It works by binding to and blocking the effects of CD3 expressed on T-lymphocytes. However, its use was limited to acute cases due to reported side-effects (e.g., human anti-mouse antibody response). A major stumbling block was the fact that the production of early monoclonal antibodies was limited by whether or not there was a suitable myeloma cell line available (usually mouse or rat).

No method of generating a hybridoma that avoids the use of animals has been found. Recent in vitro techniques allow the intracellular production of antigen-binding antibody fragments, but such techniques are still experimental and have an uncertain yield, efficacy, and antibody function. It has been possible to genetically replace much of the mouse mAb producing genes with human sequences, reducing the immunogenicity of mAb destined for clinical use in humans though. Before the advent of the hybridoma method, investigators could produce only polyclonal serum antibodies; this required large numbers of immunized animals and did not immortalize the antibody-producing cells, so it required repeated animal use to obtain more antibodies. Development of the hybridoma technology has reduced the number of animals (mice, rabbits, and so on) required to produce a given antibody but with a decrease in animal welfare when the ascites method is used.

Hybridomas may be low yielding or genetically unstable. More recently, many different expression systems for monoclonal antibodies have been tested, each with contrasting effects.

Bi-Specific Monoclonal Antibodies

The MAbs in which the two arms of Fab (antigen-binding) have two different specificities for two different epitopes are referred to as bi-specific MAbs. They may be produced by fusing two different hybridoma cell lines or by genetic engineering. Bi-specific Fab MAbs theoretically, are useful for a simultaneous and combined treatment of two different diseases

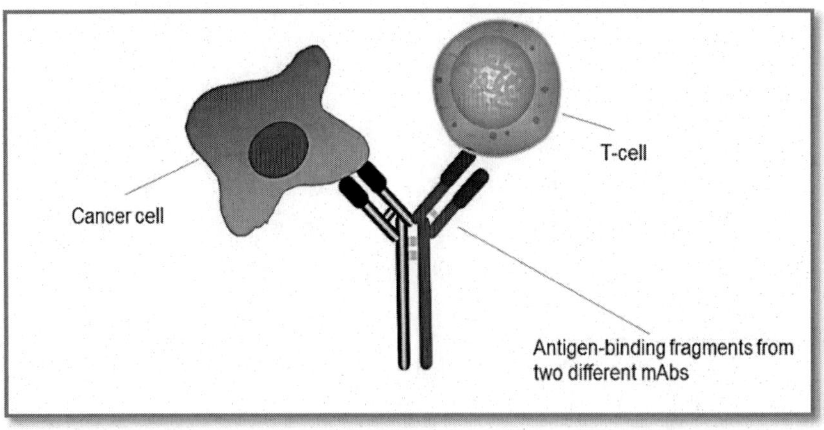

Figure 8.12. Bi-specific antibody is an engineered protein composed of antigen-binding fragments from two different monoclonal antibodies.

Production of MAbs in *E. coli*

The hybridoma technology is very laborious, expensive and time consuming. To overcome these limitation, researchers have been trying to genetically engineer bacteria, plants and animals. The objective is to develop bioreactors for the large scale production of monoclonal antibodies.

E. coli was found to be an excellent system for expression of antibody fragments such as single-chain variable fragments (scFv) and antigen-binding fragments (Fab). However, the synthesis of a relatively larger, full-sized antibody (i.e., consisting of 2 heavy chains and 2 light chains joined together by disulphide bridges giving a total molecular weight of ~150 kDa) may be a step too far for such a relatively small microorganism, although the lack of glycosylating enzymes in E. coli may also prove to be beneficial for antibodies whose primary role is to block protein–protein interactions as opposed to invoking downstream immune effector responses (e.g., the complement system), which can lead to potential immunogenic side-effects.

The Generation of Monoclonal Antibodies Using Phage Display

Another method of generating monoclonal antibodies is by using phage display. This involves isolating B-lymphocytes from the blood of humans and then isolating the mRNA and converting it into cDNA using PCR to amplify all the VH and VL segments. These segments can then be cloned into a vector (usually as scFv) next to the PIII protein of a bacteriophage before being used to infect E. coli in order to generate a library containing approximately 1010 cells by inoculating the library with an additional helper phage. E. coli can then secrete the bacteriophage containing the VH and VL segments as part of the bacteriophage coat. Specific VH and VL segments against the antigen can then be selected and used to re-inoculate E. coli with the bacteriophage. Cells containing the plasmid can then be isolated and sequenced (Figure 8.13).

Its advantages include once the library is made, the same library can be used to generate new antibodies and does not have to be remade, no immunisations are required as the entire process is done in vitro, antibodies can be obtained much more quickly than the traditional hybridoma technique and the library can be used to generate antibodies to toxic antigens that could not be used to immunise an animal.

Improvement of Monoclonal Antibody Efficacy

Targets for improving antibody efficacy include: immunogenicity, antigen-binding affinity, effector functions and pharmacokinetics.

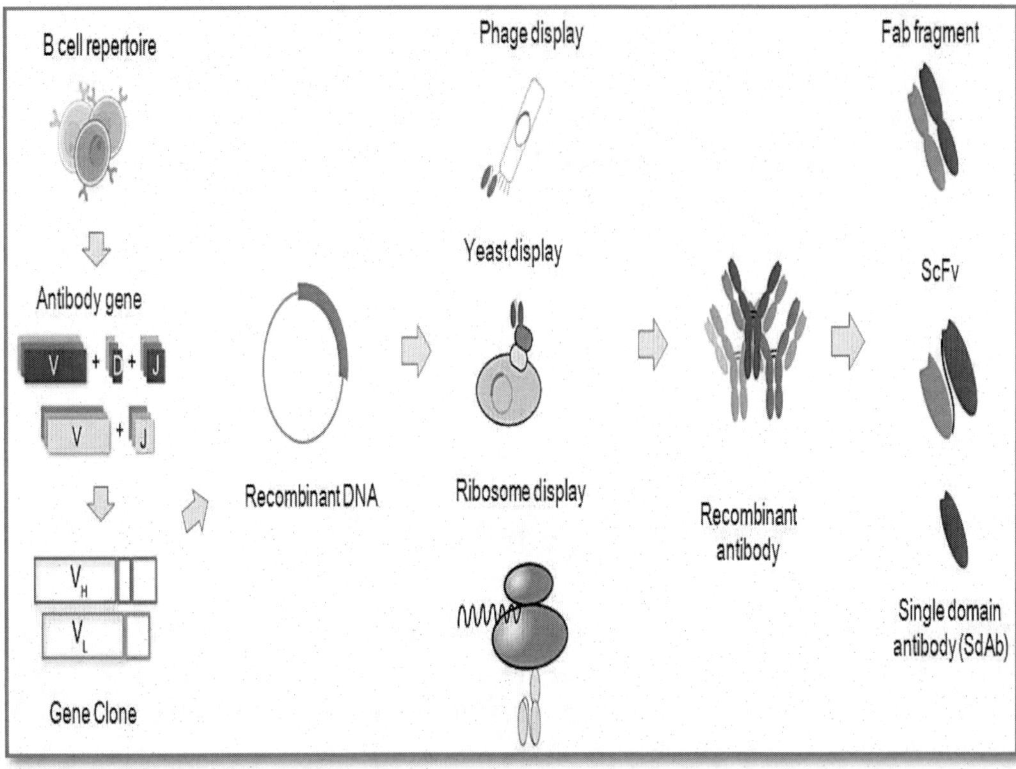

Figure 8.13. Illustrating the generation of monoclonal antibodies using phage display.

Immunogenicity involves minimising non-human sequences by creating chimeric, humanised or human versions of the antibodies with as few T-lymphocyte epitopes as possible. Antibody fragments are usually less immunogenic due to a lack of Fc domain.

Antigen-binding affinity can be improved by using phage display libraries to isolate antibodies with strong affinities for the antigen. However, sometimes antibodies with a lower affinity for the antigen may be required to allow better penetration of a tumour.

Effector functions can be improved by genetically engineering the Fc region to contain point mutations or glycan modifications. Yamane-Ohnuki and Satoh review and discuss the development of de-fucosylated antibodies which have increased affinities for the FcγRIIIa receptor and enhanced antibody-dependent cell-mediated cytotoxicity (ADCC).

A particularly interesting aspect of antibody efficacy is its unique *pharmacokinetic* characteristics once inside the body. For example, it has been noted that the bioavailability of IgG in plasma is partially dependent on its interaction with the neonatal Fc/Brambell receptor (FcRn). The FcRn functions as a salvage receptor that leads to the rescue of IgG internalised in cells from degradation by lysosomes and causes recycling of antibodies into the plasma, thus prolonging its half-life. The plasma half-life of IgG can also be increased by developing antibodies (e.g., through phage display) with increased affinity for FcRn. Antibody fragments treated with polyethylenegycol (PEGylation) have also been shown to have an increased plasma half-life.

Advantages of Monoclonal Antibodies

Monoclonal antibodies truly represent a homogeneous state of a single molecular species. Each mAb is specific to a given antigenic determinant. This is in contrast to the conventional antiserum that contains polyclonal antibodies. Following are some of the applications of monoclonal antibodies.

Diagnostic Tests

Once monoclonal antibodies for a given substance have been produced, they can be used to detect the presence of this substance. Proteins can be detected using the Western blotting technique and immuno dot blot tests. In immunohistochemistry, monoclonal antibodies can be used to detect antigens in fixed tissue sections, and similarly, immunofluorescence can be used to detect a substance in either frozen tissue section or live cells.

Analytic and Chemical Uses

Antibodies can also be used to purify their target compounds from mixtures, using the method of immunoprecipitation.

Cancer Treatment

One possible treatment for cancer involves monoclonal antibodies that bind only to cancer cell-specific antigens and induce an immune response against the target cancer cell. Such mAbs can be modified for delivery of a toxin, radioisotope, cytokine or other active conjugate or to design bispecific antibodies that can bind with their Fab regions both to target antigen and to a conjugate or effector cell. Every intact antibody can bind to cell receptors or other proteins with its Fc region.

Autoimmune Diseases

Monoclonal antibodies used for autoimmune diseases include infliximab and adalimumab, which are effective in rheumatoid arthritis, Crohn's disease, ulcerative colitis and ankylosing spondylitis by their ability to bind to and inhibit TNF-α. Basiliximab and daclizumab inhibit IL-2 on activated T cells and thereby help prevent acute rejection of kidney transplants. Omalizumab inhibits human immunoglobulin E (IgE) and is useful in treating moderate-to-severe allergic asthma.

Limitations of Monoclonal Antibodies

Hybridoma technology is laborious and time consuming. MAbs are produced against a single antigenic determinant; therefore, they cannot differentiate the molecule as a whole. Sometimes, they may be incapable of distinguishing groups of different molecules also.

The presence of retroviruses as a part of the mammalian chromosomes is a common occurrence. Mice used in MAb production carry several viruses (adenovirus, hepatic virus, retrovirus, reovirus, cytomegalovirus, thymic virus). The presence of some of these viruses has been detected in the hybridomas.

This poses a great danger, since there is no guarantee that MAb produced is totally virus-free, despite the purification. For this reason, US Food and Drug Administration insists that MAb for human use should be totally free from all pathogenic organisms, including viruses.

Commercial Issues Surrounding Monoclonal Antibodies

Some of the relative lack of commercial success of monoclonal antibodies may be attributed to the high costs of its administration. For example, in leukaemia treatment, it costs approximately £37,000 for a year's supply of alemtuzumab. Furthermore, in cancer treatment, antibodies are rarely, if ever, curative. For example, bevacizumab (avastin) only extends median survival times by 30% (20.3 vs. 15.6 months compared with standard chemotherapy). Side-effects include first-infusion reactions (e.g., fever-like symptoms). Thus, issues surrounding the cost of administration, a need for greater clinical efficacy and side-effects still need to be addressed in order for monoclonal antibodies to become more commercially viable. Other factors that will influence the growth in commercialisation of monoclonal antibodies in the long-term include: quality control, patient compliance, a competitive pharmaceutical market and incentives for antibody development (e.g., infrastructure reimbursement programmes, etc.).

CONCLUSION

Hybridoma technology is the method in which a large quantity of identical antibodies are produced. It is done by the administration of antigen in mice which produces an immune response. The B-cells producing antibodies are then harvested from the injected mouse. The harvested B-cells are then fused with B cancer cells which remain immortal. This produces hybrid cell line called hybridoma which possesses the antibody-producing ability of the B-cell. The hybridomas can be grown in culture with one viable cell which produces cultures having genetically identical hybridomas.

FREQUENTLY ASKED QUESTIONS

Q1. What is Monoclonal antibody? How is it different from a polyclonal antibody?

Answer: Monoclonal antibody (MAb) is a single type of antibody that is directed against a specific antigenic determinant (epitope). Polyclonal antibodies consist of many (poly) different antibodies that bind to many different epitopes on the target antigen.

Q2. What is a hybridoma?

Answer: Hybridoma is a somatic hybrid cell which is produced as a result of fusion of the myeloma cell and an antibody producing plasma cell.

Q3. What do you understand by hybridoma technology?

Answer: Hybridoma technology is the method of producing hybrid cells by fusing an antibody secreting lymphocyte cell with a cancerous cell and growing in a specific medium.

Q4. Who discovered Hybridoma technology? Describe the principle behind this technology?

Answer: Hybridoma technology was developed by Kohler and Milstein during 1975. The hybridoma cell has the capacity of antibody production derived from B cells and at the same time it can divide continuously, a quality of myeloma cells. By combining the desired qualities of both myeloma cells and B cells, hybridoma technology ensures large scale production of single specificity antibodies called as monoclonal antibodies.

Q5. How are monoclonal antibodies produced?

Answer: Monoclonal antibody generation begins in exactly the same manner as polyclonal antibody generation, with the creation of a robust immune response. However, rather than collecting host serum to recover a population of polyclonal antibodies, monoclonal antibody generation requires collection of the cells that make antibodies, lymphocytes. Once harvested, lymphocytes are immortalized, established as clonal through limiting dilutions, screened for appropriate expression. The process involves the following steps:

Step 1: Immunization of Mice and Selection of Mouse Donors for Generation of Hybridoma Cells

Step 2: Screening of Mice for Antibody Production

Step 3: Preparation of Myeloma Cells

Step 4: Fusion of Myeloma Cells with Immune Spleen Cells

Step 5: Cloning of Hybridoma Cell Lines by Limiting Dilution or Expansion and Stabilization of Clones by Ascites Production

Q6. What is HAT medium?

Answer: HAT Medium (hypoxanthine-aminopterin-thymidine medium) is a selection medium for mammalian cell culture, which relies on the combination of aminopterin, a drug that acts as a powerful folate metabolism inhibitor by inhibiting dihydrofolate reductase, with hypoxanthine (a purine derivative) and thymidine (a deoxynucleoside) which are intermediates in DNA synthesis.

Q7. What is the use of HAT medium in Hybridoma technology?

Answer: HAT medium is used for the selection of hybrid cells. It is used to isolate monoclonal-antibody-producing cells, i.e., hybrids made from fusion of myeloma cell mutants that lack hypoxanthine-guanine phosphoribosyl transferase, and consequently cannot utilize hypoxanthine, with spleen cells, which do not proliferate in vitro.

Q8. What are a few advantages of monoclonal antibodies?

Answer: The key advantages of MAbs are listed below:

1. MAbs are homogenous and consistent.
2. They can be renewably generated once a suitable hybridoma is developed.
3. The purity and concentration of a specific antibody is higher in MAbs as compared to polyclonal antibodies.
4. MAbs are highly sensitive to small changes in both salt concentration and pH.
5. They can be easily tested for cross-reactivity.

Q9. Give a few disadvantages of mAbs.

Answer: Some disadvantages of MAbs are listed below:

1. MAbs' mono-specificity also limits their applications
2. Minor changes in antigen epitope structure affect the function of MAbs
3. MAb production should be very specific to the antigen to which it needs to bind.
4. They are not suitable for use in assays such as hemagglutination involving antigen cross-linking; slight modifications affect the binding site of the antibody
5. Though these limitations can be overcome by pooling in multiple MAbs of required specificities, the identification of such MAbs can prove expensive, laborious, and time consuming.

Q10. List a few applications of mAbs.

Answer: The use of monoclonal antibodies to treat diseases is called immunotherapy therapy because each type of monoclonal antibody will target a specific targeted antigen in the body. Uses for monoclonal antibodies includes treatment of:

1. Cancer
2. Infectious diseases
3. Hormonal disorders etc.
4. Mabs are also used in Biochemical Analysis, Diagnostic Imaging etc.

Q11. What do you understand by passive immunotherapy?

Answer: Passive immunotherapy strategies do not rely on the human body's immune system to fight diseases as they don't require an active response of the immune system to foreign body cells. Passive immunotherapy includes the administration of immune system components, such as antibodies, to target foreign body cells like cancer cells.

Monoclonal antibodies (mAb's) still represent the largest class of commercialized cancer immunotherapies and are directed to a single target on a cancer cell. This type of therapy is therefore considered targeted passive immunotherapy.

Q12. How are Mabs used in passive immune therapy?

Answer: Antibodies are a key component of the adaptive immune response. Once produced by the immune system, they can bind to a target antigen or a receptor site on cell surfaces, including those on cancer cells. After binding to the cancer cell, the immune response is stimulated to recruit immune cells that can lyse the cancer cells.

Passive immunotherapies in the form of antibody treatment usually involve the following steps:

1. Monoclonal antibody production in the lab
2. Immunization of animals by antigen injection
3. Antibody-forming cells are isolated from the spleen
4. Formation of hybridomas of antibody-forming cells with cultured tumor cells

5. Hybridoma's are screened for antibody production and the best is selected
6. Hybridomas are cloned and expanded to produce mAB
7. Antibody therapy given to the patient
8. The antibody will work as a binder between the cancer cell and the immune cell
9. The immune cell will excrete granzymes and other proteins that will kill the cancer cell

Q13. Describe what "a humanised monoclonal antibody" is?

Answer: Humanized antibodies are antibodies from non-human species whose protein sequences have been modified to increase their similarity to antibody variants produced naturally in humans. The process of "humanization" is usually applied to monoclonal antibodies developed for administration to humans (for example, antibodies developed as anti-cancer drugs). Humanization can be necessary when the process of developing a specific antibody involves generation in a non-human immune system (such as that in mice). The protein sequences of antibodies produced in this way are partially distinct from homologous antibodies occurring naturally in humans, and are therefore potentially immunogenic when administered to human patients

Q14. Name 3 FDA approved monoclonal antibodies and their uses.

Answer:

1. Bezlotoxumab sold under the trade name Zinplava™. Manufactured by Merck Sharp & Dohme Limited targets C. difficile human toxin B. It is used to treat Enterocolitis; pseudomembranous.
2. Olaratumab sold under the trade name Lartruvo is a Mab used against sarcoma. It is manufactured by Eli Lilly.
3. Rituximab sold under the trade names Rituxan® MabThera®, manufactured by Biogen Idec, Genentech (Roche) is a Chimeric IgG1 that is used against Non-Hodgkin's lymphoma; chronic lymphocytic leukemia; rheumatoid arthritis.

Q15. What are chimeric monoclonal antibodies?

Answer: A chimeric antibody is an antibody made by combining genetic material from a nonhuman source, like a mouse, with genetic material from a human being. Chimeric antibodies are generally around two thirds human, reducing the risk of a reaction to foreign antibodies from a non-human animal when they are used in therapeutic treatments. Work on the development of chimeric antibodies began in the 1980s. A chimeric antibody contains antibodies developed with animal cells in culture, with sections of the genetic code replaced with human genes in order to address concerns about a potential reaction with the animal's genetic material.

Multiple Choice Questions

Q1. Hybridoma technology is a new biotechnological approach for commercial production of:

A. Monoclonal antibodies
B. Interferons
C. Alcohol
D. Antibiotics

Q2. Which of the following is incorrect?

A. MAbs can be used to diagnose or treat diseases
B. MAbs can be used in tumour scanning
C. MAbs can be used in cancer diagnosis
D. MAbs cannot be used in such non-infectious diseases as those of endocrine system

Q3. The hybrid cells can be propagated

A. In tissue culture
B. As ascites in peritoneal cavity of mice
C. Both (a) and (b)
D. None of these

Q4. In monoclonal antibody technology, tumour cells that can replicate endlessly are fused with mammalian cells that produce an antibody. The result of this cell fusion is a

A. Hybridoma
B. Myeloma
C. Natural killer cell
D. Lymphoblast

Q5. Monoclonal antibodies are:

A. Heterogeneous antibodies produced from a single clone of plasma cells
B. Homogeneous antibodies produced from single clone of plasma cells
C. Both
D. None

Q6. mAbs are:

A. Specific towards a paratope
B. Specific towards an epitope
C. Specific towards an antigen
D. None of the above

Q7. All are Mabs except

A. Rituximab
B. Transtuzumab
C. Infliximab
D. Tamoxifen

Q8. In Hybridoma technology, hybrid cells are selected in:

A. MS medium
B. HAT medium
C. X-gal medium
D. Whites medium

Q9. Which of the following cells is made deficient of HGPRT

A. B cells
B. Hybrid cells
C. Myeloma cells
D. None of these

Q10. HGPRT stands for

A. Hypoxanthine-Guanine Phosphoribosyl Transferase
B. Hypoxanthine-Guanine Phosphorous Transferase
C. Hypoxanthine-Guanine Phosphoribosyl Transductase
D. Hypoxanthine-Guanosyl Phosphoryl Transferase

Q11. Which of the following statements is incorrect regarding HAT medium?

A. HAT medium is a selective medium
B. Aminopterin in the HAT medium blocks de novo pathway of nucleotide synthesis
C. Salvage pathway requires aminopterin and thymidine
D. Hypoxanthine is converted to guanine by HGPRT

Q12. What is the first step in producing monoclonal antibodies?

A. Injecting a mouse with an antibody
B. Injecting a mouse with an antigen
C. Removing spleen cells from a mouse
D. Inducing tumour formation

Q13. What type of cells are myeloma cells?

A. Cancerous white blood cells
B. Cancerous spleen cells
C. Cancerous red blood cells
D. Nerve cells

Q14. What cells are fused to make hybridoma cells?

A. A spleen cell and a myeloma cell
B. B cells and T cells
C. Spleen cell and a lymphocyte
D. A nerve cell and a lymphocyte

Q15. The most direct source of monoclonal antibodies are ________ cells.

A. T cells
B. Myeloma
C. Hybridoma
D. Spleen

Answer Key

1(A), 2(D), 3(C), 4(A), 5(B), 6(B), 7(D), 8(B), 9(C), 10(A), 11(C), 12(B), 13(A), 14(A), 15(C)

Assignments

Q1. Illustrate all the typical steps that are followed in any monoclonal antibody production.
Q2. Monoclonal antibodies are of great commercial and medical value. Explain.
Q3. Write in brief about the history of Mabs
Q4. What are some of the fields in which monoclonal antibodies are being used?
Q5. How has hybridoma technology helped us in treating cancer?

References

A Glukhova X, V Prusakova O, A Trizna J, M Zaripov M, Afanas128 V, S Glukhov A, A Poltavtseva R, A Ivanov A, Avila-Rodriguez M, E Barreto G, Aliev G. Updates on the production of therapeutic antibodies using human hybridoma technique. *Current Pharmaceutical Design*. 2016 Feb 1; 22(7):870-8.

An Z. Monoclonal antibodies—a proven and rapidly expanding therapeutic modality for human diseases. *Protein & Cell.* 2010 Apr 1; 1(4):319-30.

Bartal AH, Hardhat Y, editors. *Methods of Hybridoma Formation.* Springer Science & Business Media; 2012 Dec 6.

Bartal AH, Hirshaut Y. Enhancement of Hybridoma Formation. In *Methods of Hybridoma Formation* 1987 (pp. 151-161). Humana Press.

Birch JR, Racher AJ. Antibody production. *Advanced Drug Delivery Reviews.* 2006 Aug 7; 58(5-6):671-85.

Butler, M., 2004. *Animal Cell Culture and Technology*. Taylor & Francis.

Campbell AM. *Monoclonal Antibody Technology: the Production and Characterization of Rodent and Human Hybridomas*. Elsevier; 2000 Apr 1.

Conroy PJ, Hearty S, Leonard P, O'Kennedy RJ. Antibody production, design and use for biosensor-based applications. In *Seminars in Cell & Developmental Biology* 2009 Feb 1 (Vol. 20, No. 1, pp. 10-26). Academic Press.

Croce CM. *What's New in Hybridoma Technology?*

Cuello AC, Milstein C. What's new: Hybridoma technology in immunocytochemistry. *Bio Essays*. 1984 Oct; 1(4):178-9.

Dangl JL, Herzenberg LA. Selection of hybridomas and hybridoma variants using the fluorescence activated cell sorter. *J. Immunol. Methods.* 1982 Jan 1; 52(1):1-4.

Dodge TC, Hu WS. Growth of hybridoma cells under different agitation conditions. *Biotechnology Letters.* 1986 Oct 1; 8(10):683-6.

Epstein N, Kobiler D, Epstein M. The hybridoma technology: II. Applications of hybrid cell products: monoclonal antibodies and lymphokines. *Advances in Biotechnological Processes*. 1986; 6:219.

Falkenberg FW. Monoclonal antibody production: problems and solutions. *Research in Immunology*. 1998 Jul 1; 149(6):542-7.

French D, Fischberg E, Buhl S, Scharff MD. The production of more useful monoclonal antibodies I. Modifications of the basic technology. *Immunology Today*. 1986 Nov 1; 7(11):344-6.

Gamble H. Application of hybridoma technology to problems in the. *Biotechnology for Solving Agricultural Problems*. 2012 Dec 6; 10:39.

Ganguly S, Wachau R. Hybridoma technology: a brief review on its diagnostic and clinical significance. *Pharm. Biol. Evaluations*. 2016; 3(6):554-5.

Goldsby RA, Srikumaran S, Arulanandam A, Hague B, De Leon FP, Sevoian M, Guidry AJ. The application of hybridoma technology to the study of bovine immunoglobulins. *Veterinary Immunology and Immunopathology.* 1987 Dec 1; 17(1-4):25-35.

Goldsby RA. Hybridoma Technology: The Golden Age and Beyond. In Biotechnology and Food Quality: *Proceedings of the First International Symposium* 2013 Oct 22 (p. 253). Elsevier.

Gorny MK. Human hybridoma technology. *Antibody Technol J.* 2012; 2:1-5.

Hall AK. *Monoclonal Antibodies at Work: Hybridoma Technology in the Biosciences and Medicine* Edited by TA Springer. New York: Plenum Press. (1985). 602 pp. $75.00.

Harbour C, Barford JP, Low KS. Process development for hybridoma cells. InBioprocesses Including *Animal Cell Culture* 1988 (pp. 1-40). Springer, Berlin, Heidelberg.

Karkare SB, Phillips PG, Burke DH, Dean RC. *Continuous Production of Monoclonal Antibodies by Chemostatic and Immobilized Hybridoma Culture*. Academic Press: New York; 1985 Jan 1.

Kaur H, Richardson E, Murty L. Preparation of monoclonal antibodies against human telomerase. *Hybridoma*. 2001 Jun 1; 20(3):183-8.

Kennett RH, Bechtol KB. Hybridoma Technology. (Book Reviews: Monoclonal Antibodies). *Science*. 1980 Oct; 210:317.

Kim SJ, Park Y, Hong HJ. Antibody engineering for the development of therapeutic antibodies. *Molecules & Cells (Springer Science & Business Media BV)*. 2005 Aug 1;20(1).

Lindl T. Development of human monoclonal antibodies: *A review. Cytotechnology*. 1996 Jan 1; 21(3):183-93.

Ling NR. Properties of the monoclonal antibodies produced by hybridoma technology and their application to the study of diseases. *Immunology*. 1983 May; 49(1):202.

Marx U, Embleton MJ, Fischer R, Gruber FP, Hansson U, Heuer J, De Leeuw WA, Logtenberg T, Merz W, Portetelle D, Romette JL. *Monoclonal Antibody Production*. Atla-Nottingham-. 1997; 25:121-38.

Maynard J, Georgiou G. Antibody engineering. *Annual Review of Biomedical Engineering*. 2000 Aug; 2(1):339-76.

Nguyen H, Sandhu J, Hozumi N. Production of human monoclonal antibodies in SCID mouse. *Microbiology and Immunology*. 1997; 41(12):901-7.

Olsson L, Brams P. Human—Human Hybridoma Technology. In *Human Hybridomas and Monoclonal Antibodies* 1985 (pp. 227-244). Springer, Boston, MA.

Olsson L. Monoclonal antibodies in clinical immunobiology: derivation, potential, and limitations. *Allergy*. 1983 Apr; 38(3):145-54.

Osterhaus AD, Uytdehaag AG. Current developments in hybridoma technology. *Tijdschrift voor diergeneeskunde*. 1985 Oct; 110(20):835-9.

Pandey A, Takarkhede S, Gokhale S, Patil K, Srivastava S, Gupta V. *Hybridoma Technology: A New Innovation and Application*.

Pandey S. Hybridoma technology for production of monoclonal antibodies. *Hybridoma*. 2010 Mar;1(2):017.

Pörtner R, Schwabe JO, Frahm B. Evaluation of selected control strategies for fed-batch cultures of a hybridoma cell line. *Biotechnology and Applied Biochemistry*. 2004 Aug;40(1):47-55.

Randerson DH. Large-scale cultivation of hybridoma cells. *Journal of Biotechnology*. 1985 Jul 1;2(5):241-55.

Ronald, A. and Stimson, W. H., 1999. The evolution of immunoassay technology. *Parasitology*, *117*(7), pp. 13-27.

Roque AC, Lowe CR, Taipa MÂ. Antibodies and genetically engineered related molecules: production and purification. *Biotechnology Progress*. 2004 Jan 1; 20(3):639-54.

Samaranayake H, Wirth T, Schenkwein D, Räty JK, Ylä-Herttuala S. Challenges in monoclonal antibody-based therapies. *Annals of Medicine.* 2009 Jan 1; 41(5):322-31.

Schwaber J. Hybridoma Technology. *Science.* 1982 May 21; 216(4548): 798-.

Self CH, Cook DB. Advances in immunoassay technology. *Current Opinion in Biotechnology.* 1996 Feb 1;7(1):60-5.

Sell S, Wahren B, editors. *Human cancer markers.* Springer Science & Business Media; 2012 Dec 6.

Serova MA, Rybal'skiĭ NG. Hybridoma technology: review of patent information. *Zhurnal Mikrobiologii, Epidemiologii, i Immunobiologii.* 1989 Jun(6):114-24.

Sevier ED, David GS, Martinis J, Desmond WJ, Bartholomew RM, Wang R. Monoclonal antibodies in clinical immunology. *Clinical Chemistry.* 1981 Nov 1;27(11):1797-806.

Shay JW. Mechanisms of cell fusion and selection in the generation of hybridomas. *In Methods of Hybridoma Formation* 1987 (pp. 63-75). Humana Press.

Siddiqui MZ. Monoclonal antibodies as diagnostics; an appraisal. *Indian Journal of Pharmaceutical Sciences.* 2010 Jan;72(1):12.

Siegel DL. Recombinant monoclonal antibody technology. *Transfusion Clinique et Biologique.* 2002 Jan 1; 9(1):15-22.

Springer, T. ed., 2013. *Hybridoma technology in the biosciences and medicine.* Springer Science & Business Media.

Tomita M, Tsumoto K. Hybridoma technologies for antibody production. *Immunotherapy.* 2011 Mar; 3(3):371-80.

Tyagi S, Sharma PK, Kumar N, Visht S. Hybridoma technique in pharmaceutical science. *International Journal of Pharm Tech Research.* 2011 Jan; 3(1):459-63.

Wiggenhauser A, Peters JH, Baumgarten H. Preconditions for Hybridoma Technology. *In Monoclonal Antibodies* 1992 (pp. 18-38). Springer, Berlin, Heidelberg.

Yagami H, Kato H, Tsumoto K, Tomita M. Monoclonal antibodies based on hybridoma technology. *Pharmaceutical Patent Analyst.* 2013 Mar; 2(2):249-63.

In: Immunoglobulins
Editor: Manzoor Ahmad Mir ISBN: 978-1-53616-903-4

Chapter 9

RADIOIMMUNOASSAY (RIA)

Manzoor A. Mir**, *Umar Mehraj, Hina Qayoom and Safura Nisar
Department of Bioresources, School of Biological Sciences,
University of Kashmir, Srinagar J&K, India

ABSTRACT

Radioimmunoassay (RIA), a highly sensitive laboratory technique used to measure minute amounts of substances including antigens, hormones, and drugs present in the body. The substance or antigen (a foreign substance or pathogen in the body that causes antibody production by the B lymphocytes of the body) to be measured is injected into an animal, causing it to produce antibodies. Serum containing the antibodies is withdrawn and treated with a radioactive antigen and later with a nonradioactive antigen. Measurements of the amount of radioactivity are then used to determine the amount of antigen present. The technique was developed by Solomon Berson and Rosalyn Yalow. Yalow was awarded the 1977 Nobel Prize in Physiology or Medicine for her work.

Radioimmunoassay is an *in vitro* assay. It calculates the presence of an antigen with very high sensitivity. Mainly, any biological product or substance for which a specific antibody exists can be measured, even in minute concentrations. Radioimmunoassay has been the first immune-assay procedure developed to analyse picomolar and nanomolar concentrations of hormones in biological fluids. The radioimmunoassay is perhaps the oldest types of immunoassays. Here, a radioisotope is attached to an antigen of interest and bound with its complementary antibody. Then a sample with the antigen to be measured is added. It competes with the radioactive antigen, kicks it out of the binding spot and replaces it. After washing away unbound antigens the radioactivity of the sample is measured. The amount of radioactive signal is inversely related to the amount of target antigen.

Keywords: antigen, antibody, RIA, radioactivity, Ag-Ab complex

---

* Corresponding author: Dr Manzoor Ahmad Mir (Ph. D), Department of Bioresources, School of Biological Sciences, University of Kashmir, Srinagar J&K, India-190006, Contact:-+919622901319. E-mail: drmanzoor@kashmiruniversity.ac.in.

Objectives

- Understanding the concept of radioimmunoassay
- Learn about the history of RIA
- Briefly describe the principle and procedure of radioimmunoassay
- Discuss the concept of immunoassay
- Describe the principle and types of immunoassays

Radioimmunoassay

The technique of radioimmunoassay was developed by Rosalyn S. Yalow (Fig.1) and Late Salamon A. Berson (Fig. 2) during the late 1960s and it represents a milestone in the history of the application of Radio nuclide methodology to biology and to medical investigation and practice. In 1977, some years after Berson's death, the significance of the technique was acknowledged by the award of a Nobel prize to Yalow.

While attempting to study the metabolism of radioactivity labelled insulin, Berson, Yalow and their colleagues noted a curious phenomenon. Individuals who had received insulin treatment displayed slower disappearance of labelled insulin than did person who had not been exposed to insulin injection.

They considered that this was the result of binding to the antibodies produced by prior exposure to insulin. They subsequently developed highly sensitive radio-isotopic technique to detect the insulin-Ab complex.

Figure 1. Picture of Rosalyn Yalow (Noble Prize 1977).

Figure 2. Image of Salamon A. Berson.

The culmination of their effort was the development of the RIA. It has occupied a key position in biological science. Its continuing value depends partly on the high sensitivity, specificity and ease of performance and partly on the wide applicability for elimination of various biomolecules in biological fluids and tissue. Major gain for development of RIA have been given to the discipline of endocrinology. For the discipline of endocrinology, RIA led to the easy detection in molecular anomalies and exploration of a number of pathways involved in the homeostasis of human beings.

What Is RIA?

Radioimmunoassay belongs to the class of technique called, "Binder Assay". RIA involves the combination of two substances-ligand, the substance to be measured and binder (the substance which binds it). The quantitation of ligand is carried out by permitting the ligand to bind the binder and subsequently determining its distribution between bound and free phase. For quantitation of the ligand in free and bound phases a small amount of ligand is labelled with a radioactive isotope and is incorporated in the reaction mixture. RIA is based upon the competition between labelled and unlabeled antigen for specific antibody sites forming antigen, antibody complexes.

Principle and Procedure of RIA

Radioimmunoassay combines the principles of radioactivity of isotopes and immunological reactions of antigen and antibody, hence the name. The principle of RIA is primarily based on the competition between the labelled and unlabeled antigens to bind with antibody to form antigen-antibody complexes (either labelled or unlabelled). The unlabeled antigen is the substance (say some hormone) to be determined. The antibody to it is produced by injecting the antigen to a goat or a rabbit. The specific antibody (Ab) is then subjected to react with unlabeled antigen in the presence of excess amounts of isotopically labelled (^{125}I) antigen (Ag^{+}) with known radioactivity. There occurs a competition between the antigens (Ag^{+} and Ag) to bind the antibody. Certainly, the labelled Ag^{+} will have an upper hand due to its excess presence.

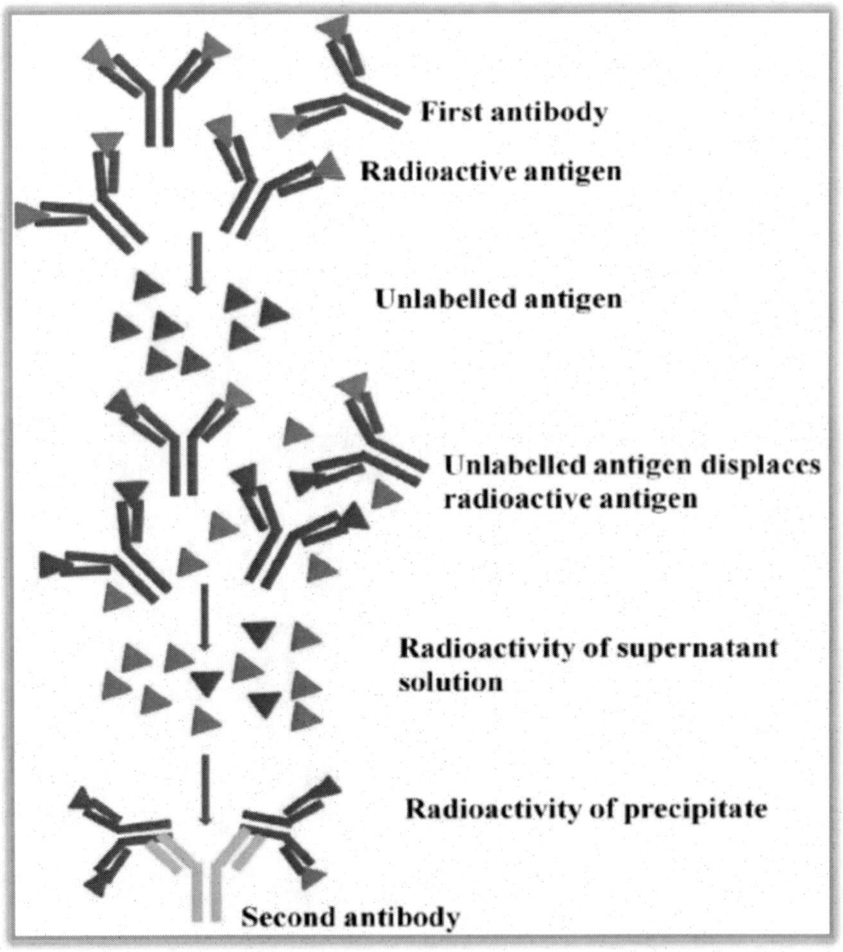

Figure 3. Diagram illustrating principle of RIA.

As the concentration of unlabeled antigen (Ag) increases the amount of labelled antigen-antibody complex (Ag^+-Ab) decreases. Thus, the concentration of Ag^+- Ab is inversely related to the concentration of unlabeled Ag i.e., the substance to be determined. This relation is almost linear. The labelled antigen-antibody (Ag^+-Ab) complex is separated by precipitation. The radioactivity of ^{125}I present in Ag^+-Ab is determined. To determine the amount of labeled antigen bound, the Ag-Ab complex is precipitated to separate it from free antigen (antigen not bound to Ab), and the radioactivity in the precipitate is measured (Figure 3). A standard curve can be generated using unlabeled antigen samples of known concentration (in place of the test sample), and from this plot the amount of antigen in the test mixture may be precisely determined (Figure 4).

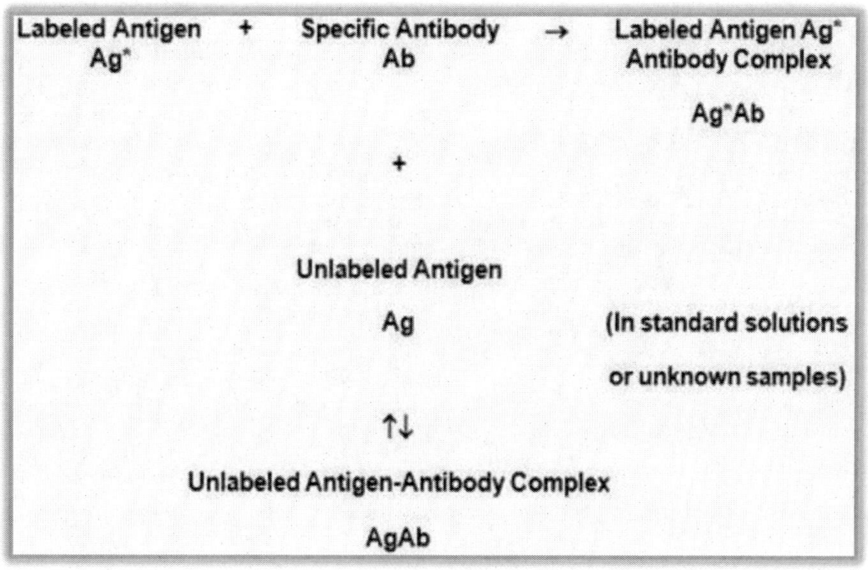

Figure 4. Basic reaction on which RIA is based. (Source: https://www. perkinelmer.com).

Several methods have been developed for separating the bound antigen from the free antigen in RIA (Figure 5). One method involves precipitating the Ag-Ab complex with a secondary anti-isotype antiserum. For example, if the Ag-Ab complex contains rabbit IgG antibody, then goat anti-rabbit IgG will bind to the rabbit IgG and precipitate the complex. Another method makes use of the fact that protein A of Staphylococcus aureus has high affinity for IgG. If the Ag-Ab complex contains an IgG antibody, the complex can be precipitated by mixing with formalin-killed S. aureus. After removal of the complex by either of these methods, the amount of free labeled antigen remaining in the supernatant can be measured in a radiation counter; subtracting this value from the total amount of labeled antigen added yields the amount of labeled antigen bound. Various solid-phase RIAs have been developed that make it easier to separate the Ag-Ab complex from the unbound antigen. In some cases, the antibody is covalently crosslinked to sepharose beads.

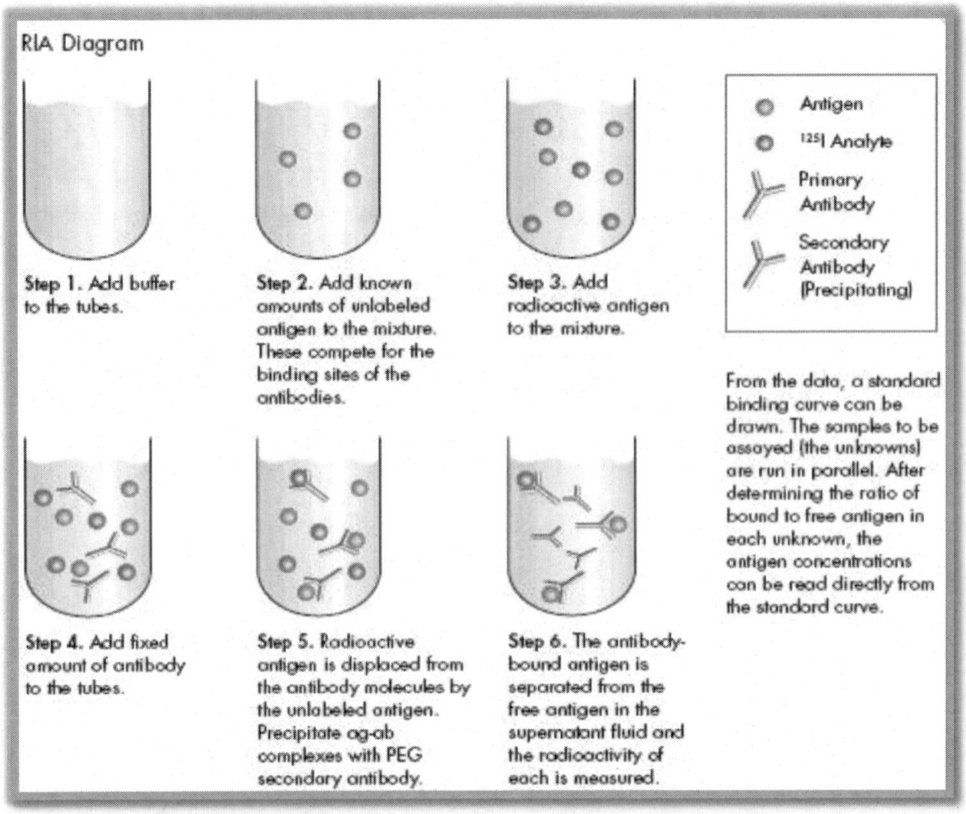

Figure 5. Showing several steps in RIA (Image credit: Georgia state university).

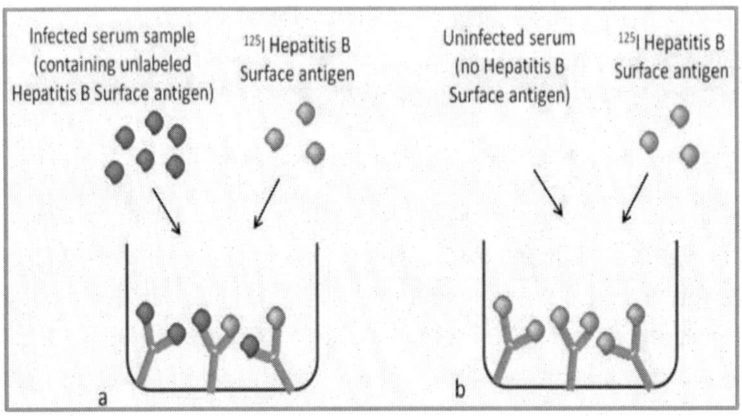

Figure 6. a) Sample containing a high amount of antigen. The unlabelled antigen competes for binding to the antibody in the tube or well. b) Sample containing no or low amounts of antigen. Antibody is bound by the radiolabeled antigen (tracer). (Source: https: //www.perkinelmer.com).

The amount of radiolabeled antigen bound to the beads can be measured after the beads have been centrifuged and washed. Alternatively, the antibody can be immobilized on polystyrene or polyvinylchloride wells and the amount of free labeled antigen in the supernatant can be determined in a radiation counter. In another approach, the antibody is immobilized on the walls of microtiter wells and the amount of bound antigen is determined. Because the procedure requires only small amounts of sample and can be conducted in small 96-well microtiter plates (slightly larger than a 3×5 card), this procedure is well suited for determining the concentration of a particular antigen in large numbers of samples. For example, a microtiter RIA has been widely used to screen for the presence of the hepatitis B virus (Figure 6). RIA screening of donor blood has sharply reduced the incidence of hepatitis B infections in recipients of blood transfusions.

Methodology of the RIA

This technique mainly is divided into two parts. Part I deals with components and is subdivided into four sections (primarily because there are four basic steps in setting up an RIA procedure): a) sample or standard; b) label or labeled antigen; c) primary binder; and d) separation step. Part II deals with the reactions occurring in the test tube when these components from Part I are combined. The principles to be discussed are RIA, competitive protein binding (CPB), sequential saturation analysis, immunoradiometric assays (IRMA), and radiolabeled binding assays.

PART I-THE COMPONENTS

Sample or Standard-More often than not, the first pipetting step in any RIA procedure is to add either the patient sample or standards into their respective test tubes. Each will be discussed individually.

Sample

The sample is the unknown and in virtually all radio-assay procedures, with rare exception, the only variable. It shares the term of unlabeled antigen with the standard, the difference of course being that the standard has known concentrations whereas the sample concentration is unknown. Just about any sample type can be analyzed in an RIA procedure, but most kits are generally designed for serum or plasma only. One must be aware of the protein dependency of the kit's components. Not just any specimen type may be analyzed. It must contain a protein concentration similar to that found in serum. Other biological fluids on which radio-assay procedures are frequently ordered are urine, spinal fluid, and most body fluids. There are, of course, the receptor assays on tissue biopsies and actual tumor tissue. In routine RIA procedures, spinal fluid and urine specimens are frequently submitted for analysis. These are major problems because their protein content is roughly 300-1,000 times less than that of serum. One approach to compensate for the protein content of these specimens is to prepare a dilution of the body fluid, generally 1:2, using the blank serum or a pooled serum. Blank serum means that it does not contain any of the constituent being measured. The result will need to be multiplied by two. Most reagent or kit manufacturers have not established reference ranges (normal) for body fluids other than for serum or plasma. The exception, of course, would be if the kit is designed specifically for measurement of a specific body fluid. A myelin basic protein kit is designed for cerebrospinal fluid only.

Standard

A primary standard is prepared from a constituent that can be extracted and purified from a tissue source or can be synthesized. It is a standard that can be weighed out and an appropriate dilution made and used in any assay. Some primary standards that are readily available in purified form (and relatively inexpensive) are: the thyroid hormones, T3 and T4, cortisol, and other adrenal and most gonadal steroids. A secondary standard is used when the primary standard needed is not easily synthesized or purified from tissue extracts and not readily available. It generally consists of a pool containing the desired analyte which is analyzed against the primary standard. Then, the value that is obtained from the primary standard curve is assigned to that pool. This pool then becomes a secondary standard and is typical of what is included in most kits. Examples of compounds that are not readily available in purified form, against which the secondary standards are calibrated, include the peptide hormones: FSH, ACTH, TSH, HCG, prolactin, and growth hormone. Kit manufacturers will maintain a supply of this primary standard by which they can calibrate their secondary standards. Most kits of this type contain secondary standards. When discussing standards, the matrix should be considered. Matrix refers to the diluent used to dilute the standard. Standards, whether primary or secondary, are diluted in some form of protein base or matrix. (Most RIA reactions are protein dependent.) Ideally, the matrix to use would be that of human serum for assays where direct pipetting is done from an unextracted sample. This matrix must be relatively free of the analyte to be measured. Human-free serum is frequently prepared by literally dumping Norit charcoal into a hepatitis free serum pool. It is mixed overnight, followed by filtration of the pool to remove the charcoal. This process, referred to as "stripping," will remove most components or analytes or effectively reduce their concentration so that they may be used as a free diluent or a free serum. The most common serum matrix is human serum. Second to this are human serum components, such as human serum albumin. Other popular, but less desirable, matrices are bovine serum albumin, horse serum, and bovine gamma globulin. The matrix may simply be a buffer. What is important is that the matrix closely approximate the binding characteristics, constituents, and pH of the patient sample. As an example, for the assays that require an extraction step and reconstitution in a buffer, the standards will be diluted in that same buffer. If the matrix does not approximate that of the patient's sample, an accuracy problem may arise. The concentration of the standard may not bind the same as that in the patient's sample. Thus, the level of the patient's serum may not be as accurate as it should be. Manufacturers, however, will generally compensate for this problem by adjusting the concentration of the standards to correspond to expected patient sample ranges. This could wreak havoc with an assay by changing the matrix they use. This change can markedly affect the accuracy of the analyte being measured, and more than likely result in a shift in quality control ranges.

Label

In an RIA technique, a label has many synonymous terms such as tracer, trace, labeled antigen, tagged material, the tag, the hot stuff, etc. Generally, it is a purified form of the analyte that one attempts to measure in the patient's sample. The difference is that it has gone through a labeling or tagging process by which a radioactive molecule has been attached. There are numerous methods available for protein purification prior to tagging and/or immunization. As was mentioned earlier in the section on standards, steroids can be obtained in relatively pure form and generally do not require further purification prior to immunization or labeling. One of the more frequently used isotopes for labeling in a clinical laboratory is iodine-125. Tritium (H-3) is less frequently used. Cobalt-57 was at one time used exclusively in the vitamin B-12 assays, since the cyanocobalamine molecule, B-12, contains a cobalt molecule in which it is easy to substitute a cobalt-57 atom. Recent advances in dual labeled kits, such as FSH/LH, utilizes Co-57 attached to LH and 1-125 attached to FSH.

Iodination

Iodination of various compounds used in an RIA laboratory falls under two generalized categories. One is a direct iodination, which requires an oxidation reduction reaction that attaches the iodine-125 directly onto an aromatic group of the molecule. The second method, indirect iodination, attaches the iodine-125 to the aromatic group of a hapten or ligand such as tyrosine, and then this ligand is attached onto the analyte to be tagged. All methods for iodination require purification of the newly tagged material by removing from the mixture any untagged antigen and all of the free iodine that did not tag to the analyte. There are a number of procedures that can be used to purify the newly labeled compound, such as Sephadex filtration, chromatography, electrophoresis, and dialysis.

Tritiation

Most laboratories do not do their own tritiated labeling primarily because of the commercial availability of low cost, prelabeled tritiated compounds, mostly steroids. The labeling is accomplished by a reduction reaction involving the antigen (steroid) to be tagged with tritium gas. One aspect that must be considered with respect to purchasing labeled compounds, especially tritiated steroids, is that even though H-3 has a 12.3 yr half-life, it does not mean that the integrity of the tagging mechanism is stable for 12.3 yr on the shelf. It is always recommended that the purity of the tritiated steroid be confirmed by chromatography prior to use. Manufacturers of H-3 steroids always report the radiochemical purity at the date of manufacture and also indicate the rate of degradation. In addition, they periodically confirm the purity every 3 to 6 mo. Degradation occurs because the H-3 (pure beta particle) is in direct contact with the steroid to which it is tagged, and some radiation damage is directed to the molecule itself. With time, the H-3 will become detached from the steroid compound to which it is labeled. By performing the purity quality control on a regular basis, it has been the experience of this investigator that problems with the tracer will be detected before manufacturer notification. Further, one can monitor the rate the label is coming off of the steroid and determine when to purchase freshly tagged steroid. For some assays, such as free testosterones, the purity of the tritiated compound is checked weekly. Fresh H-3 steroids are routinely obtained at least every 6 mo.

Advantages of Iodinated Over Tritiated Labels

- Less of the purified antigen needs to be labeled and less antigen mass improves the sensitivity of an assay.
- Higher specific activities of radio iodinated steroids can be produced since more radioactivity per molecule of antigen is obtained.
- Decreased mass of antigen in the assay requires less antiserum when compared to tritiated steroids.
- Higher specific activity can also be used, thus permitting higher counts and perhaps shorter counting times.
- The ease of handling is enhanced because, for most assays, it is not necessary to decant into another tube (or vial).
- There is no need to handle hazardous organic solvents associated with liquid scintillation cocktail.
- The need for liquid scintillation quench correction is obviated with iodinated sample counting.

One disadvantage of using an Iodine-125label is, of course, its short half life (59.7 days) as compared to the half-life of 12.3 yr for tritium. As a result, radiolabeling is done more frequently when iodine-125 is used as the tag.

Purified Antigen for Labeling

The ideal labeled antigen is the same purified antigen that was used in the immunization process to produce the specific antibodies. When the antigens are identical, the RIA kinetics of the system will have the highest affinity for the antibody sites of any antigen available. This "ideal" situation is, of course, extremely rare. An unexpected low binding may be caused by the fact that the affinity of the antibody for the labeled antigen is less than acceptable. One cause could be the damage that has occurred to the antigen during tagging. Traumatization during tagging can structurally change the sites the antigen once had for the antibody. This must be considered when binding problems occur, especially with a fresh tag. The quality of the iodine used in tagging is also important. Even kit manufacturers occasionally encounter bad batches of iodine. The initial tag may be fine, but may deteriorate within 24 to 48 hr. During the assay, it will appear as though not enough tracer is binding to the antibody (low counts). Alternatively, it could also effectively "alter" the standard concentration. Because the iodine-125 is no longer attached to its respective antigen, the antigen is now "cold" or an unlabeled antigen exactly like the standard.

Primary Binder

In RIA procedures, the primary binder is a specific antibody. This antibody has antigenic sites specifically for the antigen or the analyte to be measured. Production of antibodies requires immunization of a particular animal with the purified immunogen. The more commonly used animals are rabbits, sheep, goats, monkeys, mice, rats, chickens, and donkeys. The type of antibody to be produced will govern what kind of immunization schedule is best. Not all but most, immunizations require that a second or "booster" injection follow from 4 to 6 wk after the initial injection so that an adequate titer of antibody is achieved. The RIA technique is used to determine the presence and/or titer of an antibody. A radioactive tracer for the analyte measurements is added to varying dilutions of antibody to see if the tracer binds. Then, determination as to what dilution of the antiserum will bind to approximately 50% of the working tracer used in the assay can be made. In a conventional RIA procedure, the antibody concentration should be limited so that the antibody will be saturated by one-half of the tracer in the absence of unlabeled antigen (standard). If a higher concentration of antibody will reduce the sensitivity in the lower range of your standard curve, then the converse is also applicable that a lower concentration of antibody will reduce the sensitivity in the higher range of the standard curve (Figure 1). Thus, 50% of trace binding is optimum. To create an optimum condition for the binding of the assay, adjust the antibody titer to correspond to the preferred concentration range. Diluting the antibody will improve the sensitivity in the low range, whereas increasing the concentration of the antibody will improve the sensitivity in the high range. In the process of immunization, following the injection, the immunogen seeps out from the emulsion. The host animal's immune system considers this antigen to be foreign and will respond by producing antibodies. The spectrum of antibodies released into circulation will be capable of binding the purified antigen in the assay with varying degrees of affinity. An obtained antibody titer represents a mean titer for the diversified population of antibodies produced by the host organism. Antibodies that bind the antigen most effectively are said to have the highest affinity. Those with low affinity may have a high affinity for another segment of the antigen molecule which may resemble a similar compound, and thus may have a high affinity for another similar compound (cross reactivity). When, the antibody population consists of such an aggregate of antibodies, it is called a polyclonal antiserum.

How well the antibody detects what it is designed to detect determines the specificity of the antiserum. If it will bind with other similar compounds, it is said to cross-react with these other compounds. If the antibody has high cross-reactivity to other related compounds, it is considered to have low specificity. An antibody with high cross-reactivity to structurally related compounds may not be desirable or may require that special steps be taken in the RIA prior to using this antibody to remove cross-reacting or interfering substances from the patient's sample. An example of this would be the extraction steps that are sometimes required with certain RIA procedures and/or the chromatography (purification step of the patient sample prior to analysis). Another approach to improving the specificity of an antiserum is to supersaturate the antibodies present in the antiserum with an excessive amount of the antigen with which it cross-reacts. By doing so, all of the binding sites for this antigen are occupied prior to its use in the assay, thus eliminating any observable cross reactivity. This process is referred to as absorption of the antibody.

Recent advances in the field of antibody production have virtually eliminated antibody cross-reactivity at the level of the cell production. Monoclonal antibodies have been in use in clinical labs for almost 3 years. In brief, the production of monoclonal antibodies begins with immunization of the mouse, quite similar to the immunization of a polyclonal antibody. The spleen cells of mice with the highest titer antibodies are fused with mouse myeloma cells in the presence of "polyethylene glycol". The rapidly dividing fused myeloma spleen cells are called hybridomas. Those hybridoma cells with the desired antibody purity and specificity can then be injected into the peritoneal cavity of mice who will, in turn, produce extremely large quantities of this antibody, which accumulate in the ascites fluid of the mouse. The success rate, however, is low. Approximately five of the desired hybridoma antibodies are produced out of the 200,000 fused hybrid cells that can be produced from the billions of spleen cells from a single animal. The entire culture and selection process takes about 4 mo.

The significance of monoclonal antibodies are that they are highly specific and can be maintained in cell culture indefinitely. This advantage eliminates one of the remarkable problems in RIA, that is, antibody lot changes which frequently result in differences of affinity and specificity from lot to lot. Once the monoclonal antibody is produced and cultured, it is all but guaranteed that this same antibody will be present in all kits. Even though the lots of all the other reagents change, the antibody lot will remain the same.

Binding Proteins

The primary binder in a radiolabeled assay need not to be an antibody. For certain assays, it is simply a natural protein that is known to bind with high affinity to the analyte to be detected. Historically, a number of radiolabeled assays use protein binders instead of antibodies. Assays of this type are referred to as CPB assays.

Table 1 lists some of these procedures and their corresponding protein binders.

$$[Ag] + [Ab] \leftrightarrow [Ag\text{-}Ab] + [Ag]$$

Figure 7. This equation depicts the equilibrium reaction between antigen and antibody in a conventional RIA procedure. Ag = antigen (labeled or unlabeled); Ab = 1° antibody; Ag-Ab = antigen-antibody complex formed during the reaction; K, = the rate of formation constant, association, or affinity constant; K2 = dissociation constant. An in-depth discussion of this equation can be found in ref. 37.

Of the assays listed, only vitamin B-12 and folic acid are still widely used CPB assays. The others have been replaced by RIA techniques as antibodies for each analyte became available and as the quality of antibody was improved. Intrinsic factor is used exclusively in vitamin B-12 assays as the primary binder. *In vivo*, intrinsic factor is found in the cell membranes of the gut. In order for ingested vitamin B-12 to be absorbed from the gut into the blood stream, it must first bind to intrinsic factor for passage across the gut membrane.

Table 1. Ligand binding assays: mechanisms, advantages and disadvantages of the methods

Group of assays	Assay	Mechanism	Advantages	Disadvantages
Labeled ligand-binding assays	Fluorescent ligand binding assays	Fluorescent labeled ligand binding assay is applied to detect its binding to a target.	These assays have a broad spectrum of wavelengths, therefore it can apply multiple colors.	The assays are susceptible to different fluorescence interferences. Also, the labeling of a ligand, can lead to undesirable alterations in the binding characteristics of the ligand. In addition, it is difficult to obtain good fluorescent ligands for many receptors.
	Radioligand binding assays	Radioactively labeled ligand binding assay is applied to detect its binding to a target.	The most popular assay for membrane-bound targets. Good robustness, precise determination of receptor density and distribution, ligand binding sites and affinity.	High cost and hazards of handling high levels of radioactivity.
	Bioluminescent binding assays using nanoluciferase	Bioluminescent reporter nanoluciferase (NanoLuc) is used to study interactions of protein hormones with their receptors.	NanoLuc-based assays have several advantages: high sensitivity, high stability and reproducibility.	Since NanoLuc reporter is larger than radioisotopes, it may potentially influence the receptor binding.
Label-free ligand binding assays	Surface plasmon resonance (SPR)	SPR applies light-excited surface plasmon polaritons to track the binding of ligands to the receptors bound to a gold surface.	SPR is the most popular label-free ligand binding assay. This assay can be used to determine binding kinetics.	This assay has low sensitivity to binding-induced conformational change.
	Plasmon-waveguide resonance (PWR)	PWR applies a continuous polarized wave laser exciting electromagnetic waves in a resonator made of a thin silver film with a layer of SiO2 and a glass prism. Ligand binding changes amplitude, position and width of reflected lights.	PWR examines anisotropic optical contents of receptor-ligand complexes, thus differentiating mass density changes from conformational changes.	PWR has lower sensitivity than SPR with regard to refractive index, thickness and mass parameters.
	SPR imaging for affinity-based biosensors	This assay measures the binding kinetics and is related to intensity modulation and analyzes the reflectivity of monochromatic incident p-polarized light detected at a fixed angle.	This assay can be used to monitor binding events in real time.	This assay characterizes different compounds for a specific target, it still provides low throughput.
	Nanofluidic Fluorescence Microscopy (NFM)	NFM combines nanofluidic-based biosensors and fluorescence microscopy to analyze the protein binding kinetics.	Protein binding kinetics can be analyzed in one single experiment. The time of analysis by NFM is shorter compared to other micro-fluidic assays. Statistical errors are reduced and there are no limitations on the size of the analyzed molecules.	Limitations of fluorescence microscopy.
	Whispering gallery microresonator (WGM)	WGM is based on wave sensing binding of molecules to the surface of the cavity induces changes of the resonant wavelength changes. The resonant changes of light permit multiple analyses of molecules.	WGM is a highly sensitive assay for molecule detection.	The possibility to study drug-target interactions under non-equilibrium conditions using this assay is still unclear.

Group of assays	Assay	Mechanism	Advantages	Disadvantages
	Resonant waveguide grating (RWG)	RWG uses a nanograting to couple light into the waveguide via diffraction. The light illuminates the biosensors in microplate at a nominally normal incident angle. The drug binding of the immobilized receptors results in a shift in the resonant wavelength.	RWG assay is effective for affinity screening.	This assay is not applicable to study cells due to the large size of cells and limited penetration ability of an evanescent wave.
	Biolayer Interferometry Biosensor (BIB)	This assay uses a spectrometer to detect interference patterns formed by light reflected from an optical layer and a biolayer containing proteins of interest.	Effective assay to study binding kinetics.	This assay has been validated mostly for small molecule detection.
Structure-based ligand binding assays	Nuclear magnetic resonance (NMR)	NMR analyzes the magnetic characteristics of certain atomic nuclei, which absorb electromagnetic radiation in the magnetic field.	NMR can be applied to analyze the structure of proteins and the molecular details of protein-ligand interactions and to assist structure-based drug design. Besides, NMR can be used for any class of compounds, for almost all soluble proteins, natively unstructured proteins or membrane proteins.	The high cost of the assay, time-consuming, requires a long time to analyze obtained spectra. The resolving power of NMR is lower than X-ray crystallography.
	X-ray crystallography	This assay measures the changes of the diffracted X-ray beams in order to produce a three-dimensional image of electron density within the molecule of interest. Obtained image of electron density is used to determine the structure of the molecules.	The solvent effect can be examined by X-ray crystallography. The whole 3D structure can be obtained by the systematic analysis of a crystallized material.	Solutions and behavior of molecules in solution cannot be examined. Direct determination of secondary structures including domain movements cannot be performed.
Thermodynamic binding assays	Thermal denaturation assays (TDA)	TDA detect the thermal denaturation of proteins by differential scanning fluorimetry, which applies a prob fluorophore to monitor thermal denaturation process of proteins in the presence and ligands.	These methods have found applications in chemical profiling of different protein families, identifying novel ligands, and investigating the stability of proteins in buffers in solution	These methods may not be suitable for analysis of very small proteins that do not aggregate upon denaturation.
	Isothermal titration calorimetry (ITC)	ITC measures the binding enthalpy variation by sensing the heat caused by the binding reaction	This method is useful for detecting highly potent and entropy-driven ligands, compared to others.	This method shows low throughput and sensitivity, and requires large sample volumes.
Whole cell ligand-binding assays	Surface acoustic wave (SAW) biosensor	The surface of the biosensor is excited to oscillate with a high frequency. The phase of the generated acoustic wave is shifted upon mass changes. Conformational characteristics are indicated by a change in amplitude.	Using this assay, the sequential binding of several distinct ligands, targeting different cell surface molecules, to the whole cells can be detected directly in real time.	SAW biosensors for analysis of small molecules using living cells have not developed.
	RWG biosensor	RWG biosensor can transform drug-receptor interactions in cells into cell phenotypic signatures named dynamic mass redistribution.	This assay measures the consequence of binding, and the stimulus-response interpretation of the initial interaction of drug and receptor. Also, RWG can analyze the functional consequences of ligand binding and dissociation.	RWG biosensor may not be used to detect ligand binding directly.

Around 1979, controversy surrounding the impurity of the intrinsic factor used in vitamin B-12 kits sparked an intense study by the National Committee for Clinical Laboratory Standards (NCCLS). It was determined that the impurities in the intrinsic factor (called "R" proteins) would bind other B-12 analogs present in serum, in addition to vitamin B-12. As a result, binding this could mask the true vitamin B-12 status of a patient. All vitamin B-12 kits currently on the market contain purified intrinsic factor, or at least have the "R" protein sites blocked. Folic acid, historically, was first measured by microbiological techniques (as was vitamin B-12). This method for folate was replaced by a CPB technique after it was found that a purified protein in milk called beta lactoglobulin would act as a binder. This has been employed in folate radioassays ever since.

It is important to understand the relationships between the three components, sample, label, and primary binder, involved in the reaction mixture. In a conventional RIA procedure, the sample or standard competes with the tracer for the limited number of antibody binding sites. When these components are first introduced to each other, there is a first order reaction in the direction of binding to the (until now occupied) antibody sites. This reaction continues until equilibrium is established between what is bound to the antibody (the free fraction) (Figure 7). The time it takes for equilibrium to be established varies from method to method, because of the affinity and specificity of the antibodies and the optimum conditions established for the reaction to reach equilibrium. Another way to visualize equilibrium is that the proportions of the tracer versus-patient bound to the antibody are equal to the proportions of tracer-versus-patient not bound to the antibody. The trend is toward speed without sacrificing quality. To meet this challenge of "faster" assays, manufacturers have begun a trend toward shorter incubations. Some assays have been redesigned so that complete equilibrium is not achieved. Instead, the reaction is stopped prior to equilibrium, but supposedly without sacrificing the necessary sensitivity. The success or failure of this type of assay to maintain sensitivity can be demonstrated by setting up two to three parallel runs and extending the incubation interval two to three times beyond what is prescribed. Make note of the higher binding (zero cpm/total activity) and in most cases improved sensitivity. The basic RIA concepts can be applied to a practical situation.

If a patient has an elevated HCG concentration and this sample is analyzed by conventional RIA procedures as explained above, there is so much patient HCG present after equilibrium is established that it will displace most of the tracer from the antibody. This results in a relatively high level of patient HCG bound to the antibody; thus, there will not be much tracer bound to the antibody (Figure 3). If you were to remove the antibody from the reaction system and count the radioactivity, the radioactive counts would be low. The converse is also true; if you have a patient with little or no HCG, after the equilibrium is established, what will be bound to the antibody will be mostly, or all, tracer. Thus, when the bound fraction is counted, the radioactive counts will be high (Figure 4). The inverse relationship between a patient concentration and radioactive bound counts has been clearly explained. The key point is that the inverse relationship exists only when the antibody bound phase is counted in a conventional RIA procedure.

Other modifications of the above procedures will be discussed in the section on binding reactions.

Separation Step

The final step in an RIA procedure is removing the antibody bound fraction from the unbound or free fraction so that the radioactivity can be counted. This is referred to as a separation step. There are two general categories under which separation techniques can be classified: a) liquid phase separation; and b) solid phase separation.

The designation liquid or solid phase separation refers to the location of the primary antibody after the separation step is completed. In the liquid phase separation, the primary antibody is liquid or part of the supernatant. Solid phase means that the primary antibody is in some sort of solid form at the completion of the assay. There are two subgroups in the solid phase category: liquid reagents, which result in a pellet, and large solid matrices, which require only a wash and/or decantation.

LIQUID PHASE SEPARATION

Dextran Coated Charcoal

Charcoal, by itself, is a clarifying, decolorizing, filtering agent (30). Alone, it is used as a stripping agent for serum. If you were to add it to the assay tubes, it would absorb everything-both the bound and free fraction. When the charcoal particles are coated with dextran (a polysaccharide), it forms a mesh around the charcoal particles, thus, limiting the size of the molecules that can be adsorbed by the charcoal. When added to the RIA reaction mixture, it will adsorb only the tracer or patient analyte not bound to antibody. The antibody bound fraction is too large to be adsorbed by the dextran-coated charcoal (DCC), thus, following a centrifugation step, the primary antibody is located in the liquid phase or supernatant. The charcoal (free phase) is in the pellet. Once DCC is added to the reaction mixture, it is generally allowed to incubate 10 min at 4 oc for optimum adsorption. Centrifugation is then carried out, also at 4 oc. If not carried out at this temperature, the charcoal will not pack well in the bottom of the tube, with some of it remaining on the sides. A word of caution: Do not allow the DCC to remain in the reaction mixture any longer than specified. Prolonged exposure to DCC can result in the DCC stripping the antigen (labeled or unlabeled) from the antibody

Resin

The action of resin is very similar to that of DCC, but it has been used less frequently. T3 resin uptakes are one of the few tests still utilizing the substance.

SOLID PHASE SEPARATION-LIQUID REAGENTS

Polyethylene Glycol

PEG is a water soluble polymer with a basic formula H(OCH2CH2)nOH, where n~4. PEG is known to precipitate proteins, although the exact mechanism is not fully understood. It is thought to be related to a breakdown of the hydroxy group on the PEG in the formation of complexes. Factors that control effective use of PEG are: 1) percent of PEG; 2) protein content in the assay tube; 3) pH; and 4) the salt content of the buffer. The optimum conditions will vary with the analyte being measured. Alone it is a fast, clean separation of bound and free in which it quickly precipitates the bound fractions. Most procedures require that it be pi petted and centrifuged at 4 oc. Its disadvantage is its non-specificity. Depending on the nature of the proteins in the assay tube, it can precipitate most large proteins in the reaction vessels, not just the antibody in question.

Second Antibody (Precipitating Antibody)

The second antibody is simply an antibody against the primary antibody, i.e., if the first antibody was produced in rabbits, this 1 ° antibody will be a rabbit gamma globulin. To produce a second antibody to this first antibody, ordinary rabbit gamma globulin is used as an immunogen and injected into another species, i.e., sheep, goats, etc. This second animal will then produce an antibody to rabbit gamma globulin. When this antirabbit gamma globulin is reacted with the primary antibody (being a rabbit gamma globulin), it forms a huge complex with the primary antibody, which then renders itself insoluble and precipitates out of solution. The advantages are that it is the cleanest, most sensitive form of separation available. However, it is a slow process. Depending on the quality and affinity of the second antibody, incubations with the reaction mixture could vary from 2-24 hr.

Second Antibody/PEG

This method is a successful attempt to combine the sensitivity and precision of the second antibody with the speed of PEG. Virtually all second antibody separations now incorporate PEG. Once this reagent is added, incubations range from 1-30 min depending on the assay.

Microimmobilized Bead with Primary Antibody

In this system, the primary antibody is bonded to a microscopic bead (glass, sephadexs, or plastic). In the separation step, the immobilized microbeads containing the bound fraction are centrifuged. The free fraction is then decanted. Microimmobilized Bead with Second Antibody. In this case, the second antibody has been covalently linked to a solid particle. After incubation and centrifugation, the free fraction is discarded and the bound fraction counted.

Solid Phase Separation-Solid Reagents

Macro-immobilized Beads with Primary Antibody. In this case, the primary antibody is coated onto a large bead. One bead is added per test tube. The separation step involves washing the free antigen-labeled and unlabeled-away from the bead and counting the bead. What is bound to the bead is not disturbed by the washing process.

Coated Tubes

In this widely used technique, the primary antibody is bound to the walls of the assay tube (generally 12 x 75 mm). The advantage is thus saving centrifugation time. The free fraction is decanted and what is bound to the primary antibody on the wall of the tube is counted. One disadvantage is a reduction in precision when compared to that of a liquid separation step. This reduction is negligible, however, being only 2-3% less (higher CVs).

PART II-BINDING REACTIONS

RIA

The basic reactions in a conventional RIA procedure have been previously discussed.

CPB

This technique has been previously mentioned in the section on primary binders. It is very similar to that of RIA. There are two types of CPB systems: a) one is identical to RIA in that the sample (or standard) and tracer compete for available binding sites on the binding protein (except there is only one binding site per protein binding molecule, as opposed to two binding sites on each antibody molecule); and b) the protein binder is first saturated with the tracer. The sample (or standard) is then added, and then displaces the tracer in proportion to the relative concentration of the sample or standard.

Sequential Saturation Analysis

In an RIA procedure where the tracer may have a higher affinity for the antibody binding sites than the material to be detected in the patient sample, the technique of sequential saturation analysis may be employed. In these assays, the patient sample or standard is allowed to react with the antibody first. In this case, the concentration of antibody is high relative to the concentration of antibodies used in a routine RIA procedure. In the sequential saturation analysis, all of the patient sample or standard will bind to the antibody, leaving many unoccupied antibody sites. After a suitable incubation period, the tracer is added. It will then occupy all of the binding sites not occupied by the patient or standard. The second incubation is crucial since over-incubation could result in complete displacement of the patient sample or standard, as the tracer has a higher affinity for the antibody. Thus, the step must be initiated without delay following the tracer incubation. This approach has been used in numerous other assays since it has been determined that allowing the patient sample to have first exposure to the antibody binding sites improves sensitivity in the low ranges of the standard curve.

Immunoradiometric Assay

Immunoradiometric assay (IRMA) is the term for what was referred to as "sandwich technique." Many monoclonal assays utilize this technique because the fewer number of pipetting steps reduces hands-on time. It is most often employed in macro-bead technology where the primary antibody is coated onto the macroscopic bead. The patient sample or standard is first permitted to react with the antibody. The remaining unwanted patient sample constituents are then washed away. Another primary antibody is added, but this antibody is labeled with a radioactive isotope, generally iodine-125. This labeled antibody will bind to the patient's antigen, which is already bound to the primary antibody on the bead. Following incubation, the labeled antibody not bound is washed away. In this type of procedure, the counts are directly proportional to the patient sample concentration. More patient sample bound to the primary antibody on the bead will result in more labeled antibody bound to the patient sample sites. Recent advances in IRMA technology and monoclonal antibodies have eliminated the second wash step.

Radiolabeled Binding Assays

In these assays, the presence of an antibody or a receptor not an antigen-is detected. A radiolabeled antigen is added to the patient sample. If the antibody, receptor, or binder is present, the labeled antigen will bind to it. Following incubation and separation, the presence and quantitation of the bound fraction will determine if the antibody, binder, or receptor is present. In some of the radiolabeled binding assays, the results are expressed as a percent of the radioactivity bound. The more radioactivity present, the more antibody present. Testing for insulin antibodies, anti-DNA antibodies, and B-12 binders is generally expressed this way. Other assays, such as breast tumor receptor assays, require more precise quantitation of the exact concentration of receptor present. In these procedures, by adding known concentrations of the radiolabeled antigen, one can quantitate the concentration of the binder by using the method of Scatchard ("Scatchard's analysis").

One additional reaction type that deserves special attention because it does not fall into the above categories is T3 uptake assay, which has remained essentially the same since about 1959. In this assay, the patient sample is mixed with radiolabeled T3. The primary binder is the patient's own thyroxine binding globulin (TBG). Therefore, there are two variables: the patient's thyroid hormone concentration and the patient's TB concentration. The labeled T3 will occupy all available binding sites on the patient's TBG molecules. It will compete with the patient's own T3 on TBG, but will not displace the patient's T4 due to the T;s high affinity to bind TBG much more than that of T3. A separation step (resin) follows, which absorbs all of the labeled T3 not bound to the TBG. The bound fraction is discarded and the free fraction is counted. The results are expressed as a percent of radioactive T3 taken up by the separating agent, such as resin. Thus, an elevated T3 uptake result means either a high T3 concentration or a low TBG concentration. A low T3 uptake result would of course mean the opposite.

QUALITY CONTROL PARAMETERS IN RIA

Radioimmunoassay needs to be evaluated periodically for its reliability. The criteria employed for this are:

- Sensitivity
- Specificity
- Precision
- Accuracy

Sensitivity

Sensitivity is the minimal detection limit of the assay. It is the smallest concentration of hormone which can be distinguished from zero standard (sample containing no unlabeled hormone). Sensitivity cannot be calculated from the slope of the standard curve as the slope would vary according to parameters taken on X and Y axis (e.g., bound/free ratio vs concentration of standard will have a different slope them when free bound ratio is plotted against standard concentration).

Specificity

Specificity is the extent of freedom from interference by substances other than the one intended to be measured. Non-specificity or the extent of interference from other substances is of two types. Specific non-specificity is due to substances structurally similar to the hormone (e.g., various steroid hormones, anterior pituitary glycoprotein hormones LH, FSH and TSH, existence of the hormone in different forms). The antisera against the hormone may cross-react with interfering substance due to presence of antibodies against an epitope (portion of hormone molecule to which antibody binds) which may be present in the interfering molecule also. One must obtain the cross-reactivity table from the supplier of antiserum. The antiserum would be suitable for the estimation of a particular hormone in a biological sample only if the substances cross-reacting with the antibody are precept in very low concentrations in the sample. Non-specific non-specificity is due to presence of substances in the biological sample which do Dot possess structural similarly to the hormone but which can somehow inhibit hormone-antibody complex formation.

Precision

The precision is a measure of the 'reproducibility' of the assay. It is defined as the degree to which repeated measurements of the same sample agree with the mean concentration. In other words, the 'imprecision' is the measure of variation between repeated determinations of the same sample. Imprecision, in combination with 'inaccuracy' reflects the errors in the assay.

Accuracy

Accuracy is the degree to which the estimated value is close to the 'true' or actual concentration of hormone in the sample. In a broader sense it encompasses the specificity of the assay. If the endogenous ligand behaves in the same manner in the assay as the standards and tracer. the assay is considered to be accurate. If, on the other band, the endogenous ligand behaves differently under the assay conditions the assay would be inaccurate. The recovery of the hormone reflects the accuracy of the assay.

ADVANTAGES OF RIA

- Radioimmunoassay is very sensitive technique used to measure concentrations of antigens without the need to use a bioassay.
- It can measure one trillionth (10^{12}) of a gram of material per milliliter of blood.
- It is structurally specific as antigen antibody reaction are highly specific.
- It is indirect method of analysis,
- It is a saturation analysis as active reagents are added in smaller quantity than that of analytic

USES OF RIA

- Narcotics (drugs) detection.
- Blood bank screening for the hepatitis (a highly contagious condition) virus.
- Early cancer detection.
- Measurement of growth hormone levels.
- Tracking of the leukemia virus.
- Diagnosis and treatment of peptic ulcers.
- Research with brain chemicals called neurotransmitters.

APPLICATIONS

- RIA is used in the estimation of substances that exhibit antigenic property either as such or by chemical modification.
- Substances that are estimated by RIA are hormones, tumor markers (peptides, proteins, steroids, etc.), vitamins, drugs, etc.
- Thus, radioimmunoassay has applications in the diagnosis of hormonal disorders, cancers and therapeutic monitoring of drugs.
- It is also useful in biomedical research.

CONCLUSION

Radioimmunoassay (RIA) is sensitive method for measuring very small amounts of a substance in the blood. The presence of a radioactive tracer in radioimmunoassay greatly enhances the sensitivity of the detection when radioisotopes are used as labels to be conjugated with antigen or antibodies. These techniques of detection of the antibody antigen complex is called as radioimmunoassay (RIA).

FREQUENTLY ASKED QUESTIONS

Q1. What is radioimmunoassay (RIA)?

Answer: Radioimmunoassay (RIA) is a very sensitive *in vitro* assay technique used to measure concentration of antigens (for example hormone level in the blood by the use of antibodies) It is based upon the competition between labelled and unlabelled antigen for specific antibody sites, forming antigen-antibody complexes.

Q2. What is the common use of Radioimmunoassay (RIA)?

Answer: RIA is more excessively used in the medical field to measure relative amount of hormones or drugs in patient's sera.

Q3. What is the purpose of RIA?

Answer: RIA uses radiolabelled antibodies. The presence of largest proteins or antibodies is determined by measuring the amount of radioactivity.

Q4. Enumerate the different types of ELISA?

Answer:

1. sandwich ELISA
2. Direct ELISA
3. Competitive ELISA
4. Indirect ELISA.

Q5. Name different types of enzyme used in ELISA?

Answer: Various types of enzyme used in ELISA test are

1. Alkaline phosphatase
2. Beta galactosidase
3. Horseradish peroxidase

Q6. What do u understand by immunofluorescence?

Answer: Immunofluorescence is a method in which an antigen is detected by the use of fluorescence labelled immunoglobins. Immunofluorescence is an assay which is used primarily on biological samples and is classically defined as a procedure to detect antigens in cellular contexts using antibodies. The specificity of antibodies to their antigen is the base for immunofluorescence. The biological samples include tissue and cells. Immunofluorescence allows researchers to evaluate whether or not cells in a particular sample express the antigen in question. In cases where an immunopositive signal is found, immunofluorescence also allows researchers to determine which subcellular compartments are expressing the antigen. Immunofluorescence can be used on cultured cell lines, tissue sections, or individual cells.

Q7. Into how many types is immune fluorescence divided?

Answer: Immunofluorescence is divided into two types?

1. Direct immunofluorescence.
2. Indirect immunofluorescence.

Q8. What is direct immune fluorescence?

Answer: In this method, the species antibodies are primary antibodies, which are directly conjugated to fluorescent dye.

Q9. What is the basic principle of ELISA?

Answer: The basic principle is an enzyme conjugated to an antibody reacts with a colourless substrate to generate a coloured product.

Q10. What do u understand by quality control?

Answer: Quality control refers to the reliability of investigative service. Any error in the laboratory will jeopardize the lives of patients. It is therefore utmost important that laboratory errors are identified and rectified.

Q11. What is the use of ELISA test?

Answer: ELISA is widely used for the determination of small quantities of proteins (hormones, antigens, antibodies) and other biological substances. The most commonly used pregnancy test for the detection of human chorionic gonadotropin (HCG) in urine is based on ELISA. It is also been used for the diagnosis of AIDS.

Q12. What is an ELISA kit?

Answer: An ELISA kit is a commercially available ELISA test that usually contain pre-coated polystyrene plates, detection antibodies and usually chemicals needed to perform an ELISA test.

Q13. What is the principle of radioimmunoassay?

Answer: It involves combination of three principles.

1. An immune reaction i.e., antigen, antibody binding.
2. A competitive binding or competitive displacement reaction. (It gives specificity)
3. Measurement of radio emission. (It gives sensitivity).

Q14. What is the difference between Sandwich and competitive ELISA?

Answer: sandwich ELISA is mainly meant for detecting antigens with two antigenic epitopes capable of binding to antibody whereas competitive ELISA is used to measure the concentration of antibody in a sample by detecting interference in an expected signal output.

Q15. What is the difference between direct immunofluorescence and indirect inflorescence?

Answer: Direct immunofluorescence uses the single antibody (primary antibody) that carries a fluorophore. The antibodies recognise the target molecule and binds to it, and the fluorophore it carries can be detected with a microscopy. Indirect immunofluorescence uses two antibodies. The un label primary antibody binds to the target molecule and the secondary antibody that carries the fluorophore recognise the primary antibody and binds to it.

Q16. What is the disadvantage of direct immunofluorescence?

Answer: A separate fluorescent conjugate have to be prepared against each antigen to be tested.

Multiple Choice Questions

Q1. Which of the following counter is generally used in RIA?
A. Alpha counter
B. Gamma counter
C. Beta counter
D. Delta counter.

Q2. The radioimmunoassay technique was developed
A. Salamon A. Berson and Rosalyn S. Yalow
B. Berson
C. Louis Pasteur
D. Morgan

Q3. The radioimmunoassay came into existence while studying radiolabelled
A. Insulin
B. Thyroxine
C. Progesterone
D. Oxytocin

Q4. Why are radioimmunoassays a particularly efficient diagnostic tool?
A. They only require spit to work, so no blood is drawn
B. They only require a small sample of blood to work
C. They do not require anything to be added to the sample
D. They can be done without any laboratory equipment

Q5. Which of the following occurs first in a radioimmunoassay?
A. The radiolabeled antigens are placed in the sample
B. Free radioactive antigens are separated from bound ones
C. The radioactivity level of the sample is measured
D. The relevant antibodies are added to the blood sample

Q6. What is a radioimmunoassay intended to measure?
A. The amount of a specific antigen in a blood sample
B. The amount of a specific antibody in a blood sample
C. The number of white blood cells in a blood sample
D. How well a person's immune system is functioning

Answer Key

1(B), 2(A), 3(A), 4(A), 5(A), 6(B)

Assignments

Q1. What is an Enzyme Linked Immunosorbent Assay? What are its different types? Explain.

Q2. Explain briefly the procedure of Radio Immunoassay (RIA)?

Q3. Explain the principle of Radio Immunoassay (RIA)?

Q4. What is immunofluorescence?

Q5. What are the applications of Enzyme Linked Immunosorbent Assay?

References

Berson, SA; Yalow, RS; Aurback, GD; Potts, JT; Jr. Immunoassay of bovine and human parathyroid hormone. *Proc Natl Acad Sci USA*, 49, 613, 1963.

Berson, SA; Yalow, RS; Bauman, A; Rothschild, MA; Newerly, K. Insulin 1-131 metabolism in human subjects: demonstration of insulin binding globulin in the circulation of insulin treated subjects. *J Clin Invest*, 35, 170, 1956.

Berson, SA; Yalow, RS; Glick, SM; Roth, J. Immunoassay of protein and peptide hormones. *Metabolism*, 13, 1135, 1964.

Berson, SA; Yalow, RS. Assay of plasma insulin in human subjects by immunologic methods. *Nature*, 184, i648, 1954.

Berson, SA; Yalow, RS. Immunoassay of endogenous plasma insulin in man. *J Clin Invest*, 39, 1157, 1960.

Berson, SA; Yalow, RS. lmmuno-chemical heterogeneity of parathyroid hormone. *J Clin Endocrinol Metab*, 26, 1037, 1968.

Berson, SA; Yalow, RS. lodoinsulin used to determine specific activity of iodine-131. *Science*, 154, 205, 1966.

Berson, SA; Yalow, RS. Nature of immunoreactive gastrin extracted from tissues of gastrointestinal tract. *Gastroenterology*, 60, 215, 1971.

Berson, SA; Yalow, RS. Principles of immunoassay of peptide hormones in plasma, in Astwood EB, Cassidy CE (eds): *Clinical Endocrinology*. New York, Grune & Stratton, 1968, pp 699-720.

Berson, SA; Yalow, RS. Quantitative aspects of reaction between insulin and insulin binding antibody. *J Clin Invest*, 38, 1996, 1959.

Berson, SA; Yalow, RS. Radioimmunoassay in gastroenterology. *Gastroenterology*, 62, 1061, 1972.

Berson, SA. Introduction to symposium on radioimmunoassay, in Protein and Polypeptide Hormones, *Proceedings of International Symposium*, Liege, Belgium, 1968. Amsterdam, Excerpta Medica, International Congress Series !61, 1968, p3.

Blake, C; Gould, BJ. Use of enzymes in immunoassay techniques. A review. *Analyst.* 1984 Jan 1, 109(5), 533-47.

Collip, JB; Anderson, EM. The production of serum inhibitory to the thyrotropic hormone. *Lancet*, 226, 76, 1934.

Feldman, H; Rodbard, D. Mathematical theory of radioimmunoassay, In Odell WD, I) aughaday WH (eds): *Principles of Competitive Protein-Binding Assays*. Philadelphia, J. B. ippincott, 1971.

Franchimont, P; Burger, H; Legros, JJ. Impact of radioassay techniques on the field of sex hormones. *Metabolism*, 22, 1003, 1973.

Friesen, HG. Human prolactin in clinical endocrinology: The impact of radioimmunoassays. *Metabolism*, 22, 1039, 1973.

Glick, BM; Kagan, A. Vasopressin, in Jaffe BM, Behrman (eds): *Methods of Hormone Radioimmunoassay*. New York, Academic Press, 1974, pp 187-198.

Glick, SM; Roth, J; Yalow, RS; Berson, SA. Immunoassay of growth hormone in plasma. *Nature*, 199, 784, 1963.

Glick, SM; Roth, J; Yalow, RS; Berson, SA. The regulation of growth hormone secretion, In Pincus C (ed): *Recent Progress in Hormone Research*. New York, Academic Press, 1965, pp. 241-283.

Glick, SM; Rothenberg, SP. (ed): Festschrift in honor of Solomon A. Berson, M.D. *Metabolism*, 22, 961, 1973.

Goldsmith, SJ; Yalow, RS; Berson, SA. significance of human plasma insulin Sephadex reactions. *Diabetes*, 18, 834, 1969.

Goldsmith SJ. Basic principles of competitive radioassay, In Rothfeld B (ed): *In Vitro Nuclear Medicine*. Philadelphia, J. B. Lippincott, 1974, pp. 96-121.

Goldsmith, SJ. Radioimmunoassay: Review of basic principles. In *Seminars in Nuclear Medicine* 1975 Apr 1 (Vol. 5, No. 2, pp. 125-152). WB Saunders.

Gorden, P; Roth, J. Plasma insulin: Fluctuations in the big insulin component in man after glucose and other stimuli. *J Clin Invest*, 8, 2225, 1969.

Heding, LG. Radioimmunological deternination of pancreatic and gut glucogon in plasma. *Diabetologica*, 7, 10, 1971.

Herington, AC; Jacobs, LS; Daughaday, WH. Radioreceptor and radioimmunoassay quantitation of human growth hormone in acromegalie serum: Overestimation by immunoassay and systemic differences between antisera. *J Clin Endocrinol Metab*, 39, 257, 1974.

Hershman, JM; Pittman, JA. Utility of the radioimmunoassay of serum thyrotrophin in man. *Ann Intern Med*, 74, 481, 1971.

Hunter, WM; Greenwood, FC. Preparation of iodine-131 labeled human growth hormone of high specific activity. *Nature*, 194, i95, 1962.

Jaffe, MB; Behrman, HR. *Methods of Hormone Radioimmunoassay*. New York, Academic Press, 1974.

Kagan, A; Glick, SM. Oxytocin, In Jaffe BM, Behrman HR (eds): *Methods of Hormone Radioimmunoassay*. New York, Academic Press, 1974.

Landon, J; Carney, J; Langley, D. The measurement of enzymes by radioimmunoassay: A review. *Annals of Clinical Biochemistry*, 1977 Jan, 14(1-6), 90-9.

Lawrence, AM. Radioimmunoassayable glucogon levels in man: effects of starvation. *Proc Natl Acad Sci USA*, 55, 316, 1966.

llb. Berson, SA; Yalow, RS. Radioimmunoassay of ACTH in plasma. *J Clin Invest*, 47, 2725, 1968.

lla. Yalow, RS; Glick, SM; Roth, J; Berson, SA. Radioimmunoassay of human plasma ACTH. *J Clin Endoerinol Metab*, 24, 1219, 1964.

McGuigan, JC; Trudeau, WL. lmmunochemical measurement of elevated levels of gastrin in serum of patients with pancreatic tumors of the Zollinger-Ellison variety. *N Engl J Med*, 278, i308, 1968.

Melani, F; Ryan, WG; Rubenstein, AH; teiner, DF. Proinsulin secretion by a pancreatic eta cell adenoma: Proinsulin and C-peptide secretion. *N Engl J Meal*, 283, 713, 1970.

Palmieri, GMA; Yalow, RS; Berson, SA. Adsorbent techniques for the separation of antibody-bound from free peptide hormones in radioimmunoassay. *Horm Metab Res*, 3, 301, 1971.

Parker, CW. Nature of immunological responses and antigen-antibody interaction, In Odell WD, Daughaday WH (eds): *Principles of Competitive Protein-Binding Assays*. Philadelphia, J. B. Lippincott, 1971.

Reiss, E; Canterbury, JB. Primary hyper- parathyroidism. *N Engl J Med*, 280, 1381, 1969.

Samols, E; Tyler, J; Marri, G; Marks, V. stimulation of glucogon secretion by oral plucose. *Lancet*, 2, 1257, 1965.

Sherwood, LM. Similarities in the chemical structure of human placental lactogen and tituitary growth hormone. *Proc Natl Acad Sci IJSA*, 58, 2307, 1967.

Silverman, R; Yalow, RS. Heterogeneity ff parathyroid hormone: Clinical and physiological implications. *J Clin Invest*, 52, 1958, 1973.

Skelley, DS; Brown, LP; Besch, PK. Radioimmunoassay. *Clinical chemistry*. 1973 Feb 1, 19(2), 146-86.

Skelley, DS; Brown, LP; Besch, PK. Review: Radioimmunoassay. *Clin Chem*, 19, 146, 1973.

Spector, S. Radioimmunoassay. *Annual review of pharmacology*. 1973 Apr, 13(1), 359-70.

Steiner, DF; Cunningham, D; Spigelman Atcn, B. Insulin biosynthesis: evidence for a precursor. *Science*, 157, 697, 1967.

Unger, RH; Eisentraut, A; McCall, MS; et al., Measurements of endogenous glucagon n plasma and the influence of blood glucose: concentration upon its secretion. *J Clin Invest*, 1, 682, 1962.

Walsh, JG; Yalow, RS; Berson, SA. Detection of Australia antigen and antibody using radioimmunoassay techniques. *J. Infect Dis*, 21, 550, 1970.

Wolstenholme, GEW; Cameron, MP. (eds): *Infimunoassay of Hormones, Ciba Foundation Colloquia on Endocrinology*, vol 14. Boston, Little Brown, 1962.

Yalow, RS; Berson, SA. Effect of X-Rays on Trace-Labeled I131-Insulin and Its Relevance to Biologic Studies with I131-Labeled Proteins. *Radiology*, 66, 106, 1956.

Yalow, RS; Berson, SA. Further studies on the nature of immunoreactive gastrin in human plasma. *Gastroenterology*, 60, 203, 1971.

Yalow, RS; Berson, SA. General aspects of radioimmunoassay procedures, in: *In Vitro* Procedures with Radioisotopes in Medicine. IAEA, 1970 23. Eakins RP: Radioimmunoassay, protein binding and other saturation assay techniques, In: *The Yearbook of Nuclear Medicine*. Chicago, Yearbook Medical Publications, 1973.

Yalow, RS; Berson, SA. Immunoassay of plasma insulin in man. *Diabetes*, 10, 339, 1961.

Yalow, RS; Berson, SA. Introduction and general considerations, In Odell WD, Daughaday WH (eds): *Principles of Competitive Protein-Binding Assays*. Philadelphia, J. B. Lippincott, 1971.

Yalow, RS; Berson, SA. Labeling of proteins-problems and practice. *Trans NY Acad Sc*i, 28, 1033, 1966.

Yalow, RS; Berson, SA. Radioimmunoassay of gastrin. *Gastroenterology*, 58, 1, 1970.

In: Immunoglobulins
Editor: Manzoor Ahmad Mir

ISBN: 978-1-53616-903-4

Chapter 10

ENZYME LINKED IMMUNOSORBENT ASSAY (ELISA)

Manzoor A Mir*, Safura Nisar, Umar Mehraj and Hina Qayoom
Department of Bioresources, School of Biological Sciences,
University of Kashmir, Srinagar J&K, India

ABSTRACT

Enzyme-Linked Immunosorbent Assay (ELISA) is a common analytical biochemistry immunoassay, first described in1972 by Engvall and Perlmann. The assay is useful for the detection of antibodies, peptides, proteins, and small molecules using a multi-well plate. This assay is the preferred method to determine the titer of antisera and purified antibodies. ELISA can also be successfully employed for the quantitative assessment of an antigen in a sample, often available in convenient, easy-to-use kit formats. ELISA is used as a diagnostic tool in medicine, plant pathology, and biotechnology as well as for the quality control check in industries. In an ELISA, an antigen must be immobilized on a solid surface and then complexed with an antibody that is linked to an enzyme. Detection is accomplished by assessing the conjugated enzyme activity via incubation with a substrate to produce a measurable product. The most crucial element of the detection strategy is a highly specific antibody-antigen interaction. In this chapter we will discuss the history, procedure, different types of ELISA and the applications.

Keywords: ELISA, immunoassay, antigen, antibody, direct, indirect, sandwich, competitive, biomarkers, toxoplasmas

* Corresponding author: Dr Manzoor Ahmad Mir (Ph. D), Department of Bioresources, School of Biological Sciences, University of Kashmir, Srinagar J&K, India-190006, Contact:-+919622901319. E-mail: drmanzoor@kashmiruniversity.ac.in.

Objectives

- To understand the definition and historical background of ELISA
- To describe the principle and process of ELISA
- To discuss the types and methods associated with the ELISA
- To outline the applications and uses of ELISA

Definition of ELISA

- Collins English Dictionary describes enzyme-linked immunosorbent assay as "An immunological technique for accurately measuring the amount of a substance, for example in a blood sample".
- American Heritage English Language Dictionary defines ELISA as "A sensitive immunoassay that uses an enzyme linked to an antibody or antigen as a marker for the detection of a specific protein, especially an antigen or antibody". It is often used as a diagnostic test to determine exposure to a particular infectious agent, such as the AIDS virus, by identifying antibodies present in a blood sample.
- Random House Kernerman Webster's College Dictionary defines ELISA as "A diagnostic test for detecting exposure to an infectious agent, as the AIDS virus, by combining a blood sample with antigen of the agent and probing with an enzyme that causes a color change when antibody to the infection is present in the sample"

ELISA (enzyme-linked immunosorbent assay) is a plate-based assay technique designed for detecting and quantifying peptides, proteins, antibodies and hormones. In an ELISA, an antigen must be immobilized to a solid surface and then complexed with an antibody that is linked to an enzyme. Detection is accomplished by assessing the conjugated enzyme activity via incubation with a substrate to produce a measurable product. The most crucial element of the detection strategy is a highly specific antibody-antigen interaction. ELISAs are typically performed in 96-well (or 384-well) polystyrene plates, which will passively bind antibodies and proteins. It is this binding and immobilization of reagents that makes ELISAs so easy to design and perform. Having the reactants of the ELISA immobilized to the microplate surface makes it easy to separate bound from non-bound material during the assay. This ability to wash away nonspecifically bound materials makes the ELISA a powerful tool for measuring specific analytes within a crude preparation.

HISTORY OF ELISA

Before the development of the EIA/ELISA, the only option for conducting an immunoassay was Radioimmunoassay, a technique which used radioactively labeled antigens or antibodies for diagnostics before the advent of ELISA. Rosalyn Sussman Yalow and Solomon Berson published a paper in 1960 on Radioimmunoassay. Radioactivity was the reporter providing the signal which indicated the presence of analyte being sought in the sample. However, radioactivity posed a potential health threat, so safer reporter alternatives were sought.

The enzyme linked immunosorbent assay (ELIZA) is golden standard clinical diagnostic tool for detection and quantification of protein biomarkers.

Van Weeman and Schurs (1971) in Holland and Engulla and Perlman (1971-72) in, Sweden for the first time, independently introduced the idea of using the enzyme conjugated and then complexed with an antibody that is linked to enzyme. Detection is accomplished by assessing the conjugated enzyme actively via incubation with a substrate to produce a measurable product. The most crucial element of the detection strategy is a highly specific antibody-antigen interaction.

PRINCIPLE

ELISA combines the specificity of antibodies with the sensitivity of simple enzyme assays, by using antibodies or antigens coupled to an easily assayed enzyme that possesses a high turnover number. Simply put for ELISA an antigen is adsorbed/fixed to the surface of a solid matrix which may be a well on a micro titration plate/well strip, beads or may be a special type of paper made of Nitrocellulose. Then, a specific antibody is applied over the surface so it can bind to the antigen. This antibody is linked to an enzyme, and, in the final step, a substance containing the specific substrate for the enzyme is added. The subsequent reaction produces a detectable signal, a color change in the substrate. ELISA technology has advanced a lot. To make this immunoassay more and more sensitive as well as specific various advances have been made and now, we have different types of ELISAs. There are slight differences in the technology though basically ELISA remains an antigen antibody reaction performed as wet test using different reporter and signal systems. A number of enzymes have been used for ELISA such as (Alkaline phosphatase (Ap) lectopreoxidase and Tetramethylbenzidine) Horseradish peroxidase (HRP), β-galactosidase.

PROCEDURE

Although many variants of ELISA have been developed and used in different situations, they all depend on the same basic elements:

- **Coating/Capture:** direct or indirect immobilization of antigens to the surface of polystyrene microplate wells.
- **Plate Blocking:** addition of irrelevant protein or other molecule to cover all unsaturated surface-binding sites of the microplate wells.
- **Probing/Detection:** incubation with antigen-specific antibodies that affinity bind to the antigens.
- **Signal Measurement:** detection of the signal generated via the direct or secondary tag on the specific antibody (Figure 10.1).

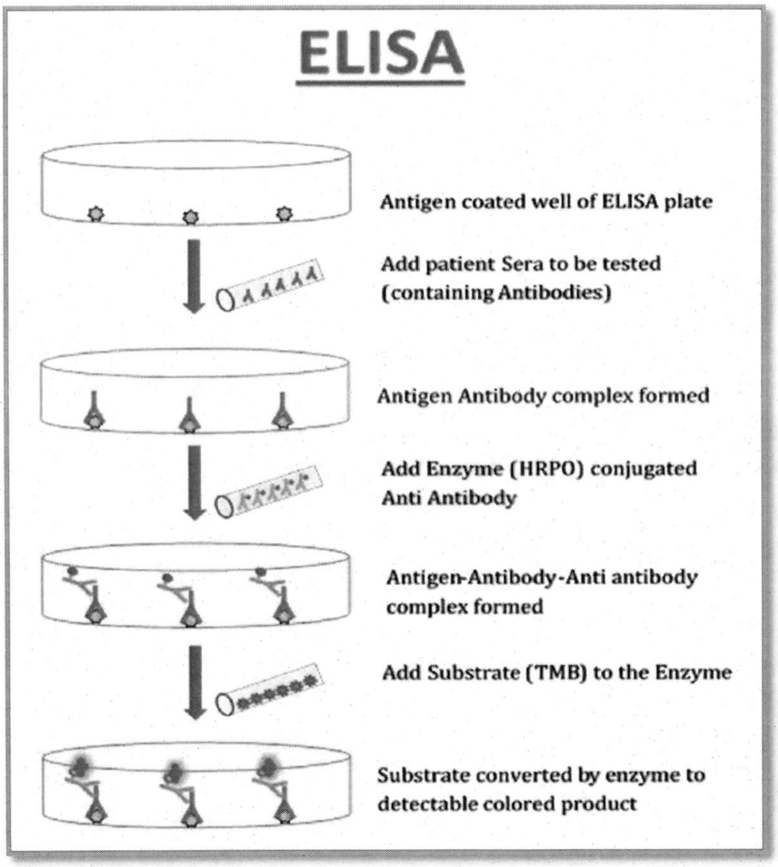

Figure 10.1. Representing the process involved in ELISA.

In a typical assay designed to detect an antigen in a sample, either the antigen is immobilized by direct adsorption or first antibody is adsorbed and then antigen is adsorbed on the well surface of the ELISA plate. The plate is blocked with albumin. The antigen is probed with a specific detection antibody. The detection antibody may be directly labeled with a signal-generating enzyme or fluorophore or it may be secondarily probed with an enzyme- or fluor-labeled secondary antibody. For enzymatic detection, the appropriate enzyme substrate is added. The signal observed is proportional to the amount of antigen in the sample. Every step of test is followed by washing to remove the un-reacted reactants, only specific (high-affinity) binding remains that causes the detection signal at the final step.

TYPES AND METHODS OF ELISA

ELISAs can be divided into the following categories based on the principle used (Figure 10.2):

- Direct
- Indirect
- Sandwich
- Competitive
- Antigen and antibody capture ELISA

DIRECT ELISA

It is the simplest configuration in which the antigen is bound by passive adsorption to the solid phase, washed to remove any unbound molecules and then directly incubated with a conjugated antibody. Following the incubation period and additional washing, substrate is added to produce signal that is allowed to develop. After certain time, the substrate reaction is stopped and the resulting signal quantified. It is commonly used for tittering conjugated secondary antibodies and very useful to estimate antigen cross-reactivity (Figure 10.3)

Advantages

- Quick because only one antibody and fewer steps are used.
- Cross-reactivity of secondary antibody is eliminated.

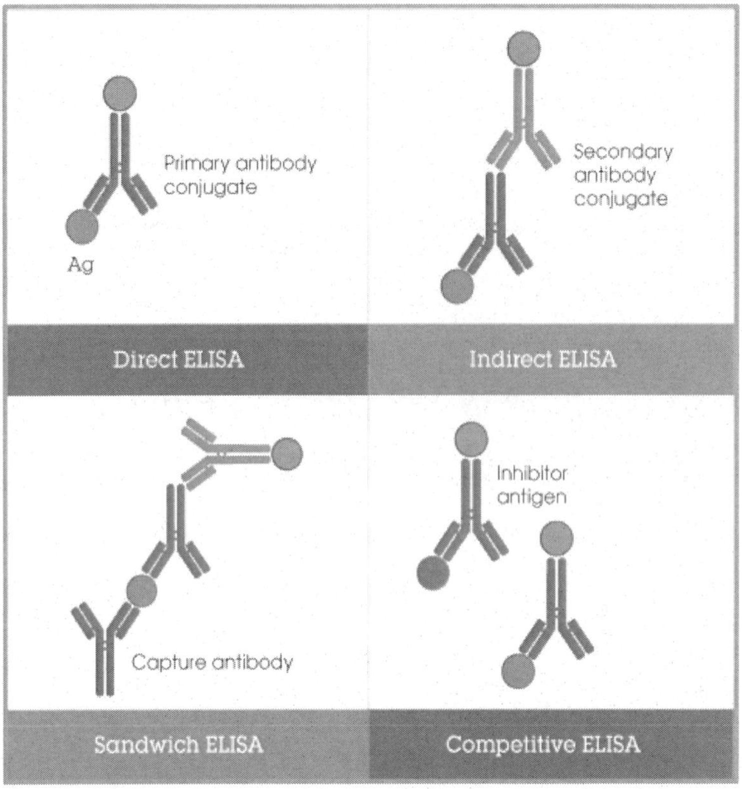

Figure 10.2. Showing various types of ELISA (Abcam).

Disadvantages

- Immunoreactivity of the primary antibody might be adversely affected by labeling with enzymes or tags.
- Labeling primary antibodies for each specific ELISA system is time-consuming and expensive.
- No flexibility in choice of primary antibody label from one experiment to another.
- Minimal signal amplification.

INDIRECT ELISA

HIV antigens are attached covalently to the solid phase support allowing corresponding/specific antibodies present in the specimen to bind, and these bound antibodies are subsequently detected by enzyme labelled anti-human immunoglobulin and specific substrate system. If the test specimen contained the antibodies specific to the antigen fixed on the solid phase color reaction will take place. The color reaction can be read with naked eye or with ELISA Reader at the specified UV light using the special filters. The result is expressed as OD (optical Density) value (Figure 10.4)

Advantages

- A wide variety of labeled secondary antibodies are available commercially.
- Versatile because many primary antibodies can be made in one species and the same labeled secondary antibody can be used for detection.
- Maximum immunoreactivity of the primary antibody is retained because it is not labeled.
- Sensitivity is increased because each primary antibody contains several epitopes that can be bound by the labeled secondary antibody, allowing for signal amplification.

Disadvantages

- Cross-reactivity might occur with the secondary antibody, resulting in nonspecific signal.
- An extra incubation step is required in the procedure.

Sandwich ELISA

One of the most useful of the immunoassays is the two antibody "sandwich" ELISA. This assay is used to determine the antigen concentration in unknown samples. This ELISA is fast and accurate, and if a purified antigen standard is available, the assay can determine the absolute amount of antigen in an unknown sample. To utilize this assay, one antibody (the "capture" antibody) is purified and bound to a solid phase typically attached to the bottom of a plate well. Specimen containing antigen is then added and allowed to complex with the bound antibody. Unbound products are then removed with a wash, and a labeled second antibody (the "detection" antibody) is allowed to bind to the antigen, thus completing the "sandwich". The assay is then quantitated by measuring the amount of labeled second antibody bound to the matrix, through the use of a colorimetric substrate. This type of ELISA has the following advantages:

- High specificity: the antigen/analyte is specifically captured and detected
- Suitable for complex (or crude/impure) samples: the antigen does not require purification prior to measurement

- Flexibility and sensitivity: both direct or indirect detection methods can be used

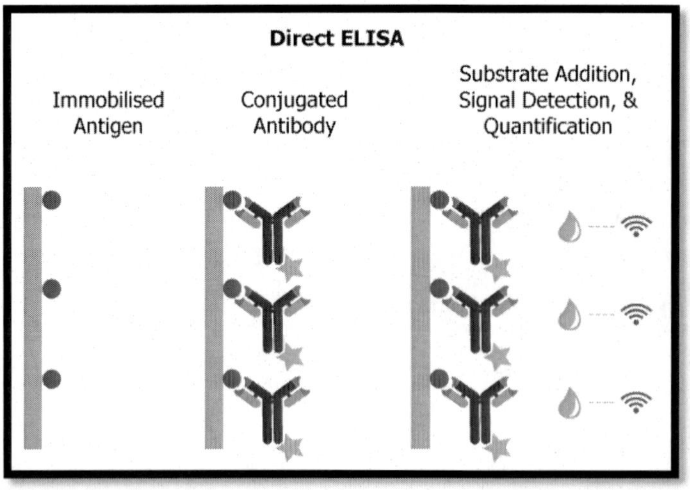

Figure 10.3. Illustrating the procedure of Direct ELISA (https://rockland-inc.com/elisa.aspx).

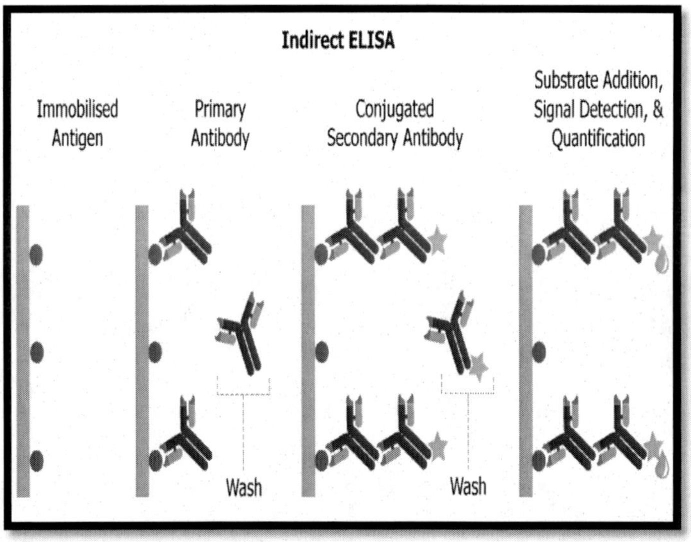

Figure 10.4. Illustrating the procedure of Indirect ELISA (https://rockland-inc.com/elisa.aspx).

Direct Sandwich

This assay uses a detection antibody conjugated to an enzyme or fluorescence tag. Following incubation with the antibody-antigen complex immobilized on the plate well, signal detection is performed upon successive addition of substrate and stopping solution or appropriate excitation/emission of the fluorescent tag (Figure 10.5).

Indirect Sandwich

This assay uses an unconjugated antibody bound to the antibody-antigen complex on the well. Following a washing step to remove unbound antibodies, the remaining antigen-bound antibodies are targeted by a conjugated secondary detection antibody. Signal detection is performed upon successive addition of substrate and stopping solution or appropriate excitation/emission of the fluorescent tag (Figure 10.6).

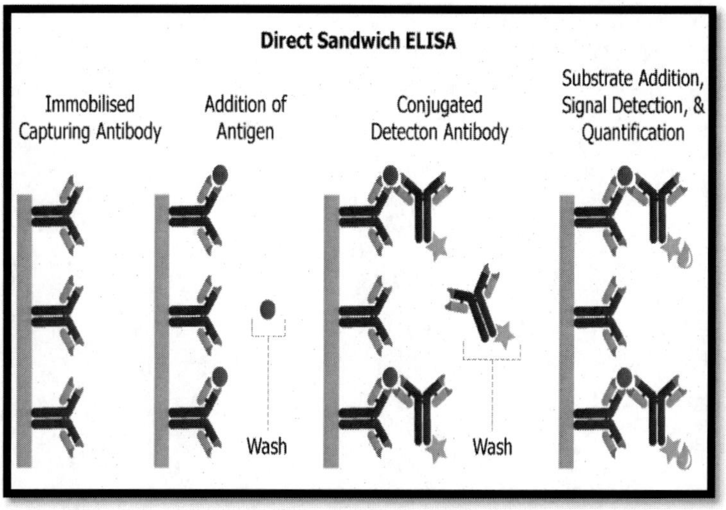

Figure 10.5. Illustrating the procedure of Direct sandwich ELISA (https://rockland-inc.com/elisa.aspx).

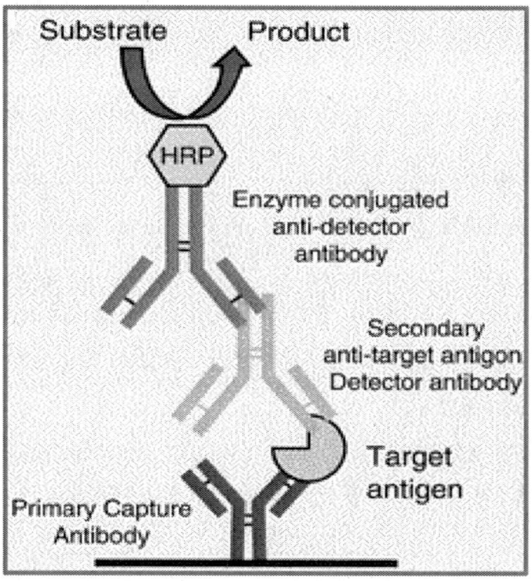

Figure 10.6. Illustrating the procedure of Indirect Sandwich ELISA (https://rockland-inc.com/elisa.aspx).

Competitive ELISA

In this assay the antibodies present in the specimen compete with the enzyme conjugated antibodies in the reagent for binding to the antigen on the solid phase. If the test specimen contains corresponding/specific antibodies, these will compete with the labelled antibodies in the reagent for binding to antigen. So that less or not labelled antibodies can attach to the solid phase. Hence, faint or no color is produced on addition of substrate if specimen contains antibodies against the antigen on the solid phase (Figure 10.7).

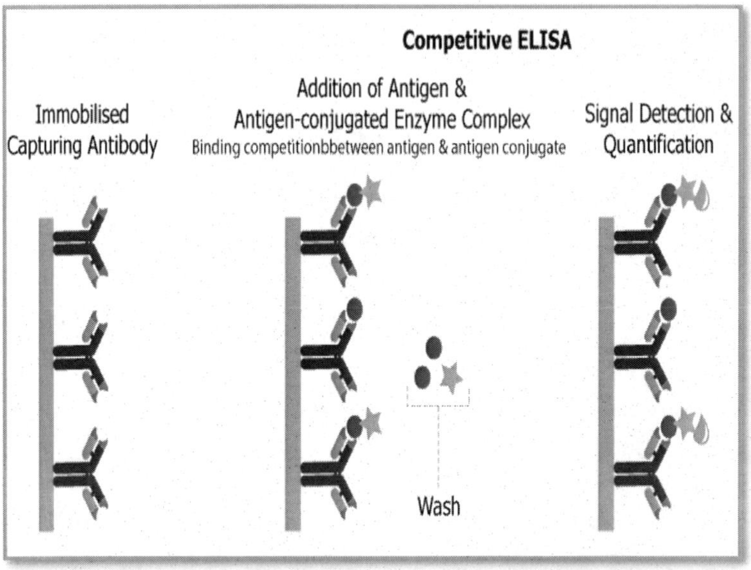

Figure 10.7. Illustrating the procedure of competitive ELISA (https://rockland-inc.com/elisa.aspx).

An example of a competition ELISA to test for antigen based on the direct detection method is shown in Figure 10.8. In this example, a known antigen is used to coat a multi-well plate. Following standard blocking and washing steps, samples containing unknown antigen are added. Labeled detection antibody is then applied for detection using relevant substrates (e.g., 3,3',5,5'-Tetramethylbenzidine or TMB). If there is a high concentration of antigen in the sample, a significant reduction in signal output will be observed. In contrast, if there is very little antigen in the sample, there will be very little reduction in the expected signal output. In the example shown in Figure 10.7, there would be a reduction in signal output.

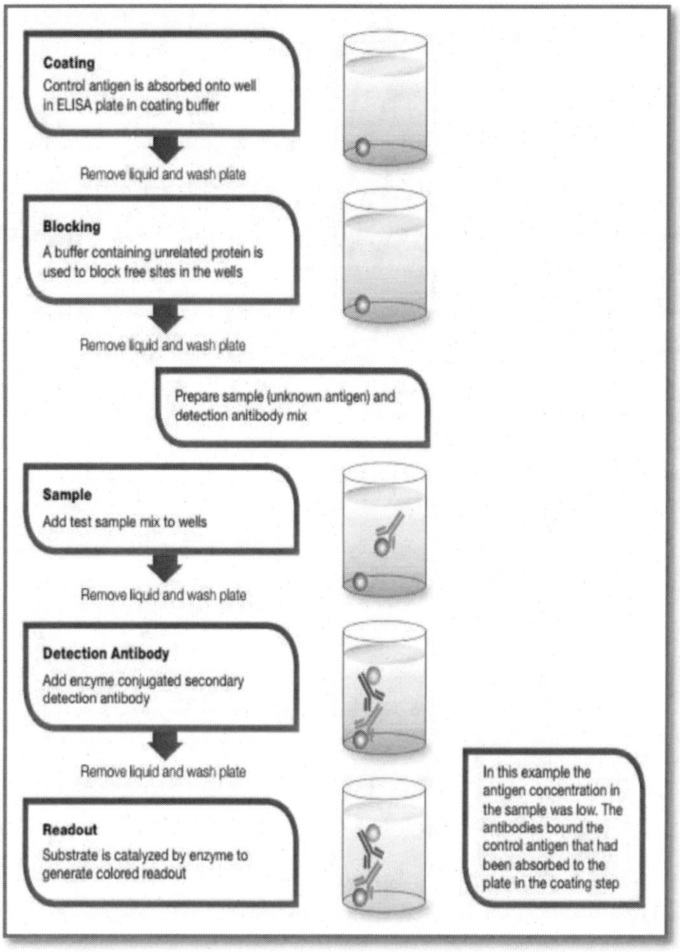

Figure 10.8. Showing the steps in competitive ELISA (https://www.bio-rad-antibodies.com/).

ANTIGEN AND ANTIBODY CAPTURE ELISA

Antigen capture ELISA can be based on principle of indirect or competitive ELISA, only difference being in the initial step of attaching antigen to the solid phase. A monoclonal antibody directed against an antigen is bound to the solid support. Next step is addition of antigen supplied as reagent in the test kit. This antigen is captured by the monoclonal antibody bound to the solid phase. Test specimen appropriately diluted is added next. Antibodies if present in the specimen bind to the antigen on solid support. Remaining principle is same as indirect ELISA Only advantage of antigen capture ELISA is that it is more specific than indirect assay.

METHODS TO PERFORM ELISA

Indirect ELISA

Indirect ELISA follow the instructions of the manufacturer of the kit which are given in the kit literature and develop a protocol/SOP. One example is given below.

Principle

Principle HIV antigens are attached covalently to the solid phase support allowing HIV antibodies present in the specimen to bind, and these bound antibodies are subsequently detected by enzyme labelled anti-human immunoglobulin and specific substrate system. If the test specimen contained antibodies color reaction will take place

Materials Required

- Single/multichannel pipettes with disposable tips: 5-50ul, 50-200ul
- Incubator (37+20C)
- ELISA reader with or without washer
- Powderless disposable gloves
- Absorbent paper
- Deionized water
- Discard jar with hypochlorite solution
- Wash bottles

Sandwich ELISA

One of the most useful of the immunoassays is the two antibody "sandwich" ELISA. This assay is used to determine the antigen concentration in unknown samples. This ELISA is fast and accurate, and if a purified antigen standard is available, the assay can determine the absolute amount of antigen in an unknown sample

Principle

To utilize this assay, one antibody (the "capture" antibody) is purified and bound to a solid phase typically attached to the bottom of a plate well. Antigen is then added and allowed to complex with the bound antibody. Unbound products are then removed with a wash, and a labeled second antibody (the "detection" antibody) is allowed to bind to the antigen, thus completing the "sandwich". The assay is then quantitated by measuring the amount of labeled second antibody bound to the matrix, through the use of a colorimetric substrate.

Materials Required

- Single/multichannel pipettes with disposable tips: 5-50ul, 50-200ul
- Incubator (37+20C)
- ELISA reader with or without washer
- Powderless disposable gloves
- Absorbent paper
- Deionized water
- Discard jar with hypochlorite solution
- Wash bottles

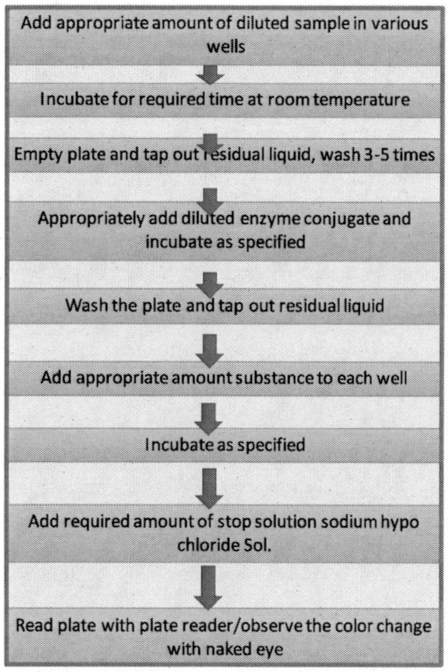

Figure 10.9. Flowchart representing the protocol of indirect ELISA.

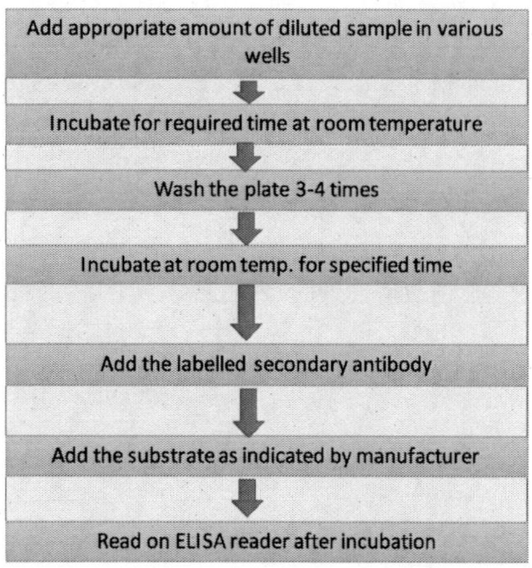

Figure 10.10. Flowchart representing protocol of sandwich ELISA.

Protocol

The following is the general protocol for a sandwich ELISA.

Competitive ELISA

Principle

In this assay the HIV-antibodies present in the specimen compete with the enzyme conjugated antibodies in the reagent for binding to the antigen on the solid phase. If the test specimen contains HIV antibodies, these will compete with the labelled antibodies in the reagent for binding to antigen. So that less or not labelled antibodies can attach to the solid phase. Hence, faint or no color is produced on addition of substrate if specimen contains HIV antibodies.

Materials Required

- Single/multichannel pipettes with disposable tips: 5-50ul, 50-200ul
- Incubator (37 + 20C)
- ELISA reader with or without washer
- Powderless disposable gloves

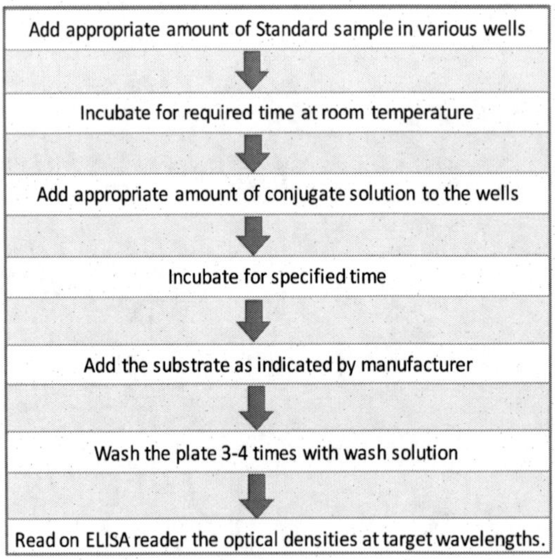

Figure 10.11. Flowchart representing the steps involved in competitive ELISA.

- Absorbent paper
- Deionized water
- Discard jar with hypochlorite solution
- Wash bottles

APPLICATION OF ELISA

Several examples for applications of this widely applied technique in the areas of food industry, vaccine development, immunology, diagnosis, toxicology, drug monitoring, pharmaceutical industry, and transplantation are briefly reviewed.

- **Food Industry:** ELISA plays a major role in food industry. It is the main platform for identifying food allergens such as those present in milk, peanuts, walnuts, almonds, and eggs. Peng et al. developed a monoclonal antibody-based sandwich ELISA for the detection of ovalbumin in food, which is the most frequent cause of food allergy, especially in children. ELISA can also be employed to corroborate the authenticity of the food products. This technique is of great help to avoid possible economic losses caused by fraudulent substitution. In the case of meat and meat-based products, ELISA has proven to be a reliable technique that provides careful monitoring of the product, especially when religious considerations in the choice of food are concerned. ELISA is also an essential technique for quality control of fish, milk (as well as their sub products), genetically modified foods, irradiated foods, or other harmful food components that can be transferred to human, such as bovine spongiform encephalopathy. Non-meat proteins such as soybean have valuable nutritional properties. Nonetheless, due to the similarity to the mean product, they are seldom added to the meat products undeclared. Careful monitoring of the products with ELISA prevents such adulteration. Unethical competitions for higher economic gain often lead to the potential health hazard through the consumed food and beverage. Production of ELISA kits for food industry applications is challenging as a selection of adequate control and standard samples is necessary to carefully calibrate the assay. Additionally, ELISA can target different types of analytes in the same food sample, thus the manufacturers should provide a complete set of kit components for the potential target biomolecules.
- **Vaccine Development:** ELISA serves as a great candidate for vaccine development. The sera sample from immunized animal or human model can be tested to detect the presence of antibodies against certain types of antigens, which were intentionally injected to the host. Normally different antigens are used to produce immune reactions in the host, among which those that elect higher protection response with less adverse effects can be selected. The main challenge in application of ELISA in vaccine development is the appropriate choice of positive and negative controls. In the experimental stage of vaccine development and when dealing with unknown samples, it is particularly difficult to achieve high analytical precision. Nonetheless, ELISA technique has proven to have a unique

position in profiling of elicited immune responses, which are widely performed for vaccine trials around the world.

- **Immunology:** The defender of the body, the immune system, can operate in cellular or humoral (innate or adaptive) modes. Measuring and monitoring the changes of the immune response underlay the foundation for understanding immune disease. Various studies have demonstrated ELISA as the gold standard method that is rapid and cost-effective for such measurements and monitoring. A great number of examples for ELISA applications in immunology are reported, while some efforts were directed to optimize ELISA protocols further and to validate/establish their accuracy, sensitivity and specificity to support the clinical practice.
- **Diagnosis:** In the area of diagnosis, ELISA has proven to be a capable platform applied worldwide for detecting variety of disease in human and animals. A number of different commercial ELISA kits are available in the market for detection of HIV, Influenza, Dengue fever, Ebola, Chagas disease, Leishmaniosis, Lyme disease, West Nile virus, among others. Even in plants pathology, ELISA technique is attracting increasing attention. ELISA has successfully overcome the drawbacks of the previous serological analyses performed in phyto-diagnosis. In this section, a brief summary on the current diagnosis applications of ELISA is provided.
- **Cancer_Detection:** Highly sensitive detection of cancer provides with the early stage diagnostic, which is crucial for patient survival. Cancer biomarkers, however, are some of the most challenging bimolecular entities as target analytes. Advancements of ELISA technique has promised its application in detection of cancer biomarkers.
- **Detection of the Infectious Disease:** ELISA-based infectious serology marks one of the most reliable means for accurate diagnosis and prognosis. There is a broad range of developed and marketed state-of-the-art assays for the detection of infectious agents. ELISA has offered a high throughput detection in three classes of infectious diseases.
- **Sexually Transmitted Diseases (STDs)** is a class of infectious diseases that has targeted adults in developing countries. A number of different ELISA platforms were designed and commercialized for sensitive and selective detection of STDs including HIV, hepatitis, syphilis, chlamydia.
- **Regional or endemic diseases**, often referred to as tropical diseases are wide spread in tropical and subtropical regions. They might appear to be mild/symptomless with serious and chronic consequences. Dengue, Chagas, borreliosis, and yellow fever are some of the examples of this class of fatal diseases among others. While existing techniques lack timely detection of such illnesses, advances in ELISA platforms have shown great promises in offering early and

effective diagnosis. **TORCH** refers to Toxoplasma, "Other infections", Rubella, Cytomegalovirus, Herpes simplex, which is a group of viral pathogens that may result in prenatal infections. This class of infectious diseases can be a potential threat to the unborn children. Illnesses such as syphilis, hepatitis B, Epstein-Barr virus, varicella-zoster virus, HIV fall under the category of "Other infections" that might also result in serious consequences for the fetus. Commercialized ELISA platforms successfully target these infectious agents in the current clinical practice.

CONCLUSION

Enzyme linked immunosorbent is one of the several methods used in the laboratory to detect and quantify specific molecules. Immunofluorescence is method in which an antigen is detected by the use of fluorescence labelled immunoglobulins. These techniques have been widely used in many important areas of pharmaceutical analysis such as diagnosis of diseases. The importance and widespread use of these techniques in pharmaceutical analysis are attributed to their inherent specificity, high sensitivity for the analysis of wide range of analytes in biological samples. ELISA combines the specificity of antibodies with the sensitivity of simple enzyme assays, by using antibodies or antigens coupled to an easily assayed enzyme that possesses a high turnover number. Simply put for ELISA an antigen is adsorbed/fixed to the surface of a solid matrix which may be a well on a micro titration plate/ well strip, beads or may be a special type of paper made of Nitrocellulose. Then, a specific antibody is applied over the surface so it can bind to the antigen. This antibody is linked to an enzyme, and, in the final step, a substance containing the specific substrate for the enzyme is added. The subsequent reaction produces a detectable signal, a color change in the substrate. ELISA technology has advanced a lot. To make this immunoassay more and more sensitive as well as specific various advances have been made and now, we have different types of ELISAs.

FREQUENTLY ASKED QUESTIONS

Q1. What is the basic principle of ELISA?

Answer: The basic principle is an enzyme conjugated to an antibody reacts with a colorless substrate to generate a colored product.

Q2. What do u understand by quality control?

Answer: Quality control refers to the reliability of investigative service. Any error in the laboratory will jeopardize the lives of patients. It is therefore utmost important that laboratory errors are identified and rectified.

Q3. What is the use of ELISA test?

Answer: ELISA is widely used for the determination of small quantities of proteins (hormones, antigens, antibodies) and other biological substances. The most commonly used pregnancy test for the detection of human chorionic gonadotropin (HCG) in urine is based on ELISA. It is also been used for the diagnosis of AIDS.

Q4. What is an ELISA kit?

Answer: An ELISA kit is a commercially available ELISA test that usually contain pre-coated polystyrene plates, detection antibodies and usually chemicals needed to perform an ELISA test.

Q5. What is the difference between Sandwich and competitive ELISA?

Answer: sandwich ELISA is mainly meant for detecting antigens with two antigenic epitopes capable of binding to antibody whereas competitive ELISA is used to measure the concentration of antibody in a sample by detecting interference in an expected signal output.

Q6. Discuss direct ELISA?

Answer: It is the simplest configuration in which the antigen is bound by passive adsorption to the solid phase, washed to remove any unbound molecules and then directly incubated with a conjugated antibody. Following the incubation period and additional washing, substrate is added to produce signal that is allowed to develop. After certain time, the substrate reaction is stopped and the resulting signal quantified. It is commonly used for tittering conjugated secondary antibodies and very useful to estimate antigen cross-reactivity and the advantages of direct ELISA are (i) Quick because only one antibody and fewer steps are used (ii) Cross-reactivity of secondary antibody is eliminated.

Q7. Describe indirect ELISA?

Answer: HIV antigens are attached covalently to the solid phase support allowing corresponding/specific antibodies present in the specimen to bind, and these bound antibodies are subsequently detected by enzyme labelled anti-human immunoglobulin and specific substrate system. If the test specimen contained the antibodies specific to the antigen fixed on the solid phase color reaction will take place. The color reaction can be read with naked eye or with ELISA Reader at the specified UV light using the special filters. The result is expressed as OD (optical Density) value. The advantages of indirect ELISA are

I. A wide variety of labeled secondary antibodies are available commercially.
II. Versatile because many primary antibodies can be made in one species and the same labeled secondary antibody can be used for detection.
III. Maximum immunoreactivity of the primary antibody is retained because it is not labeled.
IV. Sensitivity is increased because each primary antibody contains several epitopes that can be bound by the labeled secondary antibody, allowing for signal amplification.

Q8. Give an outline of competitive ELISA?

Answer: In this assay the antibodies present in the specimen compete with the enzyme conjugated antibodies in the reagent for binding to the antigen on the solid phase. If the test specimen contains corresponding/specific antibodies, these will compete with the labelled antibodies in the reagent for binding to antigen. So that less or not labelled antibodies can attach to the solid phase. Hence, faint or no color is produced on addition of substrate if specimen contains antibodies against the antigen on the solid phase.

An example of a competition ELISA to test for antigen based on the direct detection method. In this example, a known antigen is used to coat a multi-well plate. Following standard blocking and washing steps, samples containing unknown antigen are added. Labeled detection antibody is then applied for detection using relevant substrates (e.g., 3,3',5,5'-Tetramethylbenzidine or TMB). If there is a high concentration of antigen in the sample, a significant reduction in signal output will be observed. In contrast, if there is very little antigen in the sample, there will be very little reduction in the expected signal output. In the example shown in Figure 10.7, there would be a reduction in signal output.

Q9. What do you understand by the sandwich ELISA?

Answer: One of the most useful of the immunoassays is the two antibody "sandwich" ELISA. This assay is used to determine the antigen concentration in unknown samples. This ELISA is fast and accurate, and if a purified antigen standard is available, the assay can determine the absolute amount of antigen in an unknown sample. To utilize this assay, one antibody (the "capture" antibody) is purified and bound to a solid phase typically attached to the bottom of a plate well. Specimen containing antigen is then added and allowed to complex with the bound antibody. Unbound products are then removed with a wash, and a labeled second antibody (the "detection" antibody) is allowed to bind to the antigen, thus completing the "sandwich". The assay is then quantitated by measuring the amount of labeled second antibody bound to the matrix, through the use of a colorimetric substrate.

Q10. Describe the applications of ELISA test in vaccine development?

Answer: ELISA serves as a great candidate for vaccine development. The sera sample from immunized animal or human model can be tested to detect the presence of antibodies against certain types of antigens, which were intentionally injected to the host. Normally different antigens are used to produce immune reactions in the host, among which those that elect higher protection response with less adverse effects can be selected. The main challenge in application of ELISA in vaccine development is the appropriate choice of positive and negative controls. In the experimental stage of vaccine development and when dealing with unknown samples, it is particularly difficult to achieve high analytical precision. Nonetheless, ELISA technique has proven to have a unique position in profiling of elicited immune responses, which are widely performed for vaccine trials around the world.

Q11. Describe the applications of ELISA in the discipline of immunology?

Answer: The defender of the body, the immune system, can operate in cellular or humoral (innate or adaptive) modes. Measuring and monitoring the changes of the immune response underlay the foundation for understanding immune disease. Various studies have demonstrated ELISA as the gold standard method that is rapid and cost-effective for such measurements and monitoring. A great number of examples for ELISA applications in immunology are reported, while some efforts were directed to optimize ELISA protocols further and to validate/establish their accuracy, sensitivity and specificity to support the clinical practice.

Q12. Give an outline of ELISA in food industry?

Answer: ELISA plays a major role in food industry. It is the main platform for identifying food allergens such as those present in milk, peanuts, walnuts, almonds, and eggs. Peng et al. developed a monoclonal antibody-based sandwich ELISA for the detection of ovalbumin in food, which is the most frequent cause of food allergy, especially in children. ELISA can also be employed to corroborate the authenticity of the food products. This technique is of great help to avoid possible economic losses caused by fraudulent substitution. In the case of meat and meat-based products, ELISA has proven to be a reliable technique that provides careful monitoring of the product, especially when religious considerations in the choice of food are concerned. ELISA is also an essential technique for quality control of fish, milk (as well as their sub products), genetically modified foods, irradiated foods, or other harmful food components that can be transferred to human, such as bovine spongiform encephalopathy. Non-meat proteins such as soybean have valuable nutritional properties. Nonetheless, due to the similarity to the mean product, they are seldom added to the meat products undeclared. Careful monitoring of the products with ELISA prevents such adulteration. Unethical competitions for higher economic gain often lead to the potential health hazard through the consumed food and beverage.

Production of ELISA kits for food industry applications is challenging as a selection of adequate control and standard samples is necessary to carefully calibrate the assay. Additionally, ELISA can target different types of analytes in the same food sample, thus the manufacturers should provide a complete set of kit components for the potential target biomolecules.

MULTIPLE CHOICE QUESTIONS

Q1. Which of the following molecules can be detected by Enzyme linked immunosorbent assay (ELISA)?

A. Protein
B. Hormones
C. Antibody
D. All of the above.

Q2. What does a weak colour signal in competitive ELISA represents?

A. More antigens in the sample
B. Less antigen in the sample
C. Less antigen retained on the well
D. Both a and c.

Q3. which of the following is immobilized on the microtiter well in sandwich ELISA?

A. Detection antigen
B. Sample
C. Capture antibody
D. Secondary antibody conjugated to an enzyme.

Q4. What is the major advantage of ELISA in comparison to other biological quantification technique?

A. Detection of a molecule at low concentration
B. Expensive
C. Low specificity
D. Easily available.

Q5. Which of the following are commonly tested for using the ELISA method?

A. HIV
B. Lyme diseases
C. Pregnancy
D. All of the above.

Q6. Which serological techniques utilize competition between radioisotopes labelled antigen and test antigen for detection?

A. Flow cytometry
B. Immuno-electrophoresis
C. Radioimmunoassay
D. ELISA.

Q7. The most widely used serological test for antibody or antigen detection is?

A. Immunoprecipitation
B. Immunoblotting
C. ELISA
D. Flow cytometry.

Q8. Which of the following step is part of indirect enzyme linked immunosorbent assay?

A. Known antigen is absorbed in wells of the microtiter plate
B. Serum sample with known antibodies are added
C. Indicator antibody linked to enzyme binds antigen
D. Wells that remain coloured indicates positive result.

Q9. Which of the following is true of the antibody sandwich ELISA method?

A. Test antigen binds to antibody if it is complementary
B. An antibody antigen complex is formed
C. Addition of enzyme produces a visible colour change indicating positive result.
D. All of the above.

Q10. Enzyme labels often used in procedure are?

A. Alkaline phosphatase
B. Horseradish peroxidase
C. Beta-galactosidase
D. All of the above.

Q11. The following methods of diagnosis utilize labelled antibodies except?

A. ELISA
B. Haem agglutination inhibition test
C. Radioimmunoassay
D. Immunofluorescence

Q12. Immunofluorescence is a technique used for?

A. Electron microscope
B. Light microscope
C. Confocal microscope
D. Light microscopy with a fluorescence microscope.

ANSWER KEY

1(D), 2(D), 3(C), 4(A), 5(C), 6(C), 7(C), 8(A), 9(B), 10(D), 11(B), 12(D)

REFERENCES

Aydin, S. A short history, principles, and types of ELISA, and our laboratory experience with peptide/protein analyses using ELISA. *Peptides.*, 2015, Oct 1, 72, 4-15.

Butler, JE; Feldbush, TL; McGivern, PL; Stewart, N. The enzyme-linked immunosorbent assay (ELISA): a measure of antibody concentration or affinity?. *Immunochemistry.*, 1978, Feb 1, 15(2), 131-6.

Butler, JE. Enzyme-linked immunosorbent assay. *Journal of Immunoassay.*, 2000, May 1, 21(2-3), 165-209.

Chu, FS; Huang, X; Wei, RD. Enzyme-linked immunosorbent assay for microcystins in blue-green algal blooms. *Journal-Association of Official Analytical Chemists.*, 1990, 73(3), 451-6.

Clark, BR; Engvall, E. *Enzyme-Linked Immunosorbent Assay (ELISA): Theoretical and Practical Aspects*. CRC Press: Boca Raton, FL, 1980.

Clark, MF; Bar-Joseph, M. Enzyme immunosorbent assays in plant virology. In *Methods in Virology*, 1984, Jan 1, (Vol. 7, pp. 51-85). Elsevier.

Clark, MF; Lister, RM; Bar-Joseph, M. ELISA techniques. In *Methods in Enzymology*, 1986, Jan 1, (Vol. 118, pp. 742-766). Academic Press.

Dona, V. Homogeneous colorimetric enzyme inhibition immunoassay for cortisol in human serum with Fab antiglucose-6-phosphate dehydrogenase as a label modulator. *J. lmmunol. Methods*, (1985), 82, 65.

Drolet, DW; Moon-McDermott, L; Romig, TS. An enzyme-linked oligonucleotide assay. *Nature Biotechnology.*, 1996 Aug, 14(8), 1021.

Duermeyer, Wo; Wielaard, F; Van der Veen, J. A new principle for the detection of specific IgM antibodies applied in an ELISA for hepatitis A. *J. Med. Virol.*, (1979), 4, 25.

Ekins, R. Merits and disadvantages of different labels and methods of immunoassay. In: A. Voller, A. Bartlett and D. Bidwell (Eds.), *Immunoassays for the 80s*. University Park Press, Baltimore, MD, (1981b), p. 5.

Ekins, R. Toward immunoassays of greater sensitivity, specificity and speed: An overview. In: A. Albertini and R. Ekins (Eds.). *Monoclonal Antibodies and Developments in Immunoassays*. Elsevier Amsterdam, (1981a), p. 3.

Engvall, E; Perlmann, P. Enzyme-linked immunosorbent assay, ELISA: III. Quantitation of specific antibodies by enzyme-labeled anti-immunoglobulin in antigen-coated tubes. *The Journal of Immunology.*, 1972, Jul 1, 109(1), 129-35.

Engvall, E. Enzyme immunoassay ELISA and EMIT. In *Methods in Enzymology*, 1980, Jan 1, (Vol. 70, pp. 419-439). Academic Press.

Engvall, E. Quantitative enzyme immunoassay (ELISA) in microbiology. *Med. Biol.*, 1977, 55(4), 193-200.

Engvall, E; Perlmann, P. Enzyme-linked immunosorbent assay (ELISA). Quantitative assay for immunoglobulin G. *lmmunochemistry*, (1971), 8, 871.

Gan, SD; Patel, KR. Enzyme immunoassay and enzyme-linked immunosorbent assay. *J Invest Dermatol.*, 2013, Sep 1, 133(9), e12.

Gibbons, J; Skold, C; Rowley, GL; Ullman, EF. Homogeneous enzyme immunoassay for proteins employing//- galactosidase. *Anal. Biochem.*, (1980), 102, 167.

Haga, M; ItagakJ, H; Sugawara, S; Okana, T. Drug sensor: Liposome immunosensor for theophyline. *Analyt. Biochem.*, (1981), 118, 286.

Hamaguchi, Y; Phi, M; Sakakt:ra, Y; Mukojima, T. Quantitatio: J of nasal secretory lgA by enzyme-linked immunosorbant assay. *Int. Arch. Aller~ Appl. lmmunol.*, (1982), 69, 1.

Hernandez-Fuentes, MP; Warrens, AN; Lechler, RI. Immunologic monitoring. *Immunological Reviews.*, 2003 Dec, 196(1), 247-64.

Hornbeck, P. Enzyme-linked immunosorbent assays. *Current Protocols in Immunology.*, 1992, Mar, 1(1), 2-1.

Hornbeck, PV. Enzyme-linked immunosorbent assays. *Current Protocols in Immunology.*, 2015 Aug, 110(1), 2-1.

Hsu, SM; Raine, L; Fanger, H. Use of avidin-biotin-peroxidase complex (ABC) in immunoperoxidase techniques: a comparison between ABC and unlabelled antibody (PAP) procedure. *J. Histochem. Cytochem.*, (1981), 29, 577.

Huet, AC; Charlier, C; Tittlemier, SA; Singh, G; Benrejeb, S; Delahaut, P. Simultaneous determination of (fluoro) quinolone antibiotics in kidney, marine products, eggs, and muscle by enzyme-linked immunosorbent assay (ELISA). *Journal of Agricultural and Food Chemistry.*, 2006, Apr 19, 54(8), 2822-7.

IJsselmuiden, OE; Herbrink, P; Meddeus, MJM; Tank, B; Stolz, E; van Eijk, RVW. Optimizing the solid-phase immunofiltration assay. A rapid alternative to immunoassay. *J. Immunol. Methods*, (1989), 119, 35.

Jackson, T; Marshall, N; Ekins, R. Optimization of immunoradiometric (labelled antibody) assays. In: W. M. Hunter and J. E. T. Corrie (Eds.), *lmmuno-Assays for Clinical Chemistry Churchill Livingstone*, Edinburgh, (1983), 557.

Johannsson, A; Stanley, ChJ; Self, CH. A fast highly sensitive colorimetric enzyme immunoassay system demonstrating benefits of enzyme amplification in clinical chemistry. *Clin. Chim. Acta*, (1985), 148, 119.

Karawajew, L; Behrsing, O; Kaiser, G; Micheel, B. Production and ELISA application of bispecific monoclonal antibodies against fluorescein isothiocyanate (FITC) and horseradish peroxidasc (HRP). *J. Immunol. Methods*, (1988), 111, 95.

Kendall, C; Ionescu-Matiu, I; Dreesman, GR. Utilization of the biotin/avidin system to amplify the sensitivity of the enzyme-linked immunosorbent assay (ELISA). *Journal of Immunological Methods*., 1983, Jan 1, 56(3), 329-39.

Kenny, GE; Dunsmoor, F. Principles, problems and strategies in the use of antigenic mixtures for the enzyme-linked immunosorbent assay. *J. Clin. Microbiol.*, (1983), 17, 655.

King, WJ; Tomita, JT; Dowcll, BL; Dclfert, PM. Strategies for heterogeneous enzyme immunoassays for tumor markers. In: R. B. Herberman and D. W. Mercer (Eds.), *Immunodiagnosis of Cancer*. Marcel Dekker, NewYork, (1990), p. 83.

Langone, JJ. Applications of immobilized protein A in immunochemical techniques. *Journal of Immunological Methods*., 1982, Dec 30, 55(3), 277-96.

Lenz, DE. Antibody quantitation using an immunosorbem and the unlabeled antibody enzyme method. *J. Immunol. Methods*, (1976), 13, 113.

Lequin, RM. Enzyme immunoassay (EIA)/enzyme-linked immunosorbent assay (ELISA). *Clinical Chemistry*., 2005, Dec 1, 51(12), 2415-8.

Litman, DJ; Hanlon, TM; Ullman, EF. Enzyme channeling immunoassay: a new homogeneous enzyme immunoassay technique. *Analyt. Biochem.*, (1980), 106, 223.

lshikawa, E; imagawa, M; Hashid¢, S; Yoshitake, S; Hamaguchi, Y; Ueno, T. Enzyme labeling of antibodies and their fragments for enzyme immunoassay and histochemical staining. *J. Immunoassay*, (1983), 4, 209.

Maggio, ET. *Enzyme Immunoassay: 0*. Crc Press, 2018, Jan 18.

McGinlay, B; Bardsley, WG. *The Kinetics of Adsorption of Human Immunoglobulin G to Poly*, (1989).

Miles, LEM; Hales, CN. Labelled antibodies and immunological assay systems. *Nature*, (1968), 219, 186.

Morris, DL; Ellis, PB; Carrico, RJ; Jeager, FM; Schroeder, HR; Albarella, JP; Boguslaski, RC; Nakamura, RM; Voller, A; Bidwell, DE. Enzyme immunoassays: heterogeneous and homogeneous systems. In: D. M. Weir (Ed.), *Handbook of Experimental Immunology*, (1986), Vol. 1, Immunoehemistry. Blackwell, Oxford, p. 27.1.

Nakane, PK; Kawaoi, A. Peroxidase-labeled antibody. A new method of conjugation. *J. Histochem. Cytochem.*, (1974), 22, 1084.

Ngo, TT; Lcnhoff, HM. Enzyme modulators or tools for the development of homogeneous enzyme immunoassays. *FEBS Lett.*, (1980), 116, 285.

Rennard, SI; Berg, R; Martin, GR; Foidart, JM; Robey, PG. Enzyme-linked immunoassay (ELISA) for connective tissue components. *Analytical Biochemistry.*, 1980, May 1, 104(1), 205-14.

Rose, ME; Mocket, AP. Antibodies to coccidia: detection by the enzyme-linked immunosorbent assay (ELISA). *Parasite Immunology.*, 1983 Sep, 5(5), 479-89.

Schuurs, AH; Van Weemen, BK. Enzyme-immunoassay. *Clinica Chimica Acta.*, 1977, Nov 15, 81(1), 1-40.

Steward, MW; Lew, AM. The importance of antibody affinity in the performance of immunoassays for antibody. *Journal of Immunological Methods.*, 1985, Apr 22, 78(2), 173-90.

Tijssen, P. *Practice and Theory of Enzyme Immunoassays.* Elsevier, 1985, Aug 1.

Voller, A; Bartlett, A; Bidwell, DE. Enzyme immunoassays with special reference to ELISA techniques. *Journal of Clinical Pathology.*, 1978, Jun 1, 31(6), 507-20.

Voller, A; Bidwell, DE; Bartlett, A. *The Enzyme Linked Immunosorbent Assay (ELISA). A Guide with Abstracts of Microplate Applications.* Dynatech Europe, Borough House, Rue du Pre., 1979.

Walls, KW; Bullock, SL; English, DK. Use of the enzyme-linked immunosorbent assay (ELISA) and its microadaptation for the serodiagnosis of toxoplasmosis. *Journal of Clinical Microbiology.*, 1977, Mar 1, 5(3), 273-7.

Wernette, CM; Frasch, CE; Madore, D; Carlone, G; Goldblatt, D; Plikaytis, B, Benjamin, W; Quataert, SA; Hildreth, S; Sikkema, DJ; Käyhty, H. Enzyme-linked immunosorbent assay for quantitation of human antibodies to pneumococcal polysaccharides. *Clin. Diagn. Lab. Immunol.*, 2003, Jul 1, 10(4), 514-9.

Wisdom, GB. Enzyme-immunoassay. *Clinical Chemistry.*, 1976, Aug 1, 22(8), 1243-55.

Wright, PF; Nilsson, E; Van Rooij, EM; Lelenta, M; Jeggo, MH. Standardisation and validation of enzyme-linked immunosorbent assay techniques for the detection of antibody in infectious disease diagnosis. *Revue Scientifique Et Technique*-Office International Des Epizooties., 1993, Jun 1, 12, 435-.

Yolken, R; Kim, H; Clem, T; Wyatt, R; Kalica, A; Chanock, R; Kapikian, A. Enzyme-linked immunosorbent assay (ELISA) for detection of human reovirus-like agent of infantile gastroenteritis. *The Lancet.*, 1977, Aug 6, 310(8032), 263-7.

ABOUT THE EDITOR

Dr. Manzoor Ahmad Mir (MSc, PhD, PGDHE, JRF-NET)
Senior Assistant Professor
Coordinator/Head Department of Bioresources
School of Biological Sciences
University of Kashmir Srinagar
E-mail: drmanzoor@kashmiruniversity.ac.in

Dr. Manzoor Ahmad Mir holds Master's Degree in Zoology from HNBG Central University after qualifying prestigious National level CSIR-JRF-NET examination and worked jointly for his Ph. D at Jawaharlal Nehru University New Delhi and CSIR-Institute of Microbial Technology Chandigarh in the field of Immunology and Cell Biology. He currently teaches Endocrinology, Animal Physiology, Immunology and Developmental Biology at the Department of Bioresources, University of Kashmir, and has been a Research Scientist at SATCAS Stroke Research Chair Majmaah University KSA. Dr Manzoor has research and teaching experience of 13 years and has attended many courses and conferences on immunology, cancer Biology, stroke biology and endocrinology at USA, UK, Kuwait, China, UAE and Saudi Arabia. His basic research interests include molecular immunology, Tuberculosis immunology, Cancer and Stroke Biology. He has published more than 50 high-impact research papers and book chapters, in recognition of which he has received several awards and Royalties from international publishing houses. Dr. Manzoor has authored several books with international publishers like Elsevier USA and Nova Science Publisher USA. He is currently receiving Royalty from Academic Press USA for his book in the field of Immunology entitled "Developing costimulatory molecules for immunotherapy of diseases". Dr. Manzoor is on the editorial board of some prestigious journals from DOVE medical press New Zealand and Springer Plus UK and has been an invited speaker at various scientific meetings/conferences within India and abroad. He is member of many scientific organizations and societies like International Immunology Association, Indian Association of Immunology, Indian National Science Association, IMMUNOCON etc. Dr Manzoor was awarded Teachers Associate Research Exellence Fellowship (TARE) by DST Govt of India. He has been awarded Summer Research Fellowship Programme (SRFP-2019) by Indian Academy of Sciences and National Science Academy. Dr Manzoor has developed Massive Open Online Course (MOOCs) in Immunology for UG students by sanctioned by UGC- Consortium for Educational Communication (CEC) SWAYAM Ministry of HRD Govt of India. He has been awarded another MOOC Course on Endocrinology by Academic advisory committee of CEC-UGC.

About the Contributors

Mr. Umar Mehraj
Junior Research Fellow UGC-CSIR
Research Scholar
Department of Bioresources,
School of Biological Sciences, University of Kashmir Srinagar
Email: umarshk92@gmail.com

Miss Safura Nisar
Research Scholar
Department of Bioresources,
School of Biological Sciences, University of Kashmir Srinagar
Email: safooranisar99@gmail.com

Hina Qayoom
Research Scholar
Department of Bioresources,
School of Biological Sciences, University of Kashmir Srinagar
Email: hinaqkhan1@gmail.com

INDEX

A

B

C

D

E

F

G

H

I

K

L

M

Q

R

U

V

W

Y